AutoCAD®
for Electronics
A Tutorial

AutoCAD® for Electronics
A Tutorial

Second Edition

THOMAS TUMILTY

Humber College of Applied Arts and Technology
Rexdale, Ontario

Prentice Hall
Englewood Cliffs, New Jersey   Columbus, Ohio

Library of Congress Cataloging-In-Publication Data

Tumilty, Thomas 1938-
AutoCAD© for electronics: a tutorial / Thomas Tumilty. --2nd ed.
p. cm.
Includes index.
ISBN 0-13-232638-8
1. Electronics--Data processing. 2. AutoCAD (Computer file)
I. Title.
TK7835.T87 1996
621.381'0285'5369--dc20 95-34468
CIP

Editor: **Stephen Helba**
Production Editor: **Linda Hillis Bayma**
Copy Editor: **Rebecca Bobb**
Interior Design/Page Layout: **Tom Tumilty**
Cover Design: **Proof Positive/Farrowlyne Assoc., Inc.**
Production Manager: **Patricia A. Tonneman**
Marketing Manager: **Debbie Yarnell**

This book was printed and bound by Victor Graphics.
The cover was printed by Victor Graphics.

If your diskette is defective or damaged in transit, return it to Prentice Hall at the address below for a no-charge replac within 90 days of the purchase. Mail the defective diskette with your name and address.
Prentice Hall
College Operations
Englewood Cliffs, NJ 07632

TRADEMARK ACKNOWLEDGMENTS

AutoCAD® is a registered trademark of Autodesk, Inc.

Electro Sonic™ is a trademark of Electro Sonic, Inc.
IBM® is a registered trademark and PC DOS™ is a trademark of International Business Machines Corporation.
MS-DOS® is a registered trademark of Microsoft Corporation.

Printed in the United States of America
10 9 8 7 6 5 4 3 2 1

ISBN 0-13-232638-8

Prentice-Hall International (UK) Limited, *London*
Prentice-Hall of Australia Pty, Limited, *Sydney*
Prentice-Hall Canada Inc. *Toronto*
Prentice-Hall Hispanoamericana, S.A., *Mexico*
Prentice-Hall of India Private Limited, *New Delhi*
Prentice-Hall of Japan, Inc., *Tokyo*
Simon & Schuster of Asia Pte. Ltd., *Singapore*
Editora Prentice-Hall do Brasil, Ltda., *Rio de Janeiro*

To My Daughter, Leanne

1966 - 1984

Preface

THE NEED

There are many books written about AutoCAD and its applications in architecture and mechanical drafting, but one field has been left out of this maze of literature. This missing link, until now, was the electronics industry. After making a long and exhaustive search in this field and finding nothing, I decided to develop a comprehensive up-to-date tutorial on how to use AutoCAD for electronics applications. To my surprise, I found the electronics industry was also looking for such a journal.

As this second edition goes to press, this text still provides high schools, colleges and the electronics industry with the only tutorial that covers this subject with this much detail.

THE TUTORIAL

This second edition of *AutoCAD For Electronics, A Tutorial* will show you how to design a complete new electronics project, from scratch. Using a set of hand drawn electrical sketches and some help from the text, *you* will develop the manufacturing drawings for a multi-functional power supply.

This power supply is three supplies in one. The first is a fixed, +5 volt power supply rated at one-half amp which can be used for applications involving digital logic and integrated circuits. The remaining two supplies are dual-tracking, 0 to +/-15 volt supplies that are also rated at one-half amp.

These dual supplies are ideal for use with operational amplifiers and integrated circuits requiring dual voltage sources and will complement the experiments a student would use in an electronics/electrical course or in the engineering department at an electronics firm.

The project is designed around a two-part course. The first part develops the engineering drawings for the project and the second part covers the manufacturing of the project in a prototype manufacturing environment, such as in a school's electronics shop. A supplementary text, *Prototype Manufacturing Methods for Electronics, A Text and Workbook*, is used for this phase of the course.

WHAT YOU WILL GAIN FROM THIS TUTORIAL

Through the development of this electronics project, you will learn:

1. How a computer system functions, its many idiosyncrasies, and its great power.

2. How to use AutoCAD by exploring many of the commands which make this software a powerful tool.

3. The technique for translating a hand sketch into a full blown schematic/wiring interconnect drawing.

4. The process of laying out a simple printed circuit board (PCB) followed by the hands-on development of the power supply PCB board for this project.

5. Mechanical and electrical assembly methods, and why components must be oriented on the circuit board in specific ways.

6. The use of drawing standards universally accepted by industry as well as the techniques to extract data from data sheets or a parts catalog.

7. Many methods to create high-quality hard copy of computer-generated drawings.

This project and AutoCAD software make an ideal combination for learning the basic design criteria to produce the working drawings used by industry in the manufacture of an electronics/electrical product. AutoCAD is also an ideal software system to use to show the basic layout steps to produce a simple printed circuit board (PCB). It is in effect an "electronic hand taping" system.

Practice exercises provide many opportunities for you to expand your skills in dimensioning, creating text styles, and developing isometric and 3D drawing techniques, as well as developing drawing entities for use in your design.

WHAT IS RETAINED

Many versions of AutoCAD have come, and many more will go, so trying to issue a new text every time Autodesk releases a new version is a never-ending task. It also becomes very expensive to update a computer lab containing 20 to 40 computers with the latest AutoCAD software. With this in mind, I decided to choose only those basic AutoCAD commands which are needed to do the job and concentrate on showing you how to design and develop a project rather than overloading you with an explanation of the more than 170 AutoCAD commands. The basic commands I have selected will stay current no matter which version of AutoCAD you are using. You will come to realize that 20 percent of the commands are used 90 percent of the time.

WHAT IS NEW

This second edition of the tutorial contains a totally new project. It was chosen to fit into the semester time frame of a school year, as well as the budget of the student who goes on to construct the project in the next semester.

The drawing software for this text is based on AutoCAD releases 10 through 13, and I will show you much of the similarities and differences between these releases as we proceed. Release 13 has introduced some very simple but unique commands and we will play with many of these new commands as we proceed with this tutorial.

I have chosen the DOS version of release 13 to write about because implementing the Windows version requires much updating of your computer system. This would require you to have at least a 486 or Pentium based machine with sixteen megabytes or more of RAM and a CD ROM. In these days of constraint in the school system, this is usually not feasible. On the other hand, the DOS version allows the full-blown release 13 software to be run on older as well as newer computer systems. It just doesn't have all the niceties of the Windows environment.

As we progress, I will introduce you to some of the mistakes we all make and show you how to overcome them through many Tips, Tricks and Traps. These sidebars will highlight problem areas you may not be aware of, such as what happens when we draw over a line a second time. You can find the answer to this in Chapter 3.

When you have completed this tutorial, you will know the fundamental techniques of design, and you will have gained much practice with the development, placement, spacing and scaling of the building blocks used in producing technical drawings. You will see how AutoCAD excels in time-saving drawing techniques and have a solid foundation to explore more elegant AutoCAD commands found in the latest version. You will be ready to use these new commands when they come your way.

ACKNOWLEDGMENTS

I would like to thank my many colleagues who have given me a hand and encouragement in preparing the second edition of this text. A special thanks to Bruce Horin, Kelly Gray, Ken Sooknanan, Carlos Deoliveira, Bogdan Sokolowski, Ashley Daminato and Claudio Dominianni, who made available many of their drawings. To my peers Ed Vokurka and Bruce Horin whose many suggestions have fine-tuned this document and whose influence is imbedded between these lines. To my son Tom, and daughter-in-law Carol who supplied their computer for the testing of AutoCAD version 13.

A special thank you to my computer guru David Boudreau, who has spent many hours on the old 486 finding ways to incorporate the many screen captures, Tif, Gif, and DXF formats into WordPerfect.

As always, the people at Prentice Hall are great people to work with. Thanks Steve Helba for your input. I appreciate your suggestions and recommendations. A special thanks to Linda Bayma, my production editor, and Rebecca Bobb my copyeditor, for keeping me on track and for the encouragement, patience and understanding to get this book to press.

Many thanks to those of you who have reviewed the original text for your input, constructive criticism and encouragement. Thanks to Michael Doran (ITT Technical Institute, Dayton, Ohio), D. Gerry Davis (Central Piedmont Community College), J. Hirst Lederle (Old Dominion University), John B. Fitch (Indian River Community College), Donald R. Montgomery (ITT Technical Institute, Houston, Texas), Kurt Common (City College of San Francisco), and Marie Weigel (Nassau Community College).

Lastly, Thank you Marg (my wife) for putting up with all this nonsense for the last eight years and for the many hours of proof reading, consultation and coffee, for without your help and understanding I would still be in the midst of Chapter 1.

Tom Tumilty
Humber College of Applied Arts and Technology
Tumilty@Admin.HumberC.On.Ca

Contents

CHAPTER 2 Getting to Know the Computer Environment

CHAPTER 3 Getting AutoCAD Up and Running

CHAPTER 5 The Printed Circuit Board

CHAPTER 6 Main Assembly Drawing and Front Panel

CHAPTER 7 Using The Printer and Plotter

CHAPTER 8 Summary Of AutoCAD Commands

APPENDIX A AutoCAD Fonts, Linetypes, and Character Sets

APPENDIX B Function Key Designation

APPENDIX C Major Parts List

APPENDIX D Manufacturers' Data Sheets

APPENDIX E Practice Sessions

APPENDIX F Answers To Quizzes

APPENDIX G Dimension Variables and Default Settings

APPENDIX H Hatch Patterns

Notes:

List of Figures and Tables

List of Photographs

List of Figures

List of Tables

List of Charts

chapter 1

Introduction

DISCUSSION

Welcome to the world of AutoCAD. In this modern age of electronics, the computer has quickly replaced all of the manual tools we once associated with the drafting work place. Such common tools as pencils, erasers, T-squares and templates have disappeared and are now replaced with the mouse, tablet, CRT and plotter, to become an integral part of the new "draft-person's workstation."

A major player in this new age of drafting is the software system called AutoCAD. Without a doubt, AutoCAD is one of the most powerful Computer Aided Design (CAD) software packages available in this high-tech world of CAD. It allows you to create any type of line drawing using an IBM or IBM compatible computer, a Macintosh computer and recently it was exported to the Sun Work-stations. If you can draw it by hand, it can be created in AutoCAD.

TIP: To assure that AutoCAD works properly on an IBM style computer, your clone must be 100 percent compatible with IBM hardware.

With additional software, AutoCAD can change your 2D or 3D drawing into the magical world of animation, through such software as Autodesk's 3D Studio, but that's a story for another book.

Photo 1 Final Project.

Photo 2 Front and Right Side View.

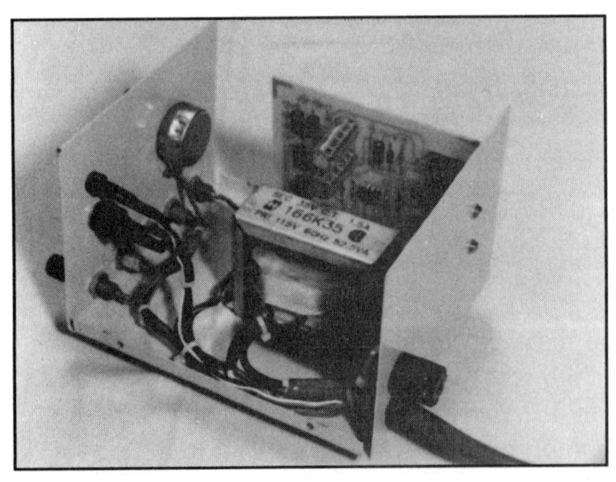

Photo 3 Back and Right Side View.

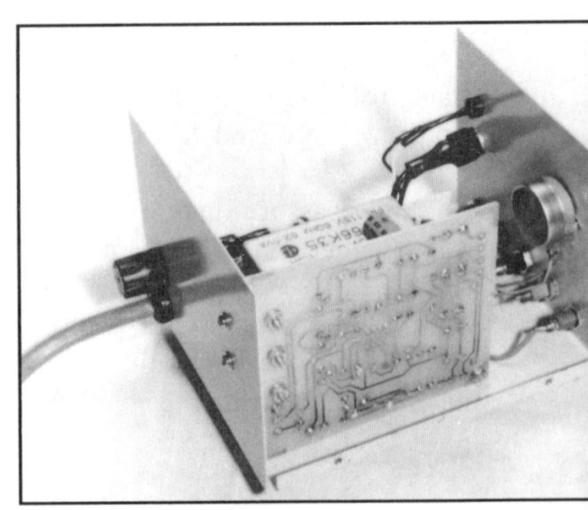

Photo 4 Left Side View.

If you are a first-time user of a CAD system, this text is for you. It is designed as a tutorial and acts as a primer to guide you through many of the basic AutoCAD commands. The text will whet your appetite for CAD by demonstrating the more frequently used commands of this powerful software. It is not my intention to overwhelm a newcomer to CAD systems with all of the possible command combinations of AutoCAD, but to provide a foundation and firm framework from which you can launch yourself into the more exciting possibilities with this software.

For the "power" user and those with experience, this tutorial includes many "**Tips** and **Tricks**" to help you shorten the time to perform specific functions. It also highlights "**Traps**" which you should avoid while using AutoCAD. I will show you how to make use of "symbol libraries", a powerful feature of AutoCAD that will save you time, give you better grades, and/or save your company money.

This tutorial is directed not only at the electronics student but at those in the workplace as well. In fact, this tutorial can be used by anyone wanting more information on CAD systems in general. I have chosen an electronics project rather than a mechanical or architectural project to highlight AutoCAD's versatility, for little has been written using CAD in this area.

OUR PROJECT

The project we will develop is a simple power supply with three outputs. The supplies are: a fixed, +5 volt supply rated at 1 amp and a dual tracking plus and minus variable supply (0 to 15 volts) rated at 1 amp. A view of the finished project is shown in photographs 1 through 4.

Together, we will work our way through this tutorial which incorporates the drafting skills needed to meet the standards set by industry, as well as develop the drawing techniques necessary to design large or small scale drawings.

The tutorial starts by showing you how to develop a simple drawing, that of the basic drawing page which includes a title block and border (Fig. 1-1). We then move to the more complex drawings of an electronic schematic, a printed circuit board assembly drawing, printed circuit artwork, and finally the mechanical assembly drawing for this project. Chapter 7 discusses several methods of transferring your drawing into hard copy, the printed page, so it can be used in the manufacturing of this project. The final chapter summarizes AutoCAD commands.

As added exercises, Appendix E contains a series of practice sessions to further your knowledge on dimensioning, modifying and creating text styles, isometric drawing techniques, a introduction to 3D drawings and modelling and playing with GRIPS.

To aid those of you who may be using a computer for the first time, Chapter 2 introduces a short tutorial on the computer's Disk Operating Systems (DOS) as well as the Windows environment. You will learn just enough about computer commands to get you started on such things as formatting a disk, the Copy and Diskcopy commands, checking your disk for corrupted files and how to erase unwanted files.

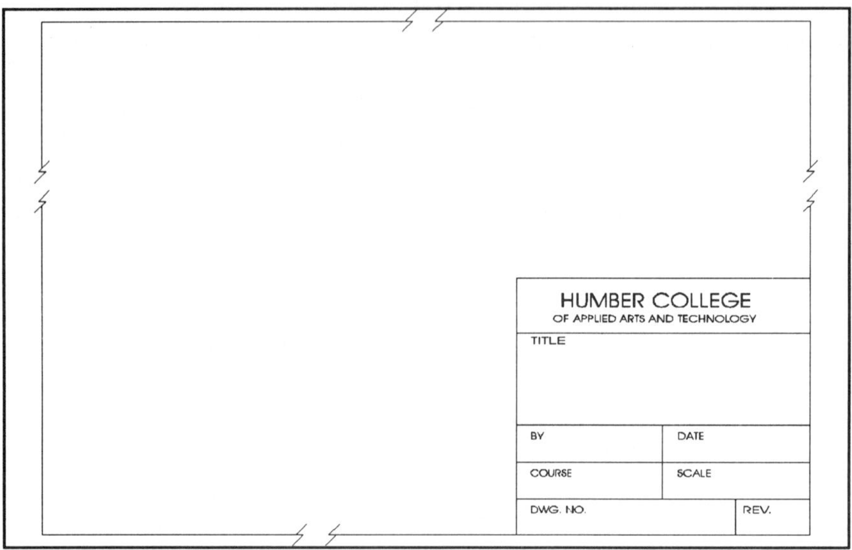

Figure 1-1 Drawing sheet and title block.

It is highly recommended that you learn the computer environment as quickly as possible. The only way to do this is by using the computer at every opportunity. In this case, **practice really does make perfect**.

COMPUTER CONFIGURATION

For the discussions used in this tutorial, the minimum computer configuration is based on the IBM computer system, although versions of AutoCAD are available for the Macintosh and Sun Work-stations. You will need:

* The IBM, or true IBM compatible clone designed with at least a 386 microprocessor and associated math-coprocessor. Since release 2.5, AutoCAD requires a hardware math-coprocessor as the software is very heavily math intensive. Don't even think of using a software coprocessor emulator as this slows down the system even further. Hardware is always faster than a software fix or work around. It is also recommended that you use the latest hardware system available such as the 486DX or Pentium based system in order to take full advantage of future software releases.

* Minimum of 8 megabytes of Random Access Memory (RAM), and 16 is recommended.

* A high capacity 5.25 inch, 1.2 megabyte floppy drive or a 3.5 inch, 1.44 megabyte floppy drive. A 2.88M drive would even be better.

* Hard disk with approximately 30 megabytes of free disk space to install all AutoCAD files.

* A color monitor running in VGA or SVGA mode.

* A mouse or digitizing tablet known as an input device.

* A plotter, laser printer, or dot-matrix printer.

* DOS operating system, Version 3.3 or later, or Microsoft Windows, Version 3.1 or Windows 95.

TIP: To protect your computer system, it is highly recommended that you use a surge protector to prevent damage from those erroneous over-voltage-spikes that periodically appear on the 115 volt power line.

STUDENT NEEDS

Each student should provide ten formatted 5 1/4- or 3 1/2-inch floppy disks. Before buying the disks, note the type of floppy drives your system is using. Half of these disks will be used to store your drawings and the other half for backup purposes. Today, many disks come preformatted for the IBM but are slightly more expensive to buy. The process of formatting a disk is discussed in detail in Chapter 2.

TIP: Make sure all your disks are formatted before starting AutoCAD.

You should provide two sheets of borderless C size paper (24 in. x 18 in.) and four sheets of D size (36 in. x 24 in.) borderless paper as shown on page 90 in Table 3-2, Civil and Architectural paper sizes. Note: Use this size of paper rather than metric, because the exercises are developed around them!

For plotting, you will need four colored pens. These should be the fine (F) or superfine (S) nib size. Check with your instructor or computer center for plotting pen requirements. Some pens come in a four-pack in the colors of black, blue, red, and green.

AutoCAD VERSIONS

Many of you may have found it too expensive to move to the latest version of AutoCAD, therefore; this text discusses many of the similarities and differences between versions 10, 11, 12 and 13. It also highlights many of the newer features found with the latest release, version 13. Remember; release 13 will run all of the commands of older versions.

TIP: You should be aware that drawings created with early versions of AutoCAD (10 and under) can be upgraded to run with version 11, 12 and 13, but once used with these later versions (11, 12 and 13), they can *no longer* be used with the earlier version software. What this means is that AutoCAD software is *not downwards compatible* after version 11. This is the price one pays for software updates. AutoCAD recognized this as a problem and has made version 12 data base compatible with version 11. Should you want to operate in the world of earlier software versions, you must keep a backup or archival copy of the drawings for each environment. Version 13 uses a conversion program to allow you to save your drawing in release 12 format.

COMPUTER HARDWARE

The Monitor (CRT) and Drawing Editor

The monitor or CRT (Cathode Ray Tube) screen, or video display, as it is sometimes called, is used to communicate drawing information to you from AutoCAD. Monitor sizes range from 12 inches to 21 inches and in resolution from 640x480 pixels for a monochrome VGA monitor to 2048x2048 for the ultra-high resolution monitors used for extended graphics applications.

TIP: The one factor you should consider most when purchasing a monitor is to get one that has a resolution that is better than you think you will need. This is because we often find ourselves *consumed* in long hours of having fun when we are working with AutoCAD. These long sessions lead to eye strain, and if we are using a poor quality monitor, greater damage can be done to the eye.

There are five areas on the screen that make up AutoCAD's "drawing editor", the term given to this screen. See Fig. 1-2a. These areas are:

1. Graphics Area (Drawing Area). The center portion of the screen is the work area for your drawing. The computer uses this area to display the drawing as you input data from the keyboard, mouse, or graphics tablet. This area is your "sheet of paper" or drawing area.

2. Command Line/Prompt Area. At the bottom of the screen is the **command line** and **prompt area**. It is here where all of your interactive communications with AutoCAD are displayed, and where prompts are requested. You should constantly monitor this area to find out the status of the command being executed and what AutoCAD is expecting next.

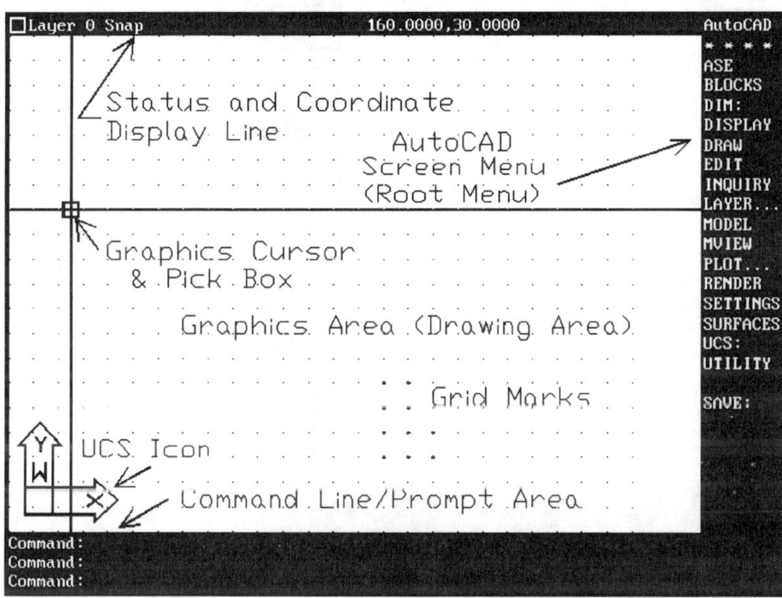

Figure 1-2a Drawing Editor Screen.

3. Screen Menu Area. On the right side of the monitor is the **screen menu** area. This area provides a "quick pick" selection of commands when you are using an input device such as a mouse or graphics tablet. The "top" choice in this menu chain is the **root menu**. The root menu is broken into submenus, which may in turn be divided into more submenus.

Since AutoCAD version 2.5, the root menu can always be reached by selecting (clicking on) the word "AutoCAD" at the top of the menu. The area marked with the "* * * *" selects the object snap mode (OSNAP option), discussed in Chapter 4.

4. Status Line. Across the top of the screen is the **status line** which gives you information about your drawing at a glance. It displays the current status of such things as the layer name, grid and snap options, AutoCAD's operating modes such as ortho or isometric, and the current coordinates of the cursor.

5. Pull-Down Menus & Menu Bar. As with most software, twenty percent of the commands are used eighty percent of the time. To quickly gain access to these more commonly used commands, AutoCAD in Release 9 introduced pull-down menus (see Fig. 1-2b). When the graphics cursor is placed at the top of the screen a Menu Bar replaces the Status Line, and "pull-down" menus appear. These allow you a quick pick of the frequently used AutoCAD commands.

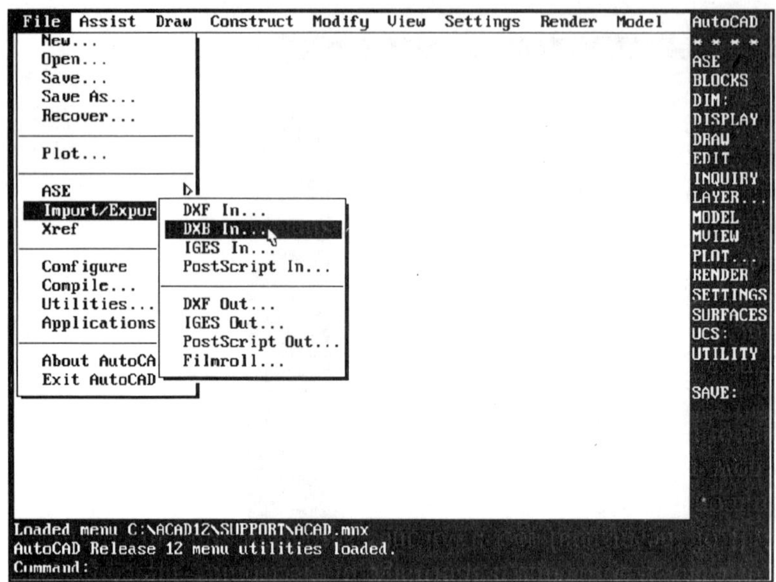

Figure 1-2b AutoCAD Screen and pull-down menus.

Graphics Cursor. The screen cursor is used to draw, pick items from the menu or dialogue box, or select entities. It may appear as crosshairs, a pickbox, or both depending upon the AutoCAD command.

The UCS Icons. In the lower left corner of the screen will be one of the two User Coordinate System (UCS) Icons, Fig. 1-2c. These icons, with their X, Y and W (Z) axis, represent the 3-dimensional world and tell you the orientation of the model in your drawing. For the drawings we will be developing in this tutorial, they are not needed, and will be turned off when we start to draw.

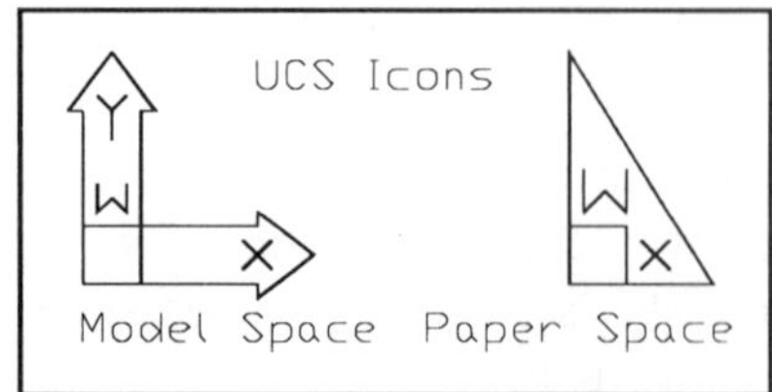

Figure 1-2c The UCS Icons.

TIP: The sooner you learn the layout, appearance, and proper use of this screen, the quicker you will master AutoCAD.

Pull-Down Windows and Dialogue Boxes

The use of pull-down windows and dialogue boxes was first introduced with the Commodore Amiga computer. Now anyone using Windows has introduced this concept into their software. It allows the user to "pick and click" their commands. Note Fig. 1-2d. Since version 9 of AutoCAD, dialogue boxes have played a much greater role in selecting options and establishing settings within AutoCAD. In fact, in version 12, the Main Menu has been completely replaced with several dialogue boxes. We will examine these in greater detail in Chapter 3.

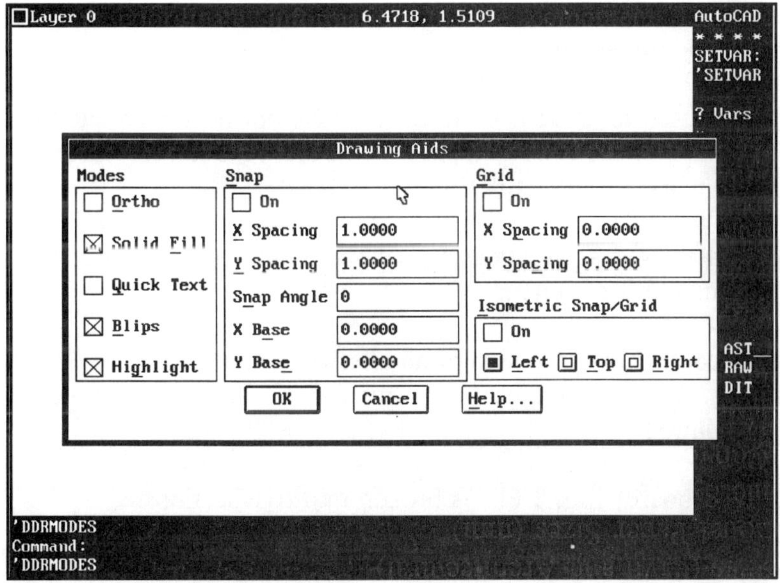

Figure 1-2d The Dialogue Box, a pick and click environment.

Keyboard

There are several styles of keyboards available for the IBM series of computers. Figure 1-3a shows the touch-typing keyboard, and Figure 1-3b the enhanced keyboard (101 type) which now comes as the standard with most computer systems. There are three basic areas to the touch-typing keyboard:

1. Function keys (10)
2. Touch-typing keypad (QWERTY style)
3. Numeric keypad/cursor control keys

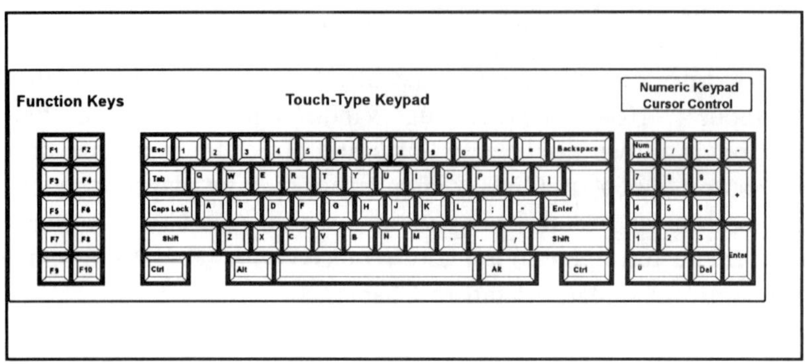

Figure 1-3a Keyboard (Standard Style).

The ten **function keys** are used to control various functions on the screen such as toggling the SNAP function ON or OFF. Appendix B provides a function key overlay to be used as an aid when working with AutoCAD.

The **touch-typing keypad** is used as an alternative input to the mouse or graphics tablet when replying to AutoCAD commands.

The **numeric keys/cursor control keys** are used as an alternative control to move the cursor around the screen. This method is slow, but can be used when precise resolution of a point is required.

The enhanced 101-style keyboard, Fig. 1-3b, is broken into five sections:

1. Function Keys (12)
2. Touch-Typing Keys (QWERTY style)
3. Independent Cursor Control Keys
4. Numeric Keypad
5. Document Edit Keys

Function Keys

Of great importance to AutoCAD are the Function Keys. Six of these keys provide instant access to basic AutoCAD commands and act in a "toggle" mode. That is, press the key once and they're activated; press them a second time and they're deactivated. The special Function Keys are:

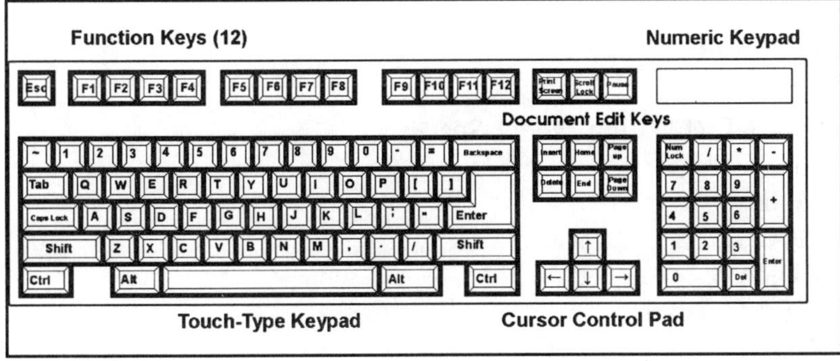

Figure 1-3b The enhanced 101-style keyboard.

F1 - Flip screen from graphics mode to text mode (toggle).
F6 - Coordinate Display On - Off (toggle).
F7 - Grid On - Off (toggle).
F8 - Ortho mode On - Off (toggle).
F9 - Snap mode On - Off (toggle).
F10 - Tablet mode On -Off (toggle).

Note Appendix B for a template which can be placed over the Function Keys for quick reference purpose.

Backspace Key. If you make a typing error, you can use the backspace key to backup to your error and then retype your command.

Hot-Keys

To speed up the selection of frequently used commands, the ACAD.PGP file (code-named the *pigpen*) allows hot-keys or Alias keys to instantly select these commands. For Example; to select the Zoom command, simply press the Z key. By editing the ACAD.PGP file, you can add or delete your own hot-key commands.

The following is a list of Alias hot-keys and the command they represent:

Table 1-1 Hot-Key, Alias Commands

Hot Key	Alias Command
A	ARC
C	CIRCLE
CP	COPY
DV	DVIEW (3D Dynamic View)
E	ERASE

L	LINE
LA	LAYER
LT	LINETYPE
M	MOVE
MS	MSPACE (Floating Model Space)
P	PAN
PL	PLINE
PS	PSPACE (Paper Space)
R	REDRAW
T	MTEXT (Text)
U	UNDO
Z	ZOOM

Input and Pointing Devices (Mice & Tablet)

AutoCAD supports many different input pointing devices such as a mouse, tablet or puck. With the introduction of Microsoft's Windows, the mouse has quickly replaced the keyboard. No matter which input device your AutoCAD system uses, each do the same basic job.

The mouse is used by sliding or dragging it over a surface where the X and Y movements are translated into electrical signals. The computer interprets this information and moves the graphics cursor to a corresponding location.

TIP: If you are a first time user, you will probably find a mouse awkward to use, as your eye-hand coordination has not yet been developed. Don't worry, I felt the same way you feel now. To overcome this, play with the mouse for a few minutes and you will quickly be up to speed with the rest of your group.

Your mouse may have two or more buttons. The left button (pick button) is used in two ways:

(A) When selecting a screen menu item, the mouse is dragged so the graphics cursor moves over the desired screen menu item and then pressed once to pick the item.

(B) To select a pull-down menu item, the mouse is dragged over the desired Menu Bar tool. The left mouse button is then "held down" which exposes the various options for that tool. With the left button continuously held down, the mouse is now dragged to the desired pull-down function. Releasing the button will signal the computer that you have made your selection.

Pressing the right mouse button is similar to pressing the "Return or Enter" key on the keyboard. The center mouse button may or may not be used, depending upon the system configuration. In release 12 of AutoCAD, the center mouse button sometimes allows pull-down menus to "float" to different locations on the screen. It can also be setup to act as a "double-click" function.

The graphics tablet is used in conjunction with a pen (stylus) or a puck. Pressing down on the pen activates a switch inside the pen and signals the computer that you have made a selection. Positioning the pen at various locations on the tablet sends X and Y coordinate information to the computer. This action translates into corresponding movement of the graphics cursor.

COMMAND TERMINATORS and CHOICE SELECTION

This tutorial directs you to use the input device to "Select" or "Pick" some of your commands. When you have made your selection, indicate your choice by pressing the input device selection button, that is, the left button on the mouse.

Most commands can be terminated and/or repeated by using the Spacebar or the Return/Enter key. The exception to using the spacebar is in the text mode, when a space is a legal text character. In this case terminate the command with the Return key.

When to "Type" (Type/Choose) and When to "Select" (Pick/Select)

Throughout this tutorial, you are asked to respond to the Command Line or a Dialogue Box by either typing (Typing/Choosing) a reply from the keyboard or by picking a selection (Pick/Select) via the input device. When asked for a typed reply, your text response is shown in **<u>BOLD UNDERLINED CAPITAL LETTERS</u>**. When asked to select or pick, select the item highlighted in **BOLD CAPITAL LETTERS** that are not underlined. Follow the text and syntax as closely as possible to assure correct response as we progress through the tutorial.

The Ellipsis (...)

Various AutoCAD variables, pull-down menu or commands maybe followed by an ellipsis (...), three periods. This signifies that there is a submenu dialogue Box or option associated with this item. Dragging the mouse over the ellipsis will display the submenu or options available.

Default Values

Many AutoCAD commands incorporate default values. The default data appears inside angle brackets "< >", and is usually the result of some previous use of the command, such as the Snap spacing. When ever a default value is displayed, you can use it by pressing the spacebar or the Return key. If a new value is typed in, this becomes the current value and the default the next time the command is used.

```
Command SNAP <2.5>:
```

FLOPPY DISK DRIVES

Over the past few years, the technology around floppy disk drives has improved to the point where we now have two sizes of disk drives, the 5 1/4 and 3 1/2 inch drive, with more on the horizon.

Likewise, the improvements in the quality of iron oxide used as medium material for the diskette has allowed us to store larger amounts of data on the disk. The result of these advancements is that each disk drive now has two storage capacities, low capacity and high capacity. Confused? Let's try to sort this out.

The 5 1/4 inch, "low capacity" disk stores 360K bytes of data and the "high capacity/density" version stores 1.2M bytes on a disk. The 3 1/2 inch "low capacity" disk stores 720K bytes and 1.44M bytes in its "high capacity" version.

TIP: With different capacity diskettes, knowing which disk is which is often quite confusing. Here is one method to remember. On the 5 1/4 inch disk, the low capacity (360K) type disk always has a "reinforcing ring" around the center hole, while the high capacity disk has the ring removed. Brand name diskettes also have the label marked as High Capacity or High Density (HD).

On the 3 1/2 inch disk, look for the "HD" or High Density label to identify the 1.44M disks. The low capacity disks (720K) do not carry this label.

Warning: Do not try to Format a Low Capacity disk as High Capacity. The lower quality iron oxide of the low capacity disk medium makes the reliability of the disk very unstable. High capacity disks may be formatted as low capacity, but remember, you lose storage capacity and pay more for the privilege. Once a high capacity disk is formatted, do not write to it using a low capacity disk drive. This will physically destroy the diskette from ever being read as a high capacity disk. This is because the low capacity track width is much wider than the high capacity track width.

<u>TIP:</u> To be able to reuse high capacity disks formatted or written to on low capacity drives, you must completely destroy the formatting and data on the disk. This can be done by simply placing a magnet over the surface of the disk and with a spiral motion rotate the magnet, assuring all areas are covered. (Who says magnets are not useful?)

There is a light at the end of the disk drive tunnel. The new "smart" drives can recognize if a disk has been formatted high or low capacity and will automatically keep this in mind when writing to it. Just make sure your system is using smart drives.

HOUSEKEEPING

As you learn more about the technique of laying out a drawing, you will find that many procedures are repeated. These procedures may establish a Layer change, set the Grid and Snap spacing, or Zoom to a new area of the drawing to work on. Instead of repeating all of the detailed command structure for each new exercise, you will be instructed to do some "Housekeeping". This is the term used when you are to call upon your past experience and knowledge to set up the drawing to a certain point, so we can continue with the exercise at hand.

KEEPING UP TO DATE

As each exercise in this tutorial lays the foundation for future exercises, it is best to work your way from start to finish of this text. If you jump around you may miss an important section critical for completing the next exercise.

Notes:

chapter 2

Getting to Know the Computer Environment

DISCUSSION

Before learning to use AutoCAD, one should become familiar with the computer, its operating system, and its basic peripherals. It is not my intention to make computer operators out of you, but to show you some of the computer's basic features as well as some of the faults you may encounter. In this Chapter we will explore the DOS operating system as well as Windows. Understanding these will help you overcome many of the simple problems you may find while running AutoCAD.

The techniques developed here may save you much time and grief, particularly if you have spent many hours on a drawing and then find the computer refuses to save or load your work for one reason or another.

SAVE YOUR WORK

One word cannot be stressed often enough: ***save, save, save.*** When developing a drawing, you should save your work at regular intervals to avoid catastrophes. In version 12 of AutoCAD the Savetime command allows you to set a time period at which AutoCAD will automatically save your drawing. Saving often will allow you to avoid losing many hours of hard work. From time to time, your drawing should also be saved under a different drawing version number. This will allow you to follow its progress. To provide additional disk space, old versions can be Erased from your disk when no longer required. Note Chapter 3 and the Erase command.

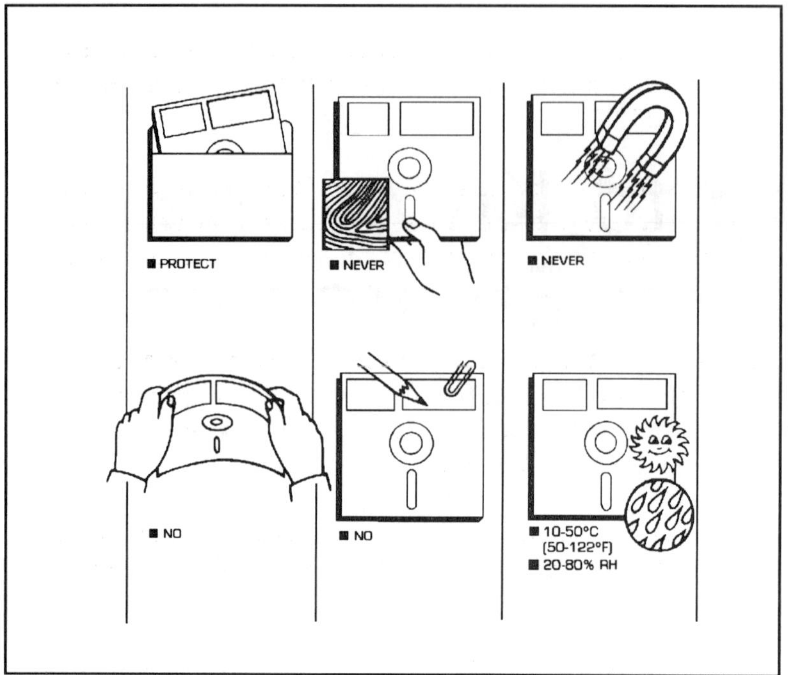

Figure 2-1 Proper care of diskette ensures an extended disk life.

FLOPPY DISKS

There are two physical sizes of floppy disks: the 5 1/4-inch disk and the 3 1/2-inch disk which is much more rugged. Both types of floppy disks use a magnetic medium to store the data. For this reason, a number of precautions should be observed when handling the disk. They are:

* Bringing the disk too close to a magnetic field will destroy the data stored on it. This means keeping your disk away from telephones, magnetic screwdrivers, the front of your CRT, and even the tablet you use as an input device with AutoCAD. Note Fig. 2-1 for disk care.

* Your disk should be treated with much care. Do not handle it by any of the exposed magnetic areas as body oils, cosmetics, and dust can contaminate it.

* Place your disk back in its protective cover when you have finished using it.

* Moisture can destroy not only your disk but also your computer. This means keeping liquids such as coffee or soft drinks away from your computer system. An accidental spill of coffee on your keyboard or disk is guaranteed

to destroy it.

* Temperature has a detrimental effect on the magnetic medium of the disk. Keep the diskette within the manufacturer's recommended temperature range of 10° to 50°C (50° to 122°F).

* Be careful when writing labels for your disk. Write the label before you stick it to the diskette jacket because pressing too hard with a pen or pencil may cause a slight indentation in the underlying magnetic medium, thus destroying it.

* When inserting the disk into the drive, do so carefully so that it does not bend in the process. Disks are flexible, to some extent, but won't last long under rough treatment. On the 5 1/4-inch disk, the label should be facing up, with the write-protect notch on the left side.

* On a periodic basis, copy your complete work disk to a blank formatted disk. This backup is your insurance policy should your work disk accidentally become corrupted.

* Never remove the disk from the drive while the red LED drive access light is on.

TIP: Remember these points as they are part of your quiz at the end of this Chapter.

WRITE-PROTECT NOTCH/TAB

Each disk has a built-in mechanical protection that can be activated to prevent the computer from writing on the disk. This device is called a *write-protect notch or tab*.

On the 5 1/4-inch disk, the write-protect notch is on the side of the disk. See Fig. 2-2a. When the notch is open, the disk may be written to. When the notch is covered, the disk is in the write-protect mode. Covering the notch prevents a beam of light from passing through this area indicating the write-protect is active. Therefore, don't use transparent write-protect tabs.

"What are write-protect tabs and where do I find them?" These are those sticky little black rectangular paper tabs that come with your disk labels. Actually anything will do as a write-protect tab, such a piece of masking tape, just so it covers the slot.

Remember: On the 5 1/4-inch disks, when the write-protect notch is left uncovered, the disk may be written to. With the stick-on write-protect tab covering the notch, it cannot be written to. These tabs as well as the labels normally come with the disks when you purchase them. Now, ... where did I put those little tabs?

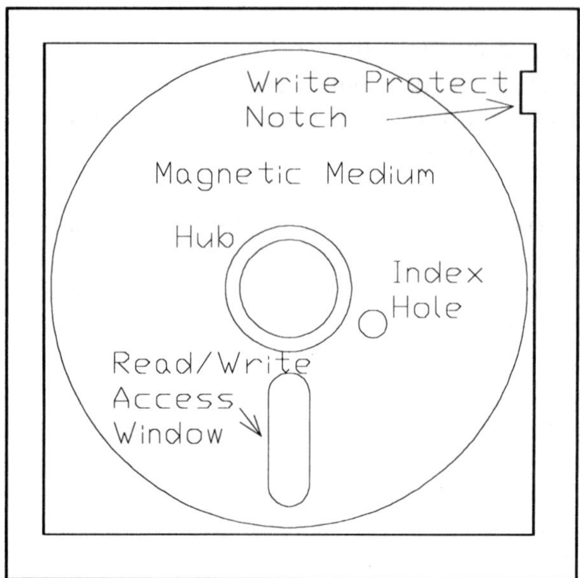

Figure 2-2a The 5 1/4-inch computer disk.

Now, let's look at the 3 1/2-inch disk. See Fig 2-2b. The construction of the 3 1/2-inch floppy is much more rugged than its big brother, the very flexible 5 1/4-inch floppy. Note that the 3 1/2-inch disk has no exposed surfaces to get your fingers on, but it is still subject to the same magnetic and temperature problems as the 5 1/4-inch floppy.

One of the great improvements of the 3 1/2-inch disk is in the write-protect tab. When using the 5 1/4-inch floppy, nine times out of ten, when you need a write-protect tab, you can't find one. On the 3 1/2-inch floppy, the tab is built in. It is that little slide tab mounted in the bottom right corner of the disk in Fig. 2-2b. Now that's handy!

Take note, on the 3 1/2-inch disk, the operation of the tab is exactly opposite to that of the 5 1/4-inch disk. That is, when the tab is open, exposing the slot, the disk is in the "write-protect" mode. When the tab is closed, not allowing light to pass through the slot, the disk can be written to. To close the slot, simply slide the tab into place.

TIP: Remember that the operation of the "write-protect" notch and tab is opposite between the 5 1/4 and 3 1/2 inch floppy disks.

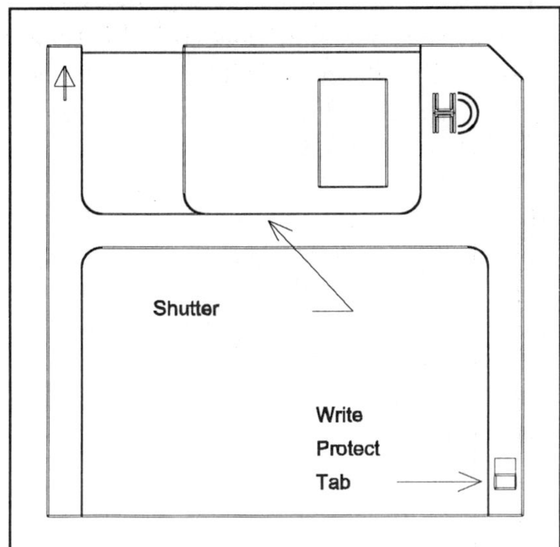

Figure 2-2b The 3 1/2 inch computer disk.

THE OPERATING SYSTEM
(How Does My Dumb Computer Get So Smart?)

Discussion

Computers, when first powered up, are usually quite dumb. That is, they only know limited functions such as how to look for disk drives or how to put a limited amount of data on the display. This is because the small amount of software resident in the computer at power up is quite limited in its scope and contains only enough instructions to "boot up" the system. This boot-up software is always stored in ROM (Read Only Memory). To improve the computer's intelligence, you must load more sophisticated software into RAM (Random Access Memory). Doing this allows the computer to become much smarter, but only in the field for which the program was written.

The control of the "behind the scenes activity" of the computer is maintained by software called the "operating system." This software controls disk drive functions, how data and graphics are displayed on the monitor, when the keyboard gets read, and the thousands of other internal functions not seen by you, the end user. In order to have control, the operating system must always be the first program to be loaded into the computer's memory.

In the early days of computers, the complete *Operating System* was small and could reside entirely in the computer's permanent memory area called ROM. As computers grew in complexity, and the operation software grew in functions, the space in ROM was quickly outgrown. The operating system thus needed to find a new home, and finally took up residence on a floppy disk. Thus, the *Operating System* became known as a *Disk Operating System* or DOS as it is now known.

Many new users to a computer feel that if they do the wrong thing, they will break it. Not true! The worst that you can do to a computer, under normal use, is to erase all the programs, or reformat the hard disk. Well, that's bad enough so be careful when you use the Erase or Format commands in DOS. By the way, should this occur, you will not be the first to accidentally do it, and I'm sure you won't be the last. If this happens, just call your local computer dealer and tell them your kids reformatted your hard drive to save you the embarrassment. You can also use two new DOS commands called Undelete and Unformat that will restore your disk back to normal.

You should remember "you" are the master of the computer, so don't become intimidated when using it. You are the boss. If you get too frustrated with what is or is not happening, you always have the option of putting it out of "your misery" by turning the power off.

WINDOWS

Microsoft introduced the concept of "picking and clicking" in an operating system called Windows, (Fig. 2-3) and with the introduction of their latest advanced operating system, code-named Chicago and now called Windows 95, you will have many more pick and click options to simplify your computing activities. We will discuss Windows in more detail near the end of this chapter.

SYNTAX AND GIGO (Garbage In = Garbage Out)

A computer, because it is a semi-intelligent machine, must be told in very precise terms what you want it to do. This requires care when typing in commands. The syntax or format of the instruction must be correct. If spaces are required in the statement, they should be there, or the computer will become confused and may take the wrong action. When this happens, it may simply just sit there saying "syntax error." If you don't want to get garbage out, you can't put garbage in. Remember; the computer is a perfect servant: it does everything you tell it, and nothing more.

A syntax error occurs when you have asked the computer to do something but have not followed the prescribed structure. This may be due to a simple spelling mistake in the command, or because the wrong command format was used. When this occurs, a syntax error message pops up on your display. Check your spelling or DOS command. If everything looks correct, then refer to the DOS reference manual that came with your computer and note how the command should be implemented.

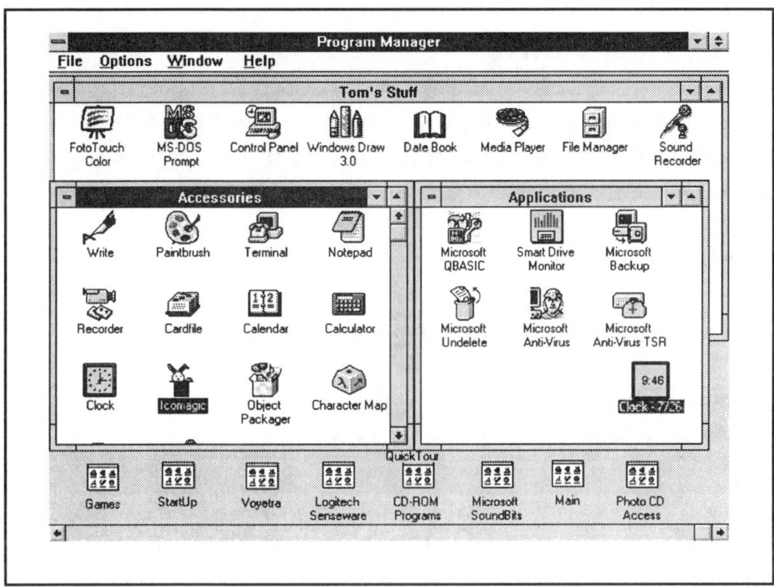

Figure 2-3 A typical Windows display at start-up.

THE DOS DISK

All commands and housekeeping routines of the operating system are found on the DOS disk. (What was DOS short for? That's right, the Disk Operating System.) These routines allow you to load and save programs, format disks, and copy and erase files, to name a few.

The DOS disk contains many useful utilities. We will briefly discuss a few of these in this chapter. For a more comprehensive understanding of all DOS commands, refer to the DOS manual that comes with your computer.

DOS and other programs can be loaded into the computer's RAM memory from a floppy disk or from the hard drive. We will first look at loading these programs from a floppy.

Floppy Disk Drive Labels

Software programs are loaded from the diskette into an area of the computer's memory called RAM (Random Access Memory). The computer reads the software using a floppy drive which is identified by a letter. The main floppy drive, always labelled "A," is the drive that has the red LED (Light Emitting Diode) light up when the power first comes on. A second floppy drive is given the letter label of "B." Almost all computer systems today use fixed or hard drives which get the labels of "C," and possibly "D."

Loading DOS from the Floppy Drive

Note: Each time the computer is powered up, the boot-up operating system always looks at the A drive to see if DOS is on the disk in that drive. If a DOS disk (System disk) is present, the Disk Operating System loads and you are on your way.

Cold and Warm Boots

There are two ways to load DOS into the computer from the A drive:

1. Place the disk that contains the DOS utilities into drive A and turn the power on. This is referred to as a "cold start" or "cold boot."

or

2. When the power is already on, place the DOS disk into drive A and press the Ctrl, Alt, and Del keys all simultaneously. This is known as a "warm boot."

In each of the above cases, the A drive will activate, as indicated by the red LED coming on, and DOS is loaded.

DOS next requests information about the current date and time. It is good practice to ensure that the correct date and time are set, as this information is appended to each file that you save and will aid in identifying files later on. This information will appear similar to the following:

```
Current date is Tue 1-01-1980
Enter new date: 08-21-95 (RETURN)

Current time is  0:00:45,76
Enter new time: 09:15:40 (RETURN)
```

You are now running the computer under a disk operating system (DOS), initiated from the A drive.

TIP: Remember, when you first power up the computer, it expects to load DOS from the disk in the A drive. If you have your work disk inserted in the drive, and the drive door closed, you will get an error message that says:

```
Non-System disk or disk error
Replace and strike any key when ready
```

This is because your work disk does not contain the DOS operating system. Should this occur, simply open the drive door, insert a disk containing DOS, and re-boot the system.

Hard Disk Drives and Loading DOS

Almost every computer that come with a hard drive has DOS installed on them, and are usually configured to automatically power up DOS when the computer is cold or warm booted. This saves having to install DOS using the floppy drive.

TRAP A note of caution when working with a computer with DOS installed on the hard drive: Remember that the start-up boot program in ROM will always look at the "A" drive initially to see if a DOS disk (System disk) exists there. If no disk is in the A drive, the boot-up program then looks at the "C" drive. If you have your work disk in drive A during this activity, you will get the "Non-Systems Disk or disk error" message because your work disk does *not* contain DOS. To overcome this, just remove your work disk and re-boot the computer by pressing the CTRL-ALT-DEL keys. The boot-up program will eventually find DOS on the hard drive and you are on your way.

Checking Programs in DOS

To see what programs are available in DOS, switch drives to the one containing DOS. If DOS is installed on the hard drive, you may need to get into the DOS subdirectory in order to see these programs.

Switching drives is done by simply typing the drive letter you want to switch to, followed by a colon (:) and then pressing the Enter or Return key. For example: If you are at the A prompt drive and want to switch to the C prompt drive, type:

```
A:\C: (RETURN)
```

Your new prompt will look similar to this:

```
C:\
```

Now to examine the programs or directory listing of that drive, type **DIR** after the prompt.

```
C:\DIR (RETURN)
```

At this point, a long list of programs similar to Table 2-1 will be displayed on the screen. One of these programs may be called FORMAT, the utility used in preparing the disk for use with AutoCAD. Did you find it?

Table 2-1 displays each program or file name, its three character extension name, the number of bytes the file contains, and the date and time the file was last saved. This is where having the correct date and time established when you first log onto the computer becomes important. At the end of the list, DOS displays the total number of files found and the amount of free bytes still remaining on the disk.

TABLE 2-1
Disk Directory

```
Volume in drive A has no label.
Directory of A:\

     COMMAND  COM  17792  1-30-84  12:00p
     ANSI     SYS   1664  1-30-84  12:00p
     FORMAT   COM   6912  1-30-84  12:00p
     CHKDSK   COM   6400  1-30-84  12:00p
     SYS      COM   1680  1-30-84  12:00p
     DISKCOPY COM   2576  1-30-84  12:00p
        .
        .
        .
     KEYBPS   COM   1596  1-30-84  12:00p
     KBPGM    COM   3364  1-30-84  12:00p

         31 FILE(S)  12288 bytes free
```

File Names and Extensions

In the DOS system, the file name is limited to a maximum of eight alphanumeric characters. If you use more, they are truncated or cut off. The extension to the file name is limited to three alphanumeric characters. When typing a file name and its extension, a period (.) is used to separate the two. The extension name usually reflects the type of file it is. For example; A picture file has the .PIC extension and a drawing .DWG. In AutoCAD, all you have to establish is the file name as AutoCAD assigns the .DWG extension automatically.

Wide Screens Versus Long Screens

When you typed "DIR", your listing may have run off the screen. For a wide listing of the files, (Table 2-2) rather than a long list, type DIR/W instead of DIR. In this case, only the file names and extensions appear. They are placed across the screen, and the number of bytes and date is omitted. Try it!

C:**DIR/W** (RETURN)

TABLE 2-2
Wide Listing of Files in DOS.

```
Volume in drive C is TUM10-89
Directory of C:\DOS

ANSI      COM   DELETE    COM   REPLACE   EXE   MORE      COM
CHKDSK    COM   ERASE     COM   BASIC     COM   TREE      COM
COMMAND   COM   FORMAT    COM   PRINT     COM   COMP      COM
COPY      COM   EDLIN     COM   VDISK     SYS   XCOPY     EXE
FDISK     COM   PARK      COM   DRIVER    SYS   FIND      EXE
```

FORMATTING A DISK

Discussion

If you have not read the section on Floppy Disk Drives in Chapter 1, do so now as it is important to know the capacity of the disk you are using.

As you progress with this tutorial, you will save your drawings on a floppy disk. All new disks are like a blank page and in order to write neatly on a blank page you should draw lines across it to act as a guide. Your new disk also requires guide lines, electronic guide lines, so that DOS knows where to place the data when writing to it. These guide lines are established when the disk is Formatted and are called "tracks" and "sectors". The DOS command and program used to format a disk and prepare it for use is **FORMAT**. FORMAT analyzes the disk for defective areas and keeps track of these in a special area on the disk called the FAT or File Allocation Table for future use.

A useful option available with the Format command is the /S option which leaves space on the disk to write the DOS system files at some time in the future. Leaving the option out provides more work space on your drawings.

Formatting a New Disk

In the following procedure, I will assume you have DOS installed on your hard drive. Place your unformatted disk in drive A. To retain as much space as possible on the disk, we will use the Format command without the /S option.

TIP: Failure to include the "**A**" in the following command will result in an attempt to format the hard disk C! This you don't want to do. Therefore, carefully type:

```
C:\ FORMAT A: (RETURN)
```

DOS responds with

```
Insert new diskette for drive A:
and strike ENTER key when ready
```

Since you have inserted your new disk in A drive, press the Enter key. The screen will display:

```
Formatting...
```

At the end of the formatting process, DOS may return with:

```
Formatting...Format complete

360256 bytes total disk space
  1024 bytes in bad sectors
359232 bytes available on disk

Format another (Y/N)?
```

Notice that 1024 sectors were detected as bad in our example. If no sectors were found defective, which is usually the case, then this line is not displayed. When errors are encountered during the formatting process, sometimes reformatting it a second time may clean up the problem.

Another disk may be formatted by typing Y (Yes) to the format question. When you have finished using Format, type N (No) and press the Return key to get back to the system prompt C:\.

To format different disk types and capacities on a smart drive, use the following DOS commands:

"Format A:", on a 360K drive will format a 360K disk.
"Format A:", on a 1.2M drive will format a 1.2M disk.
"Format A:/f:360", on a 1.2M drive will format a 360K disk.
"Format A:", on a 1.44M drive will format a 1.44M disk.
"Format A:/f:720", on a 1.44M drive will format a 720K disk.
"Format A:/s/u", will make your disk bootable by installing the DOS System on it and will format it unconditionally; that is, everything is overwritten.

Formatting an Old Disk

In addition to preparing new blank disks, Format may be used to reformat old disks for reuse. But a word of caution:

TIP: ***Always double-check the old disk to be sure that it does not contain useful files before reformatting it.***

When you are sure your old disk does not contain any useful files, only then do the reformat. Once the format process has begun, any data that existed on it becomes altered. In early versions of DOS, before version 5, the files were actually erased and the data is lost. It is therefore better to examine your old files and back them up before you reformat.

Since version 5 of DOS, the "Unformat" command is used to re-establish old file information on reformatted disks. This procedure must be used before any new data or files are written to the disk. Keep this command in mind, because sooner or later you too will make the mistake of formatting a disk with useful files. When this happens, just smile because that's Murphy's Law catching up with you. It has happened to me several times and the Unformat command saved the day.

TRAP With the introduction of the Unformat command, a security problem can occur with reformatted disks. Anyone can unformat a disk and recover your old data and files.

COPY COMMANDS

Discussion

As we progress with this tutorial, we will need to copy files from one disk to another. This is done when making a complete backup of our work disk or transferring the Electronic Symbol Template to our work disk. We use the DOS commands of Copy and Diskcopy to do this. These commands have several ways to implement their function, and we'll look at those special features most suited for our application. For an in-depth description of DOS commands, consult your DOS manual.

COPY Using A Two Drive System

COPY is designed to transfer a single file or multiple files from one disk to another "formatted" disk. It uses the syntax:

```
    COMMAND     FROM              TO

C:\COPY A:ELECTEMP.DWG B:ELECTEMP.DWG (RETURN)
```

In this example, the active drive is C. We will copy from drive A the file ELECTEMP.DWG to drive B and give it the same name.

Some points to note about the syntax:

1. There must be a space after the command **COPY**.
2. The source drive **(A):** and file name **(ELECTEMP.DWG)** are given first.
3. There is no space between **A:** and **ELECTEMP.DWG**.
4. There must be a space before the "**B**" in drive B to separate it from the source file name.
5. You must include the extension **.DWG** as it is part of the file name and is required for AutoCAD to recognize it as a drawing file.
6. Don't forget to press the Return key after every command to indicate you have completed the command.

If your default or active drive prompt is already A, then the source drive designator may be left out, such as:

```
A:\COPY ELECTEMP.DWG B:ELECTEMP.DWG (RETURN)
```

COPY allows you to copy a single file or the complete disk to a "formatted" destination disk. To copy the complete disk, use the "wild card" character of star (*) to indicate that all files and extensions are to be copied as in the following example.

```
A:\COPY *.* B: (RETURN)
```

COPY Using A Single Drive System

There are two methods of making a copy of your disk if you have only a single floppy drive system.

Method 1: Using The Hard Drive As Temporary Storage

In this method, you will use the hard drive as a temporary memory location to hold the files from the source disk, while switching to the destination disk. To do this, we make use of a subdirectory called "TEMP," as temporary storage on the hard drive.

For this to happen, we must first *make* the subdirectory TEMP, and then *change* to it. At the C:\ prompt type:

```
C:\ MD TEMP (RETURN)
```

This makes the subdirectory TEMP.
MD is the DOS "Make Directory" command.

Now to **C**hange **D**irectory (CD) by switching to the TEMP directory, type:

```
C:\ CD TEMP (RETURN)
```

You will note the prompt now looks similar to this,

```
C:\TEMP\
```

This means you are in the subdirectory Temp.

TIP: To prevent your Source disk from accidentally be written to, place a write protect tab on it now.

Now to copy all the files from the A: drive into the subdirectory TEMP, type:

```
C:\TEMP\COPY A:*.* C: (RETURN)
```

We now have copied all the files from the source disk in the A drive to the temporary subdirectory Temp. Now to write these files back to a destination disk.

Insert a new, formatted disk into drive A, and type:

```
C:\TEMP\COPY C:*.* A: (RETURN)
```

This copies all the files in the Temp storage area of the hard drive back to your new work disk in drive A.

One last thing to do before we leave the Temp storage subdirectory is to clean it up for the next time we want to use it. Following the instructions of the Erase command, with wild cards, type:

```
C:\TEMP\ERASE *.* (RETURN)
```

To get back to the root directory, type:

```
C:\CD\ (RETURN)
```

Method 2: Using a Pseudo B Drive

Note: This procedure will only work if there is physically only one floppy drive.

What is a Pseudo B Drive? A pseudo drive is one that does not physically exist. Consider a system that has only a single floppy drive. Believe it or not, we can fool the software into thinking it has two, an A drive and a pseudo B drive. How is this possible? Through the miracle of DOS!

Yes, DOS has the built-in ability to be fooled into thinking it has two floppy drives when only one actually exists. A little known feature of DOS, developed just for occasions like this.

TIP: To protect the source disk from accidentally be written to, place a write protect tab on it before proceeding further.

With your "Source Disk" in the A drive, from the A:\ prompt type:

```
A:\ COPY *.* B: (RETURN)
```

DOS activates, reads the data in the source disk, stores it in RAM and responds with:

```
Insert diskette for drive B:
strike any key when ready.
```

Remove the source disk and insert your new formatted target disk in drive A, even though DOS thinks it's drive B.

Now press a key. DOS now copies the data stored in RAM to the pseudo B drive and responds with:

```
1 File(s) copied
```

If your source disk contains many files, you may have to swap disks five or six times before the source disk gets completely copied to the destination disk.

After the Copy is complete, the next DOS message may be:

```
Insert diskette for drive A:
and strike any key when ready.

A:\
```

You have just completed copying you source disk onto a newly formatted blank disk, using one disk drive. Individual files can also be copied using the same technique as above. Just use the file name and extension in place of the wild-card characters.

DISKCOPY (BACKING UP YOUR DISK)

In the following example we will copy the entire disk in drive A to an "unformatted" disk in drive B. Note the stress on the word unformatted because Diskcopy will simultaneously format the destination disk and copy all files across to it.

```
A:\DISKCOPY A: B: (RETURN)
```

or:

```
A:\DISKCOPY B: (RETURN)
```

If you are using a *single-drive* system, type:

```
A:\DISKCOPY A: A: (RETURN)
```

or:

```
A:\DISKCOPY (RETURN)
```

COPY VERSUS DISKCOPY

Fragmentation

"A Word of Caution When You Use Diskcopy." Disks that have been used often, and may have had many files added to or erased from them, can become *fragmented.* That is, the data of each file may be spread over the entire disk, rather than neatly packed side by side. This layout consumes time when you read or write to the disk. When you use Diskcopy, the fragmentation is duplicated on the new disk. For this reason, it is better to use the Copy command instead, because Copy removes the fragmentation and neatly packs the data in each file, side by side.

CHECK DISK (CHKDSK) "A QUICK CHECKUP"

Discussion

To monitor the status of your work disk, you can use the DOS CHKDSK command. CHKDSK analyzes the disk for the total bytes of disk space, how many bytes are in hidden files, how many bytes are in user files, and how many bytes are available on the disk. It also indicates how many bytes are in the computer's total memory and how many bytes are free. One special feature of CHKDSK is that it looks for "lost files or clusters". These are files that have become faulty by having the link between the various sectors of the file being corrupted. In this next section, we will look at how to clean up these corrupted files.

Using CHKDSK

To use CHKDSK, type:

```
A:\CHKDSK (RETURN)
```

DOS responds by displaying something similar to:

```
362496 bytes total disk space
     0 bytes in 2 hidden files
278528 bytes in 16 user files
 83968 bytes available on disk
278426 bytes total memory
 89538 bytes free
```

Making Repairs with CHKDSK (Check Disk)

CHKDSK is used to check your disk for errors. It also has some repair capabilities should your disk become corrupted. There are many reasons for the disk to accumulate faults; some of them are the result of not exiting AutoCAD in the correct way, thus leaving open files on the disk. Some faults may be your error, some AutoCAD's, and some the computer's!

TIP: You should note that AutoCAD plays tricks with your disk by writing files to it that you may not be aware of. If these files remain on the disk, they create errors. They may remain there as the result of a power failure or a DISK FULL error, or you may accidentally retain them if you remove your disk from the drive before ending your session with AutoCAD. If you exit AutoCAD the correct way, through the Main Menu, or by using the END command, these files are removed automatically. Files that remain may show up as "open clusters" when CHKDSK is run. To correct this fault, we use the fix parameter "/F" when initiating the CHKDSK command.

For example; To invoke this command on a corrupted disk in the A drive, type:

```
A:\ CHKDSK /F (RETURN)
```

CHKDSK will go through a routine similar to that previously shown and will highlight the number of clusters in error. It then asks if you want the lost chains converted to files (Y/N)? If yes is replied, CHKDSK displays the number of bytes and new files made.

```
6498 bytes in 2 recovered files.
```

To observe the recovered files, do a directory listing (DIR) of the disk. You should see your old files and something similar to the following.

```
A:\ DIR (RETURN)

FILE0000 CHK    1358
FILE0001 CHK    5140
FILE0002 CHK    2660
```

The "FILE0000.CHK" are file names CHKDSK assigned to the recovered files.

TIP: CHK type files may not be useful, even though they were recovered. Programs such as Norton Utilities may help you thread the file back together, but it is a difficult task. It is like putting Humpty Dumpty back together again. AutoCAD provides a service to do this should a drawing file become corrupt. In the end, you usually wind up Erasing them to give you more disk space. This is where a back up of these files pay off. CHKDSK has many more features that you might want to investigate through your DOS manual.

ERASE AND DELETE (DEL) COMMANDS

The ERASE and DELete commands do exactly the same thing. To erase the file FILE0000.CHK, type ERASE or DEL, followed by the file name and extension.

```
A:\ ERASE FILE0000.CHK (RETURN) or A:\DEL FILE0000.CHK (RETURN)
```

To erase all CHKDSK recovered files, type:

```
A:\ ERASE *.CHK (RETURN) or A:\ DEL *.CHK (RETURN)
```

TRAP To erase a file is very serious business. *Think twice before erasing a file.* Ask yourself, "Do I really want to erase this file?" To recover from a mistake, use the "Undelete" command.

PLAYING WITH WINDOWS

For those of you who now have Windows software installed as the operating system on your computer, we will now use it to accomplish the above functions.

To get windows started, at the system prompt C:\ type:

```
C:\ WIN (RETURN)
```

When using the Windows operating system, you use a mouse and work in rectangular areas on the screen called "windows". The applications you work with, such as AutoCAD, are represented in windows by "icons", which are small graphic symbols. See Fig.2-4. It should be noted that all the DOS operations just discussed can be run through Windows by double-clicking on the MS-DOS icon. This is because Windows is an extension of DOS.

To start any application program such as AutoCAD, simply double-click the application icon.

For purposes of copying files, making backups, and erasing data, we will use the application program called "Program Manager" which is found in the application window MAIN. To activate the MAIN window move the cursor over the icon that has MAIN written under it and double-click the left mouse button. This opens the MAIN Window, see Fig. 2-5, which contains program-item icons and the icon called File Manager.

```
Double click on the Application Icon "MAIN".
```

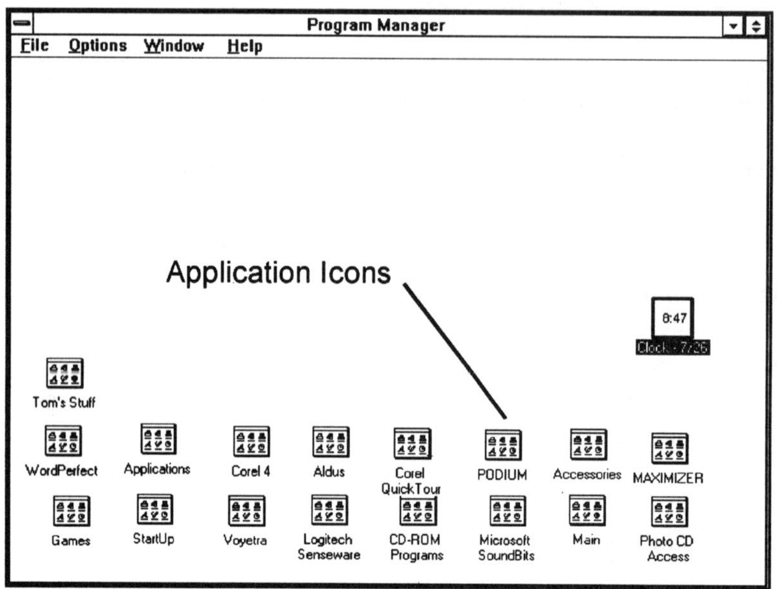

Figure 2-4 Windows Program Manager and associated application icons.

FILE MANAGER AND DIRECTORIES

When you double-click on the File Manager icon, the File Manager window opens. This allows you to view the various directories on the disk. In Fig. 2-6, the C:\ drive is shown, which is indicated with the small box around the C:\ disk drive icon. On the left side of the C:\ window are the various files, programs, directories and subdirectories on the C:\ drive. On the right side are the contents of the selected directory which may contain more files, programs and subdirectories.

To see the contents of other directories, just click on the desired directory icon.

As you have seen, across the top of the application window is a series of disk drive icons. To see the contents (Directories) of these devices, simply click on the disk drive icon and it becomes the current drive, with its contents displayed in the window.

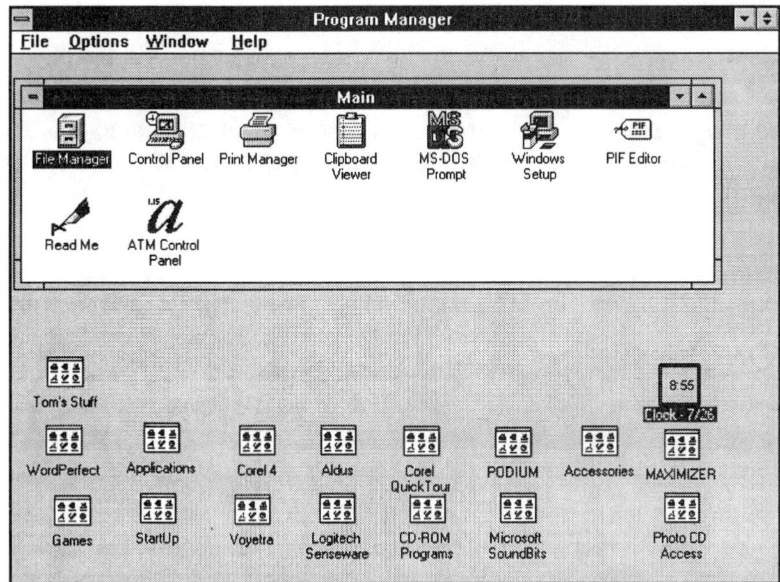

Figure 2-5 Window MAIN and the program-item icon File Manager.

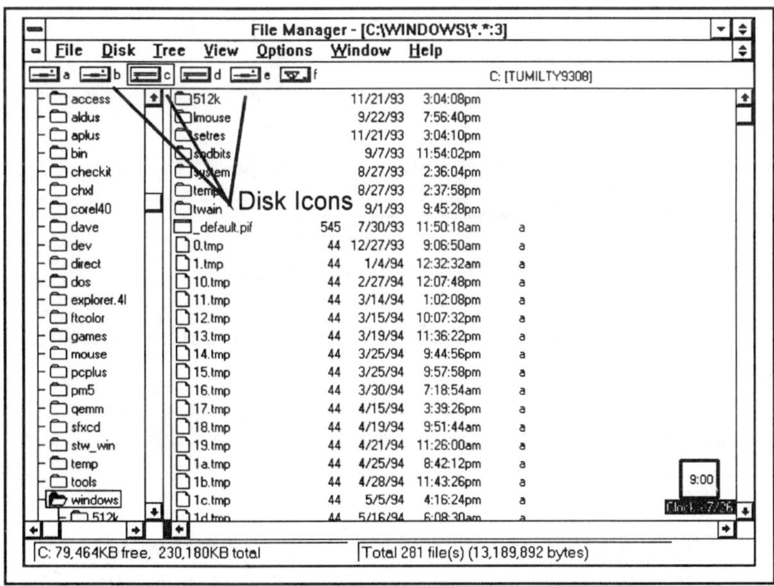

Figure 2-6 File Manager Window.

FORMATTING A DISK USING WINDOWS

Across the top of the File Manager window are the headings of File, Disk, Tree, View, Options, Window and Help. These are pull-down menus that allow you access to more commands and programs. One such program is the Format command under the Disk menu, Fig. 2-7. We will use this command to format one of your work disks by taking the following steps:

1. Insert a disk in drive A.
2. Select the Disk menu and choose "Format Disk."

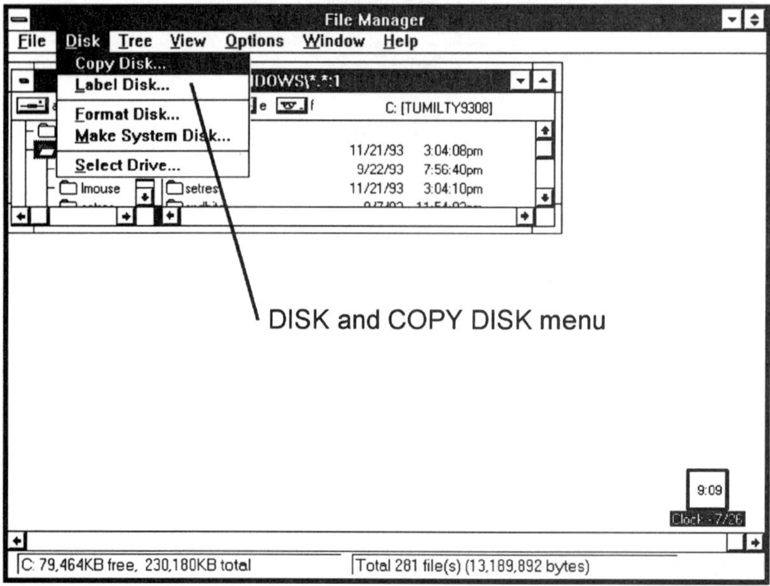

Figure 2-7 File Manager Disk menu showing Format Disk option.

This brings up the Format Disk dialogue box which allows you to change the format set up parameters. See Fig. 2-8.

TIP: Double-check the drive setting as it may be different from the drive icon selected.

3. To select the drive where the format will take place, click on the down arrow at the "Disk In" box. This allows you to choose the A or B drive. Likewise, the down arrow at the "Capacity Box" will allow you to choose either high capacity (1.44MB) or low capacity (720K) disk formats. Note that the system is smart enough to know what type of disk drive (3 1/2 or 5 1/4) you are working with, and will display the correct density combinations for you to choose from. You also have the options to add a label (11 characters total), add the necessary DOS files to make the disk a "bootable" or systems disk, or to simply do a Quick Format.

4. When you have chosen the correct parameters, click on the OK button.

This brings up a second dialogue box, the "Confirm Format Disk" dialogue box, which advises you that if you continue, all data will be erased from your disk. This is just another check to make sure you know what you are doing.

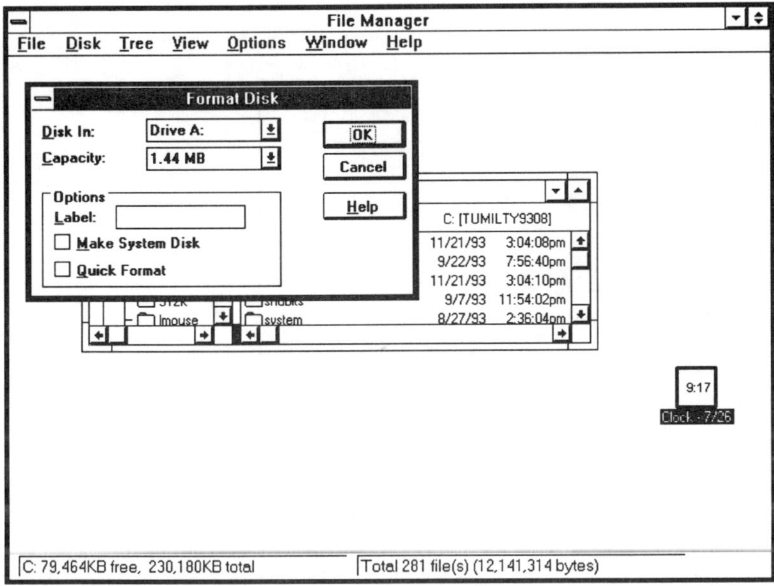

Figure 2-8 Windows "formatting parameters" dialogue box.

5. To this last request, answer Yes by clicking on the YES button, which starts the formatting process.

Windows now displays a third dialogue box (Formatting Disk) that tells you formatting has begun, and the percentage of formatting completed. You are also presented with a "Cancel" button should you decide to stop formatting, and a Label Edit Box.

When formatting is complete, you are asked to add a Label (volume label) to the disk for identification when a directory listing is requested. It's good to have a digital label built into your disk for identification purposes.

Next, another dialogue box (Fig. 2-9, Format Complete) appears which shows the total disk space and the bytes available on the disk.

6. At this time, you have the option to format another disk, and you choose "Yes" or "No".

Before version 5 of DOS, once you formatted a disk, all was lost. From version 5 on, an "Unformat" command was built into the operating system allowing you to recover from a wrongfully formatted disk.

7. To get back to the Program Manager and the application icons, select the "Control-Menu" box in the upper left corner of the window and choose "Close".

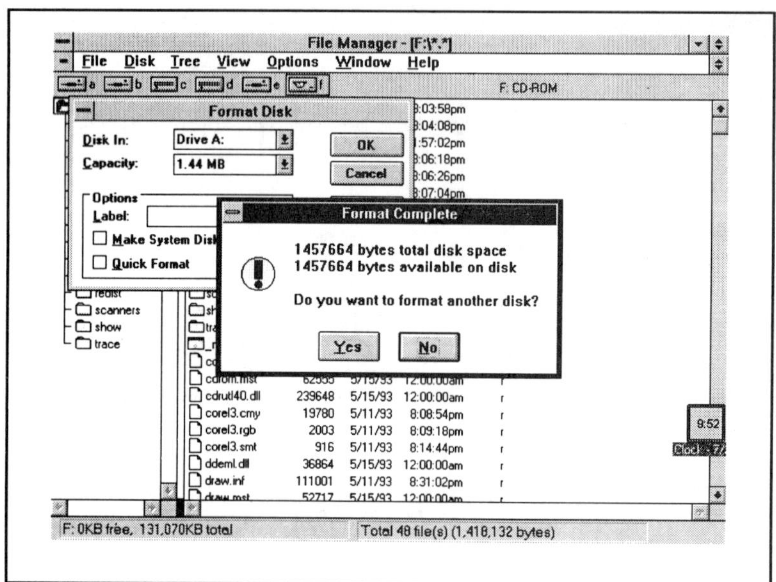

Figure 2-9 Format Complete dialogue box.

COPYING FILES

Discussion - Drag and Drop

The File Manager has a feature called "drag and drop" that will allow you to copy a file in the active drive to any other drive. We will use this feature to copy the Electemp.dwg file on the disk that came with this book to your work disk. For this exercise, I will assume you have a two-floppy drive system (A and B drives) and a hard drive labelled C. The floppy drives could be a combination of 3 1/2- and 5 1/4-inch drives.

COPY USING DRAG AND DROP

Discussion

We will copy a file from the right side of the active window, which shows the contents of the B:\ drive, to the A:\ drive which should contain your work disk. The file to be copied is ELECTEMP.DWG. To do this:

1. Insert the template disk that comes with this book into the B drive, and your work disk into the A drive.

2. From the Program Manager window, double-click the MAIN application Icon.

3. Now double-click the File Manager.

4. Double-click the B drive icon located at the top of the window.

This allows the contents of the B drive to be displayed in two windows. In the left widow is the root directory icon for the B drive, and in the right window are the three files that come with the disk, the ELECTEMP.DWG, ELECOMP.DWG and the HOUSE1.DWG. *Now to do the copy.*

5. Scan down the contents of the right window until you see ELECTEMP.DWG file.

6. Position the mouse cursor over the ELECTEMP.DWG file and press and hold the left mouse button firmly. The file now becomes highlighted and as you drag the mouse, a small file icon appears at the end of the cursor.

7. With the left mouse button held down, drag the file icon over the "A" drive icon.

8. Now, release the left mouse button to drop the file at the "A" drive and start the Copy process in motion.

A "Confirm Mouse Operation" dialogue box appears. This is to verify that the last action was correct. It's giving you a second chance to change your mind.

9. Choose the "Yes" button. We want to continue.

Now the "Copying" Dialogue box appears. See Fig. 2-10. This shows the file being copied and the "Copying From - To" disk drives associated with the copy. A "Cancel" button is present if you want to stop copying.

When the copying is complete, click on the Control-Menu box in the upper left corner of the window, and choose "Close".

TIP: *Updating The Directory Window.* If you change a floppy disk in a drive, the File Manager needs to be told this. (Windows is still kind of dumb, even after all the upgrades.) While working through the Windows File Manager Icon, you will find that after inserting a second floppy disk in a drive and clicking on the current disk icon, the system will show you the original data files from the first disk. This is because File Manager is not aware that you switched disks on it. *During the second read of the disk, the File Manager program really doesn't go out and read the disk a second time.* It assumes everything was kept the same from the last disk read. Well, everyone knows what happens when you ASS-U-ME something. It usually never happens. To overcome this, choose "Refresh" from the Windows menu at the top of the screen. This will generate a re-read of the disk and display the correct information. You can also press the F5 Key to generate a disk refresh.

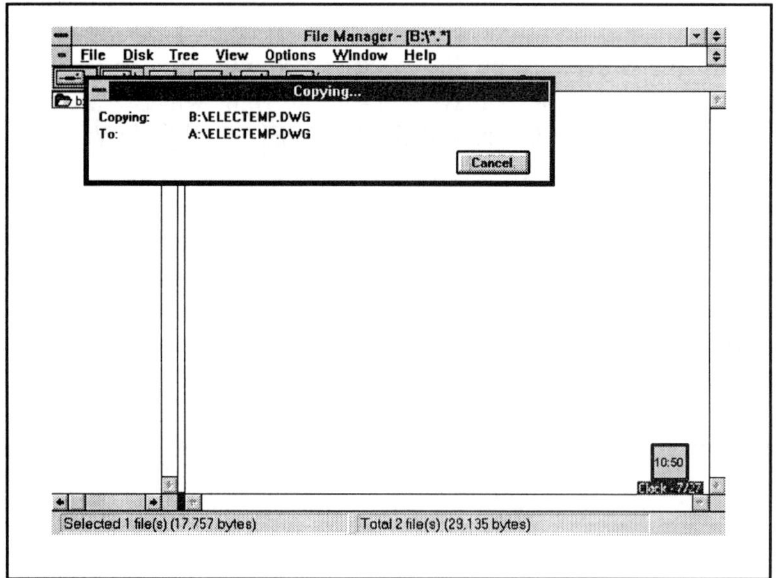

Figure 2-10 The Copying From - To dialogue box.

BULK COPYING OF FILES

Discussion

The bulk copying of files from one location to another is easy to do using the drag and drop method.

1. To accomplish this, the mouse cursor is clicked on the first file to copy, then the Shift key is held down, and the mouse is clicked on the last file to copy. This highlights all files between the selected locations.

2. Next the mouse is positioned over the selected files and the left mouse button is firmly held down.

3. Now the mouse, with its multiple file icons, is dragged to the destination location or drive. The mouse button is then released and the multiple Copy process is started.

The usual "Confirm Mouse Operation" and "Copying" dialogue boxes appear for your verification of the selected files and drives. When Yes is selected, the multiple file copy continues.

4. When the Copy is complete, click on the Control-Menu box in the upper left corner of the window and choose ‘‘Close’’.

USING "MOVE" VERSUS "COPY" FROM THE FILE MANAGER

What is the difference between the "Move" and "Copy" commands found under File Manager's File menu?

The Move process removes the file from its present location and places it at the location specified. When this happens, it no longer resides at its original location.

The Copy process causes a duplication of the original file to be located at the new location. Two files of the data now exist, the original and the copied file.

To use the Copy command:

1. Select the file you want to copy in the Files and Subdirectories window. (Right side of window.)
2. From the File Menu, choose "Copy".

The Copy dialogue box appears, with the selected file in the "From" text box.

3. At the "To" text box, the destination drive or location should be typed in the Text Box. At this stage, the file name may be changed, if necessary.

4. Next the "OK" button is clicked and the Copy process begins.

When the copy is finished, the original window is displayed.

DELETING FILES AND DIRECTORIES

While working with the File Manager, you can use several methods to delete a file, many files or even full directories. To do this:

1. Select the file, files or directory you want to delete.
2. From the File menu, choose "Delete."
3. The Delete dialogue box appears, highlighting the file or files selected for deletion.

TRAP When multiple files are being deleted, the window will show only a few of these files because of its small size. There may be files you want to keep that are not visible, so be cautious.

4. Clicking on the OK button activates the Deletion process.

A second method of deleting files is to:

1. Select the file to be deleted with the cursor
2. Press the DEL key.

Again, the Delete dialogue boxes appear. Clicking the OK button once again starts the deletion process.

THE VIRUS PROBLEM

There are people who take great joy in developing software programs which become attached to your disk and destroy random areas of data. These programs are called viruses because their sole purpose is to cause havoc, panic and frustration after you realize your disk is sick. You can get a virus by simply placing your disk into a computer that is already infected and doing a directory listing. Any disk activity will cause the virus to attach itself to your disk.

Most computers today have virus check programs built into the operating system that clean the memory at start-up. If the hard disk is write-protected, as in many school environments, the best way to avoid getting a virus is to turn the computer off and re-boot it from scratch, *every* time you use it. To prevent getting a virus, *don't let friends borrow your disks for use on their computers*. Who knows how many viruses may be lurking on their systems. Have they been practicing "Safe Computing?" What are they using for Protection?

If you get a virus, inform your instructor. He or she has access to special anti-virus programs. Unfortunately, once a virus has destroyed data it can not be recovered unless you have a backup of those files. See how important it is to have a backup!

QUIZ: THE COMPUTER ENVIRONMENT

Name ____

Circle the most correct answer or fill in the blanks.

1. Which of the following should be done at regular intervals?

   a. Save your drawing.
   b. Use CHKDSK.
   c. Make a backup of the complete disk.
   d. All of the above.

2. List four things you can do to ensure extended media life of your floppy disk.

   1. ____
      ____
   2. ____
      ____
   3. ____
      ____
   4. ____
      ____

3. Syntax refers to:

   a. How a disk is formatted.
   b. A copy program.
   c. Following the prescribed command structure.
   d. A program under DOS.

4. DOS is short for:
   Answer: ______________________________________

5. Which of the following commands will display the files on your disk?

   a. A:\DOS
   b. A:\FORMAT
   c. A:\CHKDSK
   d. A:\DIR

6. The purpose of formatting a disk is to:

   a. Prepare it to receive information.
   b. Scratch lines on it.
   c. Verify the files on the disk.
   d. Copy information from one disk to another.

7. Before formatting an old disk, what should one check for?

   Answer:____

8. In a two-drive DOS system, what is the command structure to copy the file ELECTEMP.DWG in the A drive onto a formatted disk in the B drive?

   Answer:____

9. What command is used to monitor the status of your disk?

   a. A:\ DIR
   b. A:\ DOS
   c. A:\ DISKCOPY
   d. A:\ CHKDSK

10. What is the DOS command structure to DISKCOPY the complete disk in the A drive onto a formatted disk in drive B?

    Answer:____

11. When using Windows, what is an easy method to copy a file from one drive to another?

    Answer:____

12. In the latest versions of DOS, what command is used to allow you to recover the information on a disk that has been formatted by mistake?

    Answer:____

13. What do you type at the C:\ prompt to start Windows running?

    Answer:____

14. What command is used to recover a file that was erased by mistake?

    Answer:____

15. The MOVE command:

    a. Makes a backup copy of a file.
    b. Changes places of two files.
    c. Removes a file from one place and relocates it.
    d. None of the above.

chapter 3

Getting AutoCAD Up and Running

DISCUSSION

AutoCAD has gone through many changes since it was first released in 1982. It has grown to the point where it must now be installed on the hard drive because of the numerous disks required to run the program. In this chapter, we will look at the AutoCAD start-up menu, examine dialogue boxes, have some fun by getting the itch out of our system through a quick play session, and then get serious and develop our first drawing, Fig. 3-1a. We will do this through the following exercises:

Exercise 3-1. Starting AutoCAD for the first time.

Exercise 3-2 Having fun with AutoCAD - Getting the "Itch" out.

Exercise 3-3. Setting up our housekeeping.

Exercise 3-4. Drawing our title block.

Exercise 3-5. The drawing sheets, paper sizes A, B, C, and D.

Exercise 3-6. Problems with the LINE Command and playing with the BREAK, TRIM and EXTEND Commands.

Note: It is assumed that you have used your AutoCAD manuals to install the software, and that you understand where on your system AutoCAD resides. On my system, AutoCAD resides in a subdirectory called ACAD. Our discussion will mainly deal with AutoCAD versions 12 and 13, but for those of you who have not updated to these latest releases, I will make reference to version 10 and 11 as we go.

To help you understand what is happening during each exercise, I'll discuss our game plan and how we'll proceed before we play with each command.

Don't be too concerned if you do not immediately understand the full use of each command during the exercise. As we complete each exercise, play with the command and experiment. Soon you will find these commands become second nature. You can refer to Chapter 8 and Appendix E for extra help if you need it.

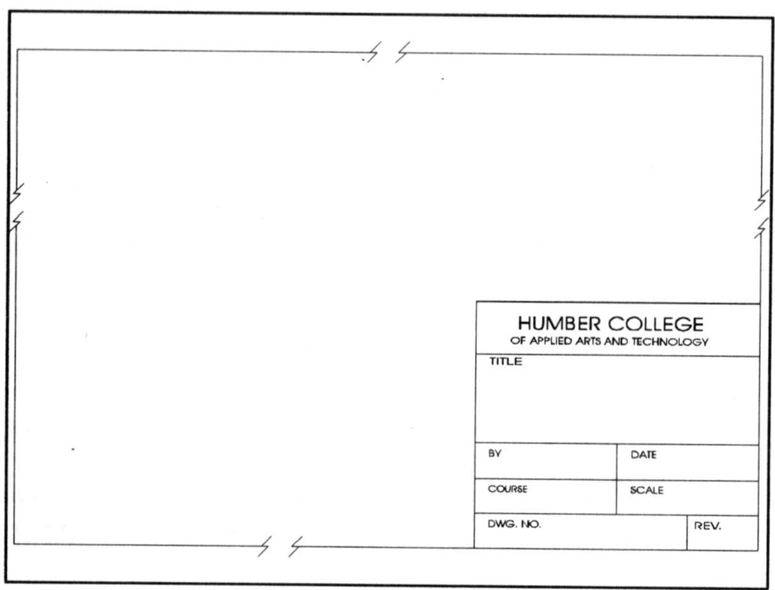

Figure 3-1a The Drawing Sheet and Title Block.

EXERCISE 3-1. STARTING AutoCAD FOR THE FIRST TIME

In this exercise, you will be introduced to the AutoCAD terms and commands of:

ACAD	CTRL-C	END	NEW
CANCEL	Dialogue Boxes	FILEDIA	
Command Line	QUIT	LINE	

Now to get started with AutoCAD.

Make sure your newly formatted disk is in the A: drive.

If your computer system at start-up is menu driven, check the menu for the AutoCAD selection prompt and follow the directions given at the menu.

If you are using version 12 or 13 of AutoCAD and have it installed with the AutoCAD batch file, you can invoke it by typing **ACADR12** or **ACADR13**.

If you do not have a batch file setup, you will probably have to switch to the AutoCAD subdirectory (ACAD in my case) using the Change Directory command. To do this type the following, replacing ACAD with your subdirectory name:

```
C:\CD ACAD (RETURN)
```

Our prompt should now look similar to this:

```
C:\ACAD\
```

To start AutoCAD running, type:

```
C:\ACAD\ACAD (RETURN)
```

The last DOS command, "ACAD," loads and runs the AutoCAD program.

For a brief moment, the copyright message, version and release information about AutoCAD is displayed, and then you are taken either directly to the Drawing Editor screen, or the Main Menu. Note that version 12 introduced a new start up procedure which eliminated the Main Menu used in earlier versions.

If you are running AutoCAD from Windows, double-click on the AutoCAD application icon. Again you will be taken to the Drawing Editor screen, after going through the standard AutoCAD start-up routine.

Is your newly formatted work disk in drive A? Make sure it is there.

Now To Have Some Fun With AutoCAD.

Note: If you are running version 12 or later, and see the Drawing Editor screen, skip directly to Exercise 3-1b, Using Version 12 and 13 At Start-Up.

If you have not upgraded to version 12 or 13, you will see the version 10 and 11 Main Menu appear, as below.

```
Main Menu:

  0.  Exit AutoCAD
  1.  Begin a NEW drawing
  2.  Edit an EXISTING drawing
  3.  Plot a drawing
  4.  Printer Plot a drawing

  5.  Configure AutoCAD
  6.  File Utilities
  7.  Compile shape/font description file
  8.  Convert old drawing file
  9.  Recover damaged drawing

Enter selection:
```

Initially we will play with AutoCAD, just to have some fun. Likewise, our drawing will be called FUN. In response to the ENTER SELECTION request, we will choose item 1, Beginning a NEW drawing. AutoCAD then requests a drawing name, and A:FUN will be our reply. The "A:" designates the drive and the "FUN" the file name. You may use upper- or lower-case letters. Let's do it:

```
Enter selection: 1 (RETURN)
Enter Name of drawing: A:FUN (RETURN)
```

AutoCAD now sets up for the new drawing and the drawing editor screen appears (Fig. 3-1b).

Having set up our drawing, now advance to Exercise 3-1c. Let's talk About AutoCAD.

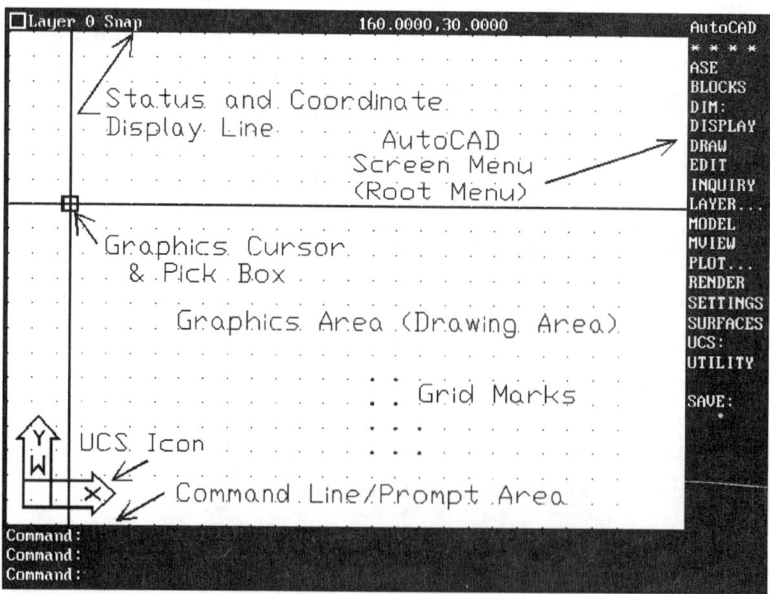

Figure 3-1b Drawing Editor Screen.

EXERCISE 3-1a. USING VERSION 12 AND 13 AT START-UP

Working on a New Drawing

In versions 12 and 13, at start up, you enter the Drawing Editor screen which displays a blank, unnamed drawing. This means you can immediately start to draw and later name the drawing when you SAVE, SAVEAS, or QSAVE it, or you can name it now by using the NEW command.

Let's give it a name now. We'll call it FUN and store it on the A drive. You can use upper- or lower-case letters. At the Command prompt type:

```
Command: NEW (RETURN)
```

This brings up the Create New Drawing dialogue box. Note Fig. 3-2a.

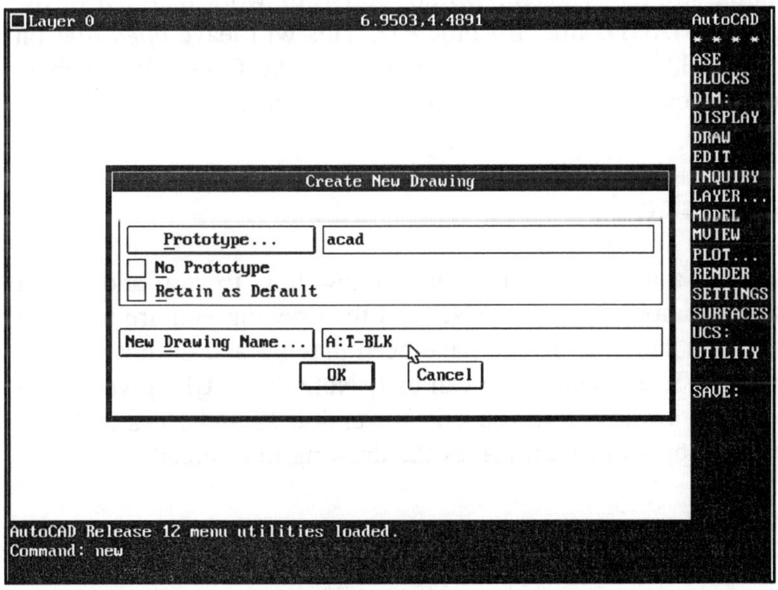

Figure 3-2a Create New Drawing dialogue box.

In the "New Drawing Name area..." edit box, type:

```
New Drawing Name... A:FUN (Don't press RETURN this time)
```

Then, with the mouse

```
Pick the OK button in the dialogue box.
```

This creates a drawing file on your work disk in the A drive called FUN. The "A" specifies the drive and "FUN" the file name.

Note: If you had been developing a drawing before using the NEW Command, AutoCAD will ask you if you want to save this drawing first.

EXERCISE 3-1b. LET'S TALK ABOUT AutoCAD FOR A FEW MOMENTS

Corrupting Your Disk

WARNING: ***Never remove your work disk from the disk drive unless you have exited AutoCAD using the "END or QUIT" command.***

I cannot stress this point strongly enough! Do not pull your work disk out of the drive while still at the COMMAND: line in AutoCAD! This will leave *open files* on your work disk, which could prevent you from using it in the future. Remember; AutoCAD stores files on your disk while you are using it and only removes them when you END or QUIT.

Ending a Drawing Session

When you are working on a drawing, it is nice to know how to stop and take a break. In AutoCAD, this is simple; just type END and the drawing you are working on will be saved. Then you can exit. If you have been playing and do not want to save the drawing, type QUIT, and again you can exit. Note that QUIT gives you a second chance to save your drawing before you exit. In version 12 and 13, a dialogue box appears for your final approval. Quit leaves the drawing unchanged.

Cancelling a Command

It's nice to know how to backtrack out of a command if all of a sudden you find yourself doing something you are not sure of. To do this, press the **CTRL** key first and while holding it, press the **C** key, shown in this tutorial as **CTRL-C**. This will **CANCEL** any command being executed and put back to the AutoCAD Command prompt. It is recommend you press the CTRL-C keys twice as some commands are two levels deep. This assures you will get back to the Command line.

DIALOGUE BOXES

Because dialogue boxes play a major role in the most recent versions of AutoCAD, we will take a few moments to discuss the basic dialogue box layout. See Fig. 3-2b. The Open Drawing dialogue box is one of the Standard File Dialogue Boxes which can be turned on or off by using the FILEDIA command and was introduced in version 12 of AutoCAD.

The FILEDIA command acts similar to a toggle switch and allows the many Standard File Dialogue Boxes to be active, when set to 1, or to remain unseen, when set to 0. The Standard File Dialogue Boxes are those dialogue boxes which write to or read from files. When the dialogue box is turned off, the Command Line is used to input the data.

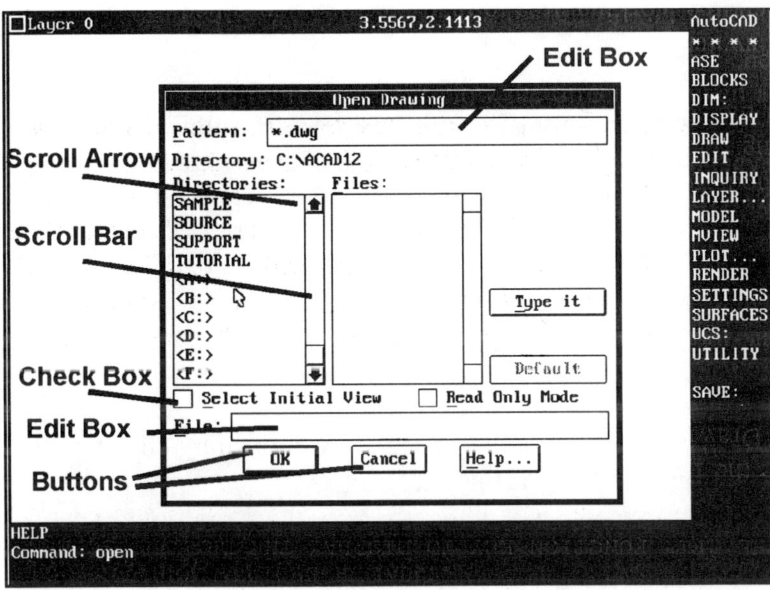

Figure 3-2b Open Drawing dialogue box.

When dialogue boxes are used, the cursor becomes a pointer. Dialogue boxes may consist of a scroll bar, buttons (Default, Type It, OK, Cancel, Help or Radio), check boxes, list boxes (List or Pop-up List), and edit boxes, to name a few.

If you want to quickly exit a dialogue box without making a change, either click on the Cancel button or press the Escape (ESC) key in the upper left corner of your keyboard.

Choosing A File Using The Scroll Bar Of The Open Drawing Dialogue Box

The scroll bar contains a slider box that allows you to scroll through a list of items such as drawing file names or directories. When the desired file is found, clicking on the file name will select it. Clicking the OK Button then confirms your selection and the command is complete. We will play with this many times before this chapter is complete.

TIP: The acceptance or approval of an entry in a dialogue box can be done in two ways: by moving the cursor arrow to the OK button and picking it, or by simply pressing the Return key.

EXERCISE 3-2. HAVING FUN BY GETTING THE "ITCH" OUT

Playing With AutoCAD

Well, you have been patient so far, and I know you are "itching" to play with AutoCAD, so here is your chance!

Many of the drawing commands such as Line, Circle, Arc, Donut, Ellipse and Erase are found under the Screen Menu Area of DRAW. With the mouse, pick DRAW and when the new menu comes up, pick LINE. Notice that the Command prompt changed to "From Point".

```
Command: PICK DRAW FROM THE SCREEN MENU.

Command: PICK LINE FROM THE DRAW MENU.

Command: From Point
```

Next, move the cursor to the screen area and anchor the Line down by clicking the left mouse button once, picking the line start point.

```
Command: From Point DRAG THE CURSOR TO THE SCREEN AND CLICK THE
                    MOUSE PICK BUTTON, TO ANCHOR THE FIRST POINT.
```

Notice the Command Line changes to "To Point." Now, drag the mouse to a new location and click the pick button again.

```
Command: To Point DRAG THE MOUSE TO ANOTHER LOCATION AND CLICK
```

Now press the RETURN key. You have just drawn your first line.

TIP: Pressing the Spacebar or Return key will terminate the command. Pressing these keys a second time will reactivate the command.

Using this technique, try to create a star pattern or the outside of a house on the screen. How creative can you be?

Creating Circles

Click on the word AutoCAD at the top of the screen menu. This brings you back to the AutoCAD Root Menu, where you can select DRAW. This time, select Circle with the CEN,RAD (Center Radius) option.

Now move the cursor to the screen edit area and click the Pick button once, and then drag the input device and you will see the circle develop. Click the Pick button again to lock the circle in place.

Plop down several circles on the drawing. Can you make a face out of them?

Don't forget, CTRL-C will cancel the command you are executing and get you back to the Command Line prompt.

Selecting the ERASE command under the DRAW menu will allow you to select screen objects to erase. To do this, place the square cursor on the objects you want to eliminate and press the pick button. This causes the object to become highlighted. Now press the Return key to terminate the command and remove the unwanted objects.

Are you having fun yet? I'm sure you are!

When you are done playing with AutoCAD, we will QUIT to assure this drawing file is *not* saved. When you are ready, at the Command line, type:

```
Command: QUIT (RETURN)
```

In version 12 and 13, AutoCAD replies with a Drawing Modification dialogue box, giving you the option to:

Save Changes, Discard Changes or Cancel Command

By picking **Discard Changes**, AutoCAD allows you to exit.

In version 10 and 11, you are asked:

```
Really want to discard all changes to drawing?: Y (RETURN)
```

Our response to this is "Y" for Yes!

Now To Get Serious and Develop Our First Real Drawing.

For those of you who are running version 12 or later, move on to Exercise 3-2b.

If you have not upgraded to version 12 or 13, you will see the old Main Menu again.

```
Main Menu:

   0.  Exit AutoCAD
   1.  Begin a NEW drawing
   2.  Edit an EXISTING drawing
   3.  Plot a drawing
   4.  Printer Plot a drawing
```

```
    5.  Configure AutoCAD
    6.  File Utilities
    7.  Compile shape/font description file
    8.  Convert old drawing file
    9.  Recover damaged drawing

Enter selection:
```

At this point you should still have your newly formatted work disk in drive A. Is it still there?

Our first drawing will be a title block called T-BLK (Fig. 3-3). In response to the ENTER SELECTION request, we will choose item 1. AutoCAD then requests a drawing name, and A:T-BLK is our reply. The "A:" designates the drive and the "T-BLK" the name. Again you may use upper- or lower-case letters.

```
Enter selection: 1 (RETURN)
Enter Name of drawing: A:T-BLK (RETURN)
```

AutoCAD now sets up for the new drawing and the drawing editor screen appears.

Having established the T-BLK drawing name, jump to Exercise 3-3.

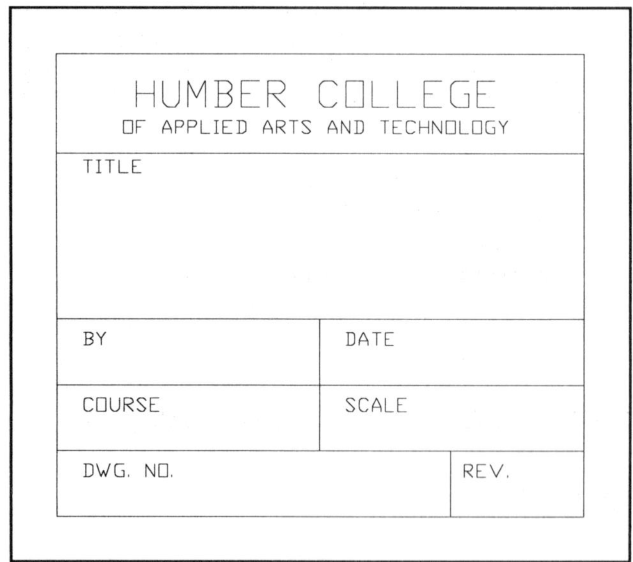

Figure 3-3 Title Block "T-BLK"

EXERCISE 3-2a. USING VERSION 12 AND 13 AT START-UP

Working on a New Drawing

You should still have your newly formatted work disk in drive A. Is it still there?

This time we'll give our new drawing the name T-BLK and store it on the A drive. At the Command prompt type:

```
Command: NEW (RETURN)
```

This brings up the "Create New Drawing" dialogue box. Note Fig. 3-2a.

Pick the "New Drawing Name ..." edit box, and type:

```
New Drawing Name... A:T-BLK (RETURN)
```

This time we did not pick the OK Button, but pressed the Return key to accept T-BLK as our new name. This action created a drawing file on our work disk in the A drive called T-BLK. When complete, this drawing will look similar to Fig. 3-3.

EXERCISE 3-3. SETTING UP OUR HOUSEKEEPING

Introduction

We are now ready to create our first drawing (Fig. 3-1). This is developed from a basic title block (Fig. 3-3) which we insert into various size borders, giving us a series of drawing sheets to develop our projects on. In this exercise, you will be introduced to the following commands:

BLOCK	Pick	Last	WINDOW
CTRL-C	Settings Menu	LIMITS	ZOOM
Editing	SNAP	OOPS	INSERT
ERASE	STATUS	OSNAP	UNITS
GRID	TEXT	UCSICON	WBLOCK

EXERCISE 3-3a. GETTING STARTED WITH OUR DRAWING

A New Type of Drawing Paper

Think of your drawing editor screen as drawing paper whose size is limited only by your imagination. You determine the size of this paper by putting limits on the outside edges of this electronic sheet. If you select a set of limits for your drawing and then later find it too large or too small, you can always change it. AutoCAD is very flexible and forgiving. Don't become too concerned with making a mistake.

First Steps (Housekeeping - Setting Up The Drawing Environment)

There are some basic "housekeeping" steps that you should do in setting up any drawing. The order in which they are done doesn't matter, but they must be considered. These include giving thought to:

1. The physical size of paper the final drawing will be plotted on.
2. The layout of the drawing and the amount of space around the outside of the basic blocks.
3. The type of units you will be working with (decimal, engineering, scientific, or architectural).
4. The scale you want your drawing in, for example, 1:1, 1:10, 5:1. Some AutoCAD advocates suggest you should always draw at one to one scale and then scale the drawing to fit the paper size later.

Once these items have been decided, you can now:

1. Set up the drawing **UNITS.**
2. Set the drawing **LIMITS.**
3. Establish a **GRID** coordinate or overlay to use as a reference.
4. Set up the **SNAP** resolution, using the input device. (SNAP will be discussed later.)

OUR FIRST DRAWING

Discussion

Our first drawing exercise will be to create a "Title Block" that is 80 drawing units (mm) wide by 70 drawing units (mm) high, as shown in Fig. 3-3 and Fig. 3-6. This will be saved under the working block (Wblock) name TITLBK1 and later inserted into four additional drawings to make the customized prototype drawing sheets for our project. These sheets are based on Table 3-2 (page 90) and shown as sheet sizes A, B, C, and D (Architectural).

To draw the title block, the AutoCAD commands of UNITS, LIMITS, GRID, and SNAP are explored. We will also take an in-depth look at the editing command ERASE and then draw the title block, using LINES, ZOOM, and the TEXT command.

The Title Block is developed using a limits setting at 100 units (mm) x 100 units (mm).

UNITS

Discussion

There are five formats of units that can be used in AutoCAD. These include Scientific, Decimal, Engineering, Architectural, and Fractional.

Each format will handle the same number in a different way. For example, the drawing unit number 18.75 will be shown as the following, using these five formats:

Scientific:	1.875E +01
Decimal:	18.75
Engineering:	1'- 6.75" (that is, 12 in. + 6.75 in. = 18.75 in.)
Architectural:	1'- 6 ¾ "
Fractional:	18 ¾

Exploring Units

For this exercise, we will select decimal units, because we want to work in metric. English decimal units and metric are handled the same way by AutoCAD. Their distinction is made at the time of plotting. To get to the Units menu, type UNITS.

```
Command: UNITS (RETURN)
```

AutoCAD responds with the following list.

```
SYSTEM OF UNITS          (EXAMPLES)

    1. Scientific        1.55E +01
    2. Decimal           15.50
    3. Engineering       1' - 3.50"
    4. Architectural     1' - 3 1/2"
    5. Fractional        15 1/2

    Enter choice, 1 to 5 <2>: 2 (RETURN)
```

Note that the default value of <2> is shown to help you in making your decision. You can either press the Return key or type in a new value in reply. In this case, we typed in the same value, 2.

Next, we are asked for the precision when using decimal units.

```
Number of digits to right of decimal point (0 to 8)<4>: (RETURN)
```

Enter your choice or press the Return key for a default value of 4.

We are next asked about the angle format to use. We'll stick with decimal degrees

(1) by pressing the Return key.

```
System of angular measure:        (Examples)
  1. Decimal degrees               45.0000
  2. Degrees/minutes/seconds       45d0'0"
  3. Grads                         50.0000g
  4. Radians                       0.7854r
  5. Surveyor's Units              N 45d0'0" E

Enter choice, 1 to 5 <1>: (RETURN)
```

Again, we are asked for the precision with which angles should be displayed:

```
Number of fractional places for display of angles (0 to 8)<0>:
(RETURN)

  Direction for angle 0:
  East   3 o'clock =   0
  North 12 o'clock =  90
  West   9 o'clock = 180
  South  6 o'clock = 270

Enter direction for angle 0<0>: (RETURN)

Do you want angles measured clockwise?<N>:(RETURN)
```

We have just setup the Units command so it works with decimal units to four decimal places. Angles are measured in degrees with 0 degrees at 3 o'clock which are measured in a counterclockwise direction.

```
Command: PRESS THE "F1" FUNCTION KEY TO DISPLAY THE DRAWING SCREEN
         AGAIN.
```

TIP: You will notice the F1 function key allows you to toggle between the Text Screen and the Drawing Editor Screen.

SETTING LIMITS

Discussion

You can use two methods to inform AutoCAD about the limits of the page size required for the drawing. One, by typing LIMITS from the keyboard, or two, by selecting LIMITS from the screen menu found under the Settings menu.

The Limits function is switchable. That is, it can be turned on or off. In the ON mode, AutoCAD restricts the working area to that which you specified. You have, in effect, an electronic fence that keeps you from straying beyond your boundaries. If you try to draw outside of the Limits range you will get an "Outside-Limits" error message.

In the Limits OFF mode, your page is now an open-sided sheet that allows you to draw anywhere outside the reference boundaries. Be careful when doing this, the world is your play ground. You should treat the Limits OFF mode with some caution, as you may spend much time working on a drawing only to find that it may not fit the paper when you plot it, or that you can not read the text because it is too small.

Exploring Limits

Let's explore the Limits function by using the AutoCAD root menu; the one on the right side of the screen. LIMITS falls under the Settings menu. Select Settings by moving the input device until the Settings menu item lights up; then press the pick button to confirm your selection.

```
Command: pick SETTINGS From The Root Screen Menu
```

This causes the Screen menu to disappear and the Settings menu replaces it. Part way down this new menu, you will find the word LIMITS:. Select LIMITS: by picking it with the mouse.

```
Command: pick LIMITS: From The SETTINGS Menu
```

The Limits menu displays:

```
Command: LIMITS
Reset Model space limits:
ON/OFF/<Lower left corner><0.0000,0.0000>:
```

You have three options to choose from:

```
1. To turn the limits ON.
2. To turn the limits OFF.
3. To set the lower left corner of the screen to 0.0000,0.0000
as the reference point.
```

For our purpose, we want item 3 and will use the default value of <0.0000, 0.0000>.

Press the Return key now.

```
ON/OFF/<Lower left corner><0.0000,0.0000>:(RETURN)
```

AutoCAD next asks for the upper right corner Limits. We will respond with 100,100 to set the X axis 100 drawing units long and the Y axis 100 drawing units high. This is the drawing size (in millimetres) of the area we will allocate for our Title Block drawing. Type:

```
Upper right corner <12.0000,9.0000>: 100,100 (RETURN)
```

TIP: The default setting of Limits is the ON mode. Should you find that your system is switched off, type ON in reply to the Limits command.

Understanding the Drawing Unit

We are using the term "drawing units" rather than millimetres, because AutoCAD really doesn't care if we are using English decimal units or metric millimeter units. They are both in decimal notation. On the other hand, we should think of these units in terms of metric, as this is how they will be plotted, and it will give us a feeling of scale.

When working with AutoCAD and its open architecture approach to scale, you quite often find yourself working in a world with limitless boundaries. Some drawings have no scale, such as a Schematic or Electrical layout drawing, thus there is no intuitive feel for how large the drawing really is. Often, you may ask yourself "What is the length of this drawing unit? Is it the thickness of the book or is it the width of the room?" Unless you can relate scale or size to some physical scale, it becomes only a number. For example, how big is big?

To overcome this feeling of random scale or size, such as the line thickness (polylines), text height, or line length, when you are using an unscaled drawing, you should try to relate the units you are working with to the scale of the page dimensions you are working on. This will bring scale and size back into perspective. The use of 100mm by 100mm is similar to 2.5 inches by 2.5 inches, and everyone has a general knowledge of how large this is.

GRID AND ZOOM

Discussion

The Grid command displays a series of reference dots across the screen area. The dots may be located any distance apart and at a different aspect ratio. That is, the X axis grid spacing may be a different spacing than the Y axis. The dots are used only as reference points and do not show up on the final plotted drawing. The grid may be turned ON or OFF using the F7 function key, or by typing GRID at the keyboard. If the grid is set to zero and the Snap option is chosen, the grid is locked to the Snap value and is thus representative of the Snap aspect.

TIP: A quick way to get rid of the distracting little marks left on a drawing after you use the Erase command is to turn the Grid OFF and ON, using the F7 function key. This quickly regenerates the drawing, establishes a new grid and removes the distractions.

To set up the grid, type GRID from the keyboard or select SETTINGS from the screen Pull-Down menus and select "Drawing Aids... ." This will open the Drawing Aids dialogue box (Fig. 3-4a) where you can then choose the aspect ratio for the X and Y spacing of the Grid. In either method, set the Aspect ratio of both the X and Y axis to 5.

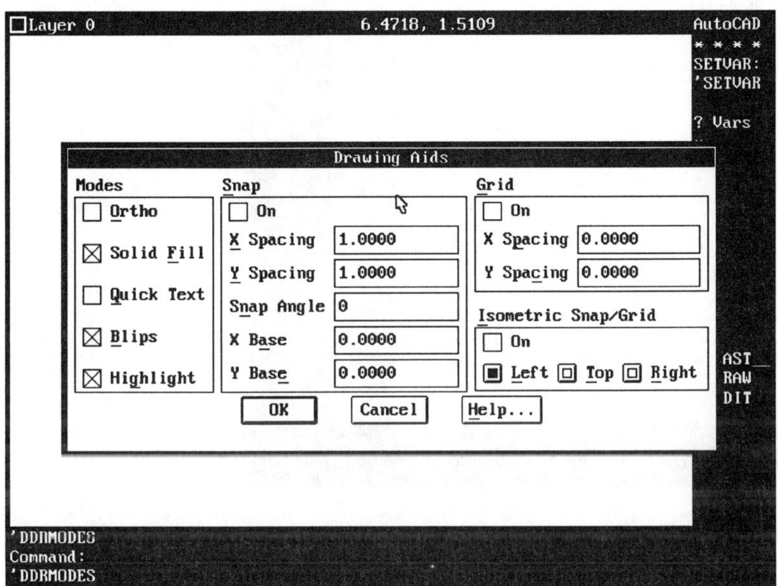

Figure 3-4a Drawing Aids dialogue box.

Type:

```
Command: GRID (RETURN)
```

Now respond with A for Aspect:

```
Grid spacing (X) or ON/OFF/Snap/Aspect<0.0000>: A (RETURN)
Horizontal spacing (X) <0.0000>: 5 (RETURN)
Vertical spacing (X) <0.0000>: 5 (RETURN)
```

A grid work (Fig. 3-4b) equivalent to 5 mm x 5 mm *may* appear on the screen if your drawing is zoomed out to its limits. If your drawing is not at its limits, let's put it there.

From the keyboard type ZOOM at the Command line.

```
Command: ZOOM (RETURN)
```

AutoCAD replies with

```
All/Center/Dynamic/Extents/Left/Previous/Vmax/
Window/<Scale(X/XP)>: A (RETURN)
```

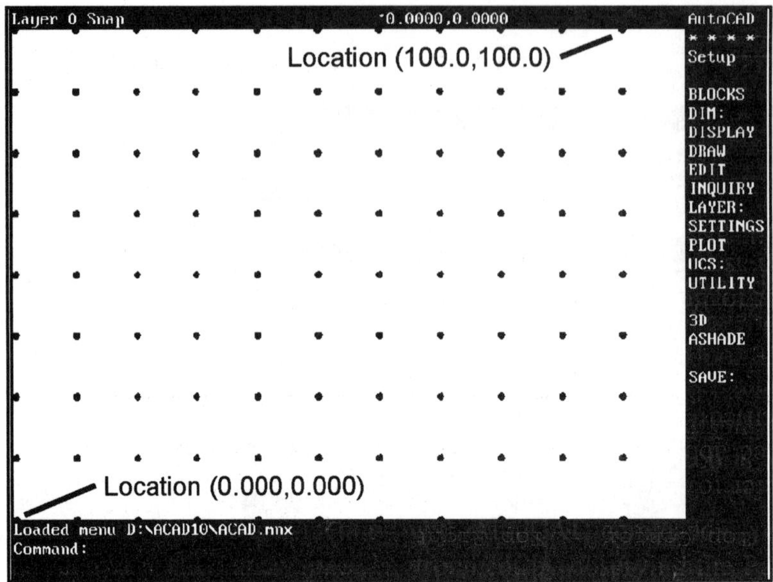

Figure 3-4b The Grid.

Choose "A" to zoom out so All of the drawing can be seen. The display (Fig. 3-4) will now show the total drawing area and your grid. For a detailed description of the Zoom options (ACDELPVWS), see ZOOM in Chapter 8.

EXERCISE 3-3b. STATUS (WHAT IS YOUR DRAWING'S STATUS?)

The Status command will display the present status of the drawing parameters which are updated after every Zoom to All. These include the Limits, Grid, Snap, and Layer status.

STATUS Command Format

The following display shows the status of the default values presently set up in AutoCAD on my system.

Using the command:

```
Command: STATUS (RETURN)
```

The following text was displayed on my system:

```
          93 entities in A:ASSEM1C
Model space limits are   X:    0.0000  Y: 800.0000 (Off) (World)
                         Y:    0.0000  Y: 500.0000
Model space uses         X:   90.0000  Y: 710.0000
                         Y:  350.0000  Y: 447.0000
Display shows            X:    0.0000  Y: 800.0000
                         Y:    0.0000  Y: 617.9757

Insertion base is        X:    0.0000  Y: 0.0000   Z: 0.0000
Snap resolution is       X:    1.0000  Y: 1.0000
Grid spacing is          X:    1.0000  Y: 1.0000

Current Space:      Model space
Current Layer:      TEXT1
Current Color:      BYLAYER  - 7 (white)
Current Linetype:   BYLAYER  - CONTINUOUS
Current elevation:  0.0000     thickness: 0.0000

Fill on Grid on Ortho off Qtext off Snap on Tablet off
Object snap modes: Endpoint, Midpoint
Free disk: 23748592 bytes
Virtual memory allocated to program: 1826K
Amount of program in physical memory/Total (virtual) program
  size: 29%
Total conventional memory: 316K   Total extension memory: 376K
Swap file size: 388KB
Page faults: 374  Swap Writes: 137  Swap reclaims: 99
```

Notice how all attributes associated with AutoCAD in my system were displayed.

Let's look at the status of your setup. Type:

Command: **STATUS** (RETURN)

You should see something similar to that displayed on my system, but the numbers will be quite different.

To display the drawing screen again, press the "F1" Function key.

AutoCAD's Lonely Entity, the Orphan

TIP: There are times when all of a sudden after regeneration of the drawing, it will shrink in size. This often happens after using the Insert command. The cause is a culprit I call an "orphan." An orphan is an entity which for one reason or another gets placed somewhere far outside the Limits of the drawing. Thus, when you Zoom to All, this orphan living in never, never land causes your drawing to shrink. I have seen full size drawings reduced to a mere dot in the left hand corner of the page. Zooming into the dot revealed that the complete drawing was still there. Orphans, depending upon their size, may or may not be seen if you view the editing screen.

So how do we get rid of these pesky little mites? First, use the Status command to highlight what coordinate is beyond its Limit. Second, re-establish the Limits well outside this expanded coordinate area. Third, using Erase with the Window option, clean the area in question of the unwanted entity. Fourth, re-establish your original Limits. Lastly, Zoom to All again.

This will get rid of the unwanted orphan. Remember, when doing this, set the Limits outside the area the orphan resides in.

SNAP AND COORDINATE DISPLAY

Discussion: The F6 Function Key - Coordinates Display

If you move the input device around, you will notice that the cursor positions itself somewhere on the screen. To see the cursor location, toggle the Coordinate Function Key (F6) to the ON position. Press F6 once, and the coordinates turn ON; press F6 again, and they turn OFF.

Now, when you move the input device, the cursor coordinates are displayed at the top right corner of the screen in the Coordinates area. These figures are relative to the limits we established for the drawing. Note how difficult it is to position the cursor cross-hair lines exactly on a dot. Try it again! Notice how the decimal portion of the coordinates value can never be set to .0000. Now to overcome this.

The Snap feature allows you to position, or snap, the cursor to one of the dots on the screen; thus you can "grab" a known location. Let's set up the Snap feature from the keyboard, type SNAP. Note: You could also use the Drawing Aids dialogue box (Fig. 3-4a).

```
Command: SNAP (RETURN)
```

We will keep the aspect ratio (that is Y distance versus the X distance) of our Snap to the same values as the GRID settings (5 units x 5 units). This time, we will enter our Snap spacing directly, rather than using the Aspect (A) option. To the following questions, type 5, and RETURN.

```
Snap spacing or ON/OFF/Aspect/Rotate/Style <1>:5 (RETURN)
Horizontal spacing <5.0000>: (RETURN)
Vertical spacing <5.0000>: (RETURN)
```

When this sequence is complete, the Snap function is turned On, as indicated by the word SNAP or letter S appearing in the Mode Area (top left corner of screen). The Snap resolution is set to 5 drawing units in both the X and Y directions. Snap can be toggled On or OFF by using the F9 function key.

Move the input device around the screen; you will see the cursor jump from grid point to grid point. This is because the Grid and Snap both have the same resolution.

Turning the Snap OFF (F9 key), will allow you to move to any location on the screen, not just to grid points.

OSNAP or Object Snap is an extension of the Snap command. It allows the cursor to have a number of options as to how it will snap, such as to the *end* of a line already drawn. We will explore this command in more detail in Chapter 4. To disable the OSNAP mode, type NONE or OFF at the OSNAP mode request.

```
Command: OSNAP (RETURN)
Object snap modes: OFF (RETURN)
```

THE UCS AND UCSICON - USER COORDINATE SYSTEM ICON

Discussion - Our Changing World

In the following discussion we will explore various coordinate systems, model space and paper space, and gain a better understanding of how to view an object from different points of view.

The recent advances in technology have helped us to redefine our thinking in many areas, and over the past few decades the world as we know it has changed greatly. In this prelude to the 21st century, science has told us of the Big Bang theory and that real space may be curved. Likewise, computer developers are weekly, making quantum leaps in the sophistication and computing power of their new machines. Today, through computer technology, we can view animated objects from any point of view and even interact with them in a world of virtual reality.

AutoCAD is no exception to this new technology of visual space. To help us in defining the drawing space or area we draw in, AutoCAD in release 10 established the User Coordinate System (UCS). We, as users, can now define how we look at a drawing and establish what is *up* and what is *down*.

Model Space: As designers, we create things on our electronic drawing sheet. Every time we draw something, we are creating a *model* of what an object should look like. We are in effect working in a "model space" environment on the drawing screen.

Prior to release 10, AutoCAD only worked in the World Coordinate System (WCS), which we know as the Cartesian Coordinate System. This is a world based on X, Y, and Z reference points which normally highlights a Top view, Front view, and Side view of an object.

Today, through the User Coordinate System, we can think of objects as three-dimensional models that can turn and rotate to allow us to see all sides and views. This allows us to view the object from a position that best suits the shape of the object.

There are two UCS icons that could appear in the lower left corner of the drawing editor screen. These are the Model Space and Paper Space icons shown in Fig. 3-5. Each allows us to visualize the viewpoint or plane our model is represented in. The Model Space icon shows the relationship between our model's world or plane, and the World Coordinate System.

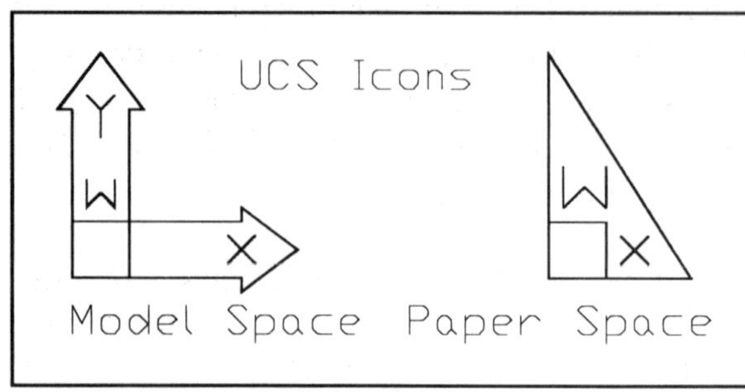

Figure 3-5 UCS Icons for Model Space and Paper Space.

Paper Space: In addition to looking at our drawing from a model space perspective, AutoCAD has established another option to allow us to have multiple views of our drawing, as if we had a series of paper windows we could look through. These paper windows are called "Paper Space" views and allow us to have such things as close up views, far away views, and views from any angle, all displayed and plotted at the same time. In the Paper Space mode, the Paper Space Icon is displayed.

Tilemode: While experimenting with AutoCAD commands, you may find yourself accidently in the Paper Space environment by observing the Paper Space UCS icon of Figure 3-5 on your drawing. Unfortunately, we shouldn't be there for this tutorial. To get back to the Model Space environment, we must turn the TILEMODE command ON again, or set it to 1. To do this, type:

```
Command: TILEMODE (RETURN)
New value for TILEMODE <0>: 1 (RETURN)
```

Before we are finished with this tutorial, we'll have a chance to play in this world. If your drawing does not require you to see objects from different viewpoints, as in two-dimensional drawings, the UCS icon is of no use and can even become a distraction. It should therefore be turned OFF.

To keep this tutorial simple, our next two drawings will only require a two-dimensional coordinate system, therefore; we will turn the UCS icon OFF. To do this, type:

```
Command: UCSICON (RETURN)
ON/OFF/All/Noorigin/ORigin/<ON>: OFF (RETURN)
```

The icon disappears until you turn it ON again with the UCSICON command.

EXERCISE 3-3c. AutoCAD's ERASE AND ITS BUDDY OOPS

Discussion

Sooner or later, you are going to make a mistake while laying out the drawing. No problem! The Erase command will help you get through what may appear to be a major crisis. You will soon realize that through editing, you can easily modify a line or object. Editing is a common, everyday event, and the Erase command is a good friend. Get to know all you can about this command. Note Appendix E, Exercise 1A, *A Practice Session Using Undo And Erase* for more help.

The Erase command is used to remove unwanted objects or entities from the drawing. There are many ways to select these objects and during this next discussion we'll look at the three most commonly used ways, as we examine the Erase command. The best method to choose will depend upon the activity at the time the error was made.

The three most common Object Selection modes allow us to:

1. Point to or (P)ick the object to erase.
2. Place a (W)indow around the object to erase.
3. Delete the (L)ast entity put on the screen.

These methods of Object Selection are universal throughout AutoCAD, and it is best to become familiar with their use. Let's take a closer look at each of them.

We get to the Erase command through Edit, found under the AutoCAD Root Menu. This time, let's just type it in:

```
Command: ERASE (RETURN)
Select objects:
```

SELECT OBJECTS USING THE PICK (P) BUTTON

Select Objects is asking us to place the cursor on the line, object, or objects we want to remove. Do this by pointing to the unwanted object, using the input device, then; press the Pick button. This will cause the object to become dotted, highlighting it. Now, pressing the Return key will cause the object to disappear.

```
Select objects: Pick the object with the mouse.
```

Then press the Return key to accept the command.

OOPS

If you accidentally erased the wrong object, place the cursor on OOPS in the menu area and press the Pick button, or type OOPS (Return). This will automatically restore the object, but it must be done before other commands are issued. You are now ready to start over again.

TIP: During Erase, make sure the cursor is not at the intersection of two objects, this will give unpredictable results. AutoCAD can't tell which object you really want to select.

SELECT OBJECTS WITH A WINDOW (W)

Select Objects using the Window (W) option by replying with "W" to the Erase command. This allows us to place a window around all objects to be erase. The mouse is the best tool to use when establishing the window as it is easy to manipulate. The window is created by selecting a corner point outside the area to capture, such as the lower left corner, then picking the opposite corner such as the upper right corner. As the mouse is dragged between the two points, the window is created.

TIP1: Any *whole entity* that falls within this window area will be erased with the "W" window opton. Note the stress on *whole entity*, because lines that are only partially encompassed by the window will remain untouched.

TIP2: To capture entities that cross into the window area, use the "C" or crossing option instead of the W option. See Appendix E, Exercise 7 for more detail.

Pressing the input device select button will cause only those entities that will be erased to appear as dotted lines. If you are satisfied with your selection, press the Return key. The Oops command will allow you to recover, should you find more was erased than you really wanted. Here is how it works:

```
Command: ERASE (RETURN)
Select objects: W (RETURN)
First Point: Pick a point outside the entities to erase.
Second Point: Drag the mouse over the objects to capture and Pick a
               second point that include all entities.
```

This causes the entities chosen to change color and become dotted. When you are satisfied,

```
Press the RETURN key to accept the entities chosen.
```

SELECT OBJECTS USING THE LAST (L) OPTION

The Select Objects Last (L) option allows you to erase the last object drawn, by either typing "L" from the keyboard or by picking "Last" as an Erase menu. When you have made your selection, the last item drawn will appear dotted. If you are satisfied, press the Return key, and the item disappears. Again, the Oops command will restore the item.

```
Select objects: L (RETURN)
```

The last item becomes dotted, now

```
Press the RETURN key to confirm your choice.
```

TIP: The entity chosen to be erased will appear dotted as well as a different color, thus making it easier to identify the items to be erased.

Multiple Selections

If more than one item is to be edited, you can chain them together by using the following commands:

Add (A) is a quick way to change the command line back to "Select Objects, or Window or Last," so you can add more objects to your selection.

Remove (R) will reverse the process that was done through the Add (A) instruction. This will remove objects so they are not affected by the Erase or Edit process.

Undo (U) removes the most recent entity or object from the screen (similar to Last). You can continue to Undo lines and work your way backwards through the drawing, erasing the lines as you go.

TRAP Remember, you must be in the original command to use Undo to remove single entities, else it removes entire blocks. The Redo command will reverse Undo problems.

CTRL-C (CANCELLING A COMMAND)

The Control C (CTRL-C), a combination of pressing the CTRL and C keys together, will cancel any command that is presently being executed. This is a quick way to bail out of a command and immediately get back to the Command line.

To gain more experience with the Erase command, see Exercise 1 in Appendix E.

EXERCISE 3-4. DRAWING OUR TITLE BLOCK

In this exercise you will be introduced to the AutoCAD commands of:

DTEXT	SAVEAS
FILEDIA	SAVETIME
QSAVE	STYLES
REDO	TEXT
SAVE	UNDO

EXERCISE 3-4a. DEVELOPING THE TITLE BLOCK LINES

LINE

Having set the Units, Limits, Grid, and Snap requirements, we are now ready to draw the title block, using the Line command. As you can see from Fig. 3-6, the title block is 80 mm wide by 70 mm high. You can think of this scale as 80 drawing units by 70 drawing units. Remember, our Grid and Snap aspect is set to a resolution of 5 units.

TRICK To save time, we will use the power of the mouse almost exclusively to make this drawing. The keyboard will only be used when it is to our advantage to do so, such as pressing the spacebar to repeat a command.

We'll use Fig. 3-7 as a reference. Now turn ON the Grid by pressing the F7 key.

From the Screen menu we will pick Draw, and from the Draw menu we'll pick Line. Note that two things happen; the Command line displays the "From Point" and the AutoCAD Root Menu area displays the special Line menu. Using the Line command, we will draw a series of straight lines to create the outline of our title block. After this, you will be given the opportunity to finish the drawing on your own, by completing the remaining lines in the center of the title block.

```
Command: Pick DRAW from the Screen Menu
```

This activates the Draw menu, now

```
Command: Pick LINE from the Draw Menu
```

The Command line displays LINE.

```
Command: LINE
```

Now to start drawing our title block.

```
From point:   SELECT POINT "A" ON FIG 3-7, BY PLACING THE CURSOR 2
              LINES UP AND 2 LINES RIGHT OF THE ZERO REFERENCE.
              (Coordinate location 10.0000, 10.0000)
```

Press the Pick button on the input device. The Command line now displays the "To Point."

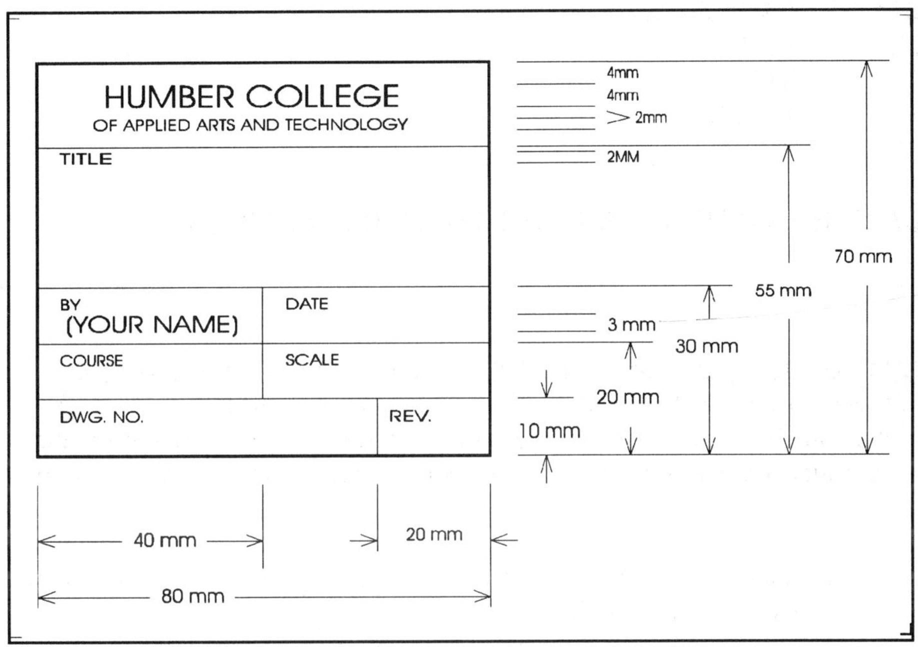

Figure 3-6 Dimensioned Title Block.

```
To point:    - MOVE CURSOR UP 14 GRID LINES AND PICK POINT B
               (70 units) (Coordinate 70.0000 <90>)

To point:    - MOVE CURSOR RIGHT 16 GRID LINES AND PICK POINT C
               (80 units) (Coordinate 80.0000 <0>)

To point:    - MOVE CURSOR DOWN 14 GRID LINES AND PICK POINT D
               (70 UNITS) (Coordinate 70.0000 <270>)

To point:    - MOVE CURSOR TO "CLOSE" in the LINE MENU
               AND PRESS THE PICK BUTTON.
```

TIP: In the last command we selected Close from the Line menu. As you saw, the Close command completed the title block box and allowed us to take a shortcut to the starting point. At the same time, it terminated the Line command.

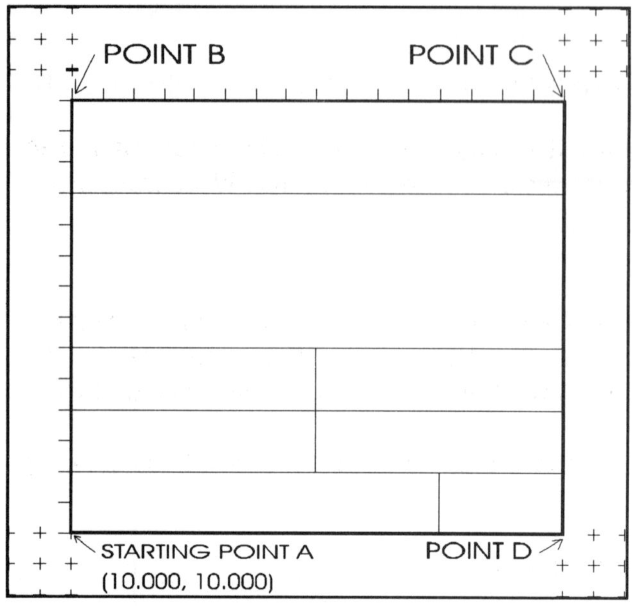

Figure 3-7 Naked Title Block.

UNDO AND REDO

Should you accidentally change something or make a mistake while drawing a line, the Undo command will erase the last command executed. If you keep Undoing, you can erase the whole drawing. To use the Undo command, type "U" (a *hot key)* and Return.

TIP: If you make a mistake and used Undo once too many times, the Redo command will redo the last undo.

TRAP Be careful; if you terminate a command such as Line and use Undo, Undo will remove the entire string of lines that were entered at that time. For Undo to work properly, you must still be in the Line command. If only one line of many is to be removed, after the command is terminated, it is better to use the Erase command with the Window or Pick option.

For more help, see Appendix E, Exercise 1, *A Practice Session Using Undo and Erase.*

TAKING THE BIG STEP!

Now it's your turn to try AutoCAD on your own!

As you can see, using the Line command is quite simple, but using Undo can create problems. Now, on your own, draw the remaining lines of the title block using Figs. 3-6 and 3-7 as references.

TIP: Remember, to regain access to the previously used command (in this case the Line command), press the **Space Bar** or the **Return** key. *Note:* The Command line again highlights the Line "From Point" when you do this.

When you have finished drawing the remaining lines of the title block, the drawing should appear similar to Fig. 3-7.

We are now ready to add the text to our drawing.

EXERCISE 3-4b. TYPING THE TEXT INTO OUR TITLE BLOCK

TEXT AND DTEXT

Discussion

There are two types of text commands available in AutoCAD. They are the original Text command and the newer Dtext (Dynamic Text) command. The basic difference between them is that Dtext places the text on the screen as you type each character, while Text enters all the typed characters after the Return key is pressed.

In this discussion we will look at the various methods to place text on the drawing. Text and Dtext give you various Styles to choose from. Text may be stretched, compressed, or mirrored, as well as positioned horizontally, vertically, or at any angle in between. To play with text styles, try Exercise 4A and 4B in Appendix E.

The Text commands may be selected from the keyboard or picked from either the AutoCAD Root Menu or a Pull-Down menu. In this exercise, we will explore all methods.

On our Title Block drawing just created (Fig. 3-7), we will now insert text into each of the sub-blocks. The syntax of the Text command uses the following format:

```
Command: TEXT (RETURN)
Justify/Styles/<Startpoint>:
Height <0.20>:
Rotation <0>:
Text: ......... (RETURN)
```

The default *startpoint* option is a request for you to show AutoCAD where you want the text to start. The text is placed above, and to the right of the point you choose. You can enter your location by typing the coordinates from the keyboard, but it is much faster to use the mouse and pick the position directly.

Height specifies the text height of capital letters, in drawing units, above the text baseline. Some special characters may extend above or below the baseline. Note Fig. 3-8.

Rotation allows you to rotate the text through 359.9 degrees, beginning at your "start point."

Text is the text string to be entered on the drawing.

When the last text character has been typed, pressing the Return key terminates the command and enters your text on the drawing.

The Styles option allows you to select various styles of fonts. The default font style is called Standard. See Appendix A for the various font styles, and Appendix E, Exercise 4, on how to use them.

Text Options

There are several options that may be chosen when entering text.

Using Justify Options When you select **Justify**, you can use one of several text alignment options. The command looks like this:

```
Command: TEXT (RETURN)
Justify/Styles/<Startpoint>: J (RETURN)
Align/Fit/Center/Middle/Right/TL/TV/TR/ML/MC/MR/BR/BC/BR:
First text line point: Pick a point
Second text line point: Pick a second point
Height<0.20>: Choose a height (RETURN)
Text:.... .... (RETURN)
```

The Align and Fit options will place the text between two specified points, shown here by the plus symbol (+). In Align, the text height and orientation are automatically adjusted to fit exactly between these points, as shown below. The Fit option allows you to specify the text height and AutoCAD adjusts the text width to fit between the two points selected.

Align Option:

```
                         HUMBER  COLLEGE
First Point Selected   +                 + Second Point Selected
```

The Center and Middle options will center the text on either side of a single point (+) you specify. See the example below. The Middle option will center text both horizontally and vertically about this point.

```
HUMBER   COLLEGE              HUMBER + COLLEGE
        +
     (Center)                     (Middle)
```

The Right option uses the point you specify as the right edge of the text and the last character of text will go no further than this point.

```
HUMBER  COLLEGE
              +
```

The use of Options TL/TC/TR/ML/MC/MR/BL/BC/BR:

TL, TC and **TR** refer to Top Left, Top Center, and Top Right. The text will be placed just below the designated start point and will be left justified, centered, or right justified with respect to it. See Fig. 3-8.

```
+            +             +
Top Left   Top Center   Top Right
```

ML, MC and **MR** refer to Middle Left, Middle Center and Middle Right. The text will be equally spaced above and below the start point, and will be left justified, centered or right justified.

```
+Middle Left    Middle+Center    Middle Right+
```

BL, BC and **BR** refer to Bottom Left, Bottom Center and Bottom Right. Text with long descenders, such as p, q and g will be placed so the descender starts along the starting point, and will be left justified, centered or right justified.

```
Bottom Left    Bottom Center    Bottom Right
+                   +                      +
```

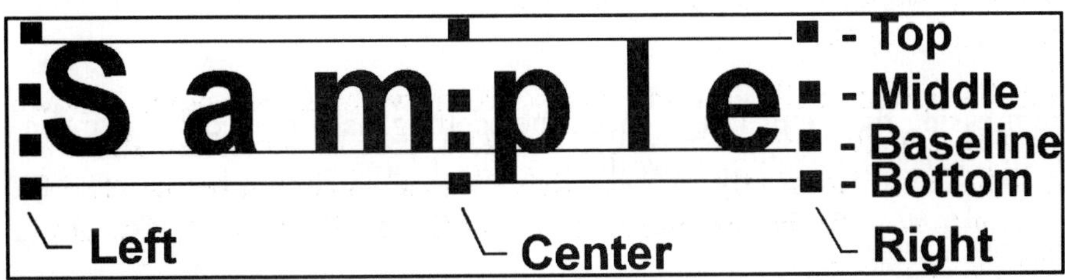

Figure 3-8 Text alignment positions.

Let's Input Some Text

To get started, we will Zoom into the title block using the Window option so that it fills the screen, similar to Fig. 3-7. Remember, your title block is 80 units wide and 70 units high on a drawing of 100 units by 100 units.

```
Command: ZOOM (RETURN)
All/Center/Dynamic/Extents/Left/Previous/Vmax/Window/
<Scale(X/XP)>: W (RETURN)
Pick your ZOOM window.
```

Next, we will change the present Grid and Snap aspect from 5 units by 5 units to 2 units by 2 units, through the Grid and Snap commands. This will give the input device a finer resolution of control when selecting points on the title block. Use Fig.3-6 for dimensions.

```
Command: GRID (RETURN)
Grid spacing(x) or ON/OFF/Snap/Aspect<5>: 2 (RETURN)
Command: SNAP (RETURN)
Snap spacing or ON/OFF/Aspect/Rotate/Style<5>: 2 (RETURN)
```

Press the F6 function key to activate the Coordinates function to On.

Now to enter the words HUMBER COLLEGE, or the name of your organization, as Text, 4 units (mm) high, into the upper area of our title block. From the keyboard, type TEXT at the Command prompt.

```
Command: TEXT (RETURN)
```

AutoCAD replies with a request for a starting point. With the input device, move the cursor to the coordinates of 22.0000, 72.0000, which is the starting point for the text in the upper part of the title block area.

```
Justify/Styles/<Start point>: PICK (22.0000, 72.0000) WITH THE MOUSE.
```

You are now asked to select a height and a rotation angle for the text. We will choose a height of 4 units and a rotation angle of 0, its default value.

```
Height <default>: 4 (RETURN)
Rotation angle <0>: 0 (RETURN)
```

Now the Text line appears. Here we type **HUMBER COLLEGE**, or your organization's name.

TIP: Before pressing the Return key, check the spelling! Is it correct? It is much easier to correct the spelling now than to edit it later.

```
Text: HUMBER COLLEGE (RETURN)
```

When you press the Return key, the text is written at location 22.0000, 72.0000.

DTEXT OPTION

We will now input the smaller text, but this time use the Dtext command. The text height is 2 units high and starts at location 20.0000, 68.0000.

At the Command prompt type:

```
Command: DTEXT (RETURN)
Justify/Styles/<Start point>: PICK (20.0000, 68.0000) WITH THE MOUSE.
Height <4.0000>: 2 (RETURN)
Rotation angle <0>: 0 (RETURN)
Text: OF APPLIED ARTS AND TECHNOLOGY (RETURN)
(RETURN)
```

Note: As you type each character, they appeared on the screen as they were entered, and two Returns were needed to terminate the command. Your drawing should now appear similar to Fig. 3-9a.

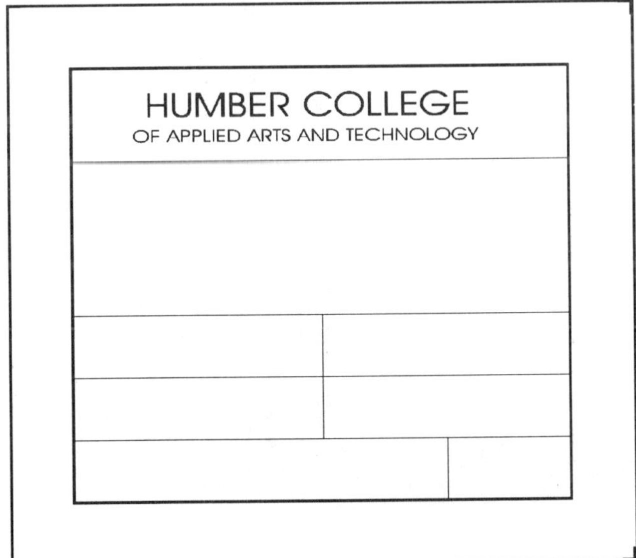

Figure 3-9a Title: HUMBER COLLEGE

TIP: Your mouse may be used to select the location for additional text and pressing the Return key will repeat the Dtext command. For more exciting ways to input text and select fonts, complete Exercises 4A and 4B in Appendix E.

Now It's Your Turn! With Fig. 3-6 as a reference, enter the remaining text on the title block using the same size text (2 units high) and font style.

EDITING TEXT - DDEDIT OPTION (Versions 12 and 13)

Discussion

In version 12 and 13, AutoCAD made available a Single Line Editor which allows you to modify a single line of text in you drawing without having to retype the complete line. Here is how it works. At the command line type:

```
Command: DDEDIT (RETURN)
<Select a TEXT or ATTDEF object>/Undo: PICK text to modify
```

The Edit Text dialogue box now appears (Fig. 3-9b). In the Text box, make your changes and click OK. This takes you back to the Drawing Editor Screen where the modified text is displayed.

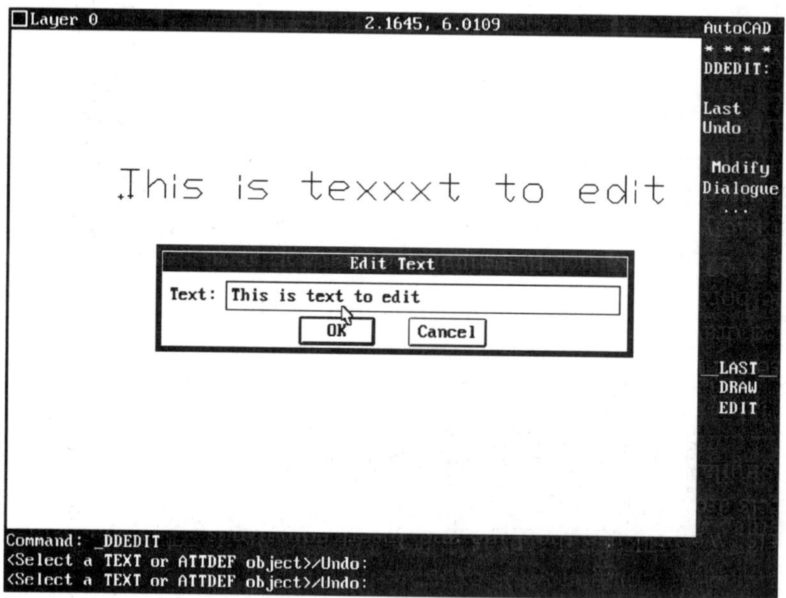

Figure 3-9b DDEDIT "Edit Text" dialogue box.

A second method of modifying the text is through the DDEDITMT command that allows you to modify the complete attributes associated with the text. This dialogue box (Fig. 3-9c) allows you to change the properties of the text such as the color and layer, text height, rotation, width factor, obliquing angle, text style, justification, upside down, and backwards, to mention a few; and of course, the text itself.

Layer 0 3.3827,8.3152 AutoCAD

THIS SSS TEXT

Modify Text

Properties

Color... BYLAYER Layer... 0

Linetype... BYLAYER Thickness: 0.0000

Text: THIS IS TEXT

Origin

Pick Point < Height: 0.4000 Justify: Left

X: 3.2304 Rotation: 0 Style: STANDARD

Y: 8.3587 Width Factor: 1.0000 Upside Down

Z: 0.0000 Obliquing: 0 Backward

Handle: None

OK Cancel Help...

Command:
Command: ddmodify
Select object to modify:

Figure 3-9c DDEDITMT "Modify Text Attributes" dialogue box.

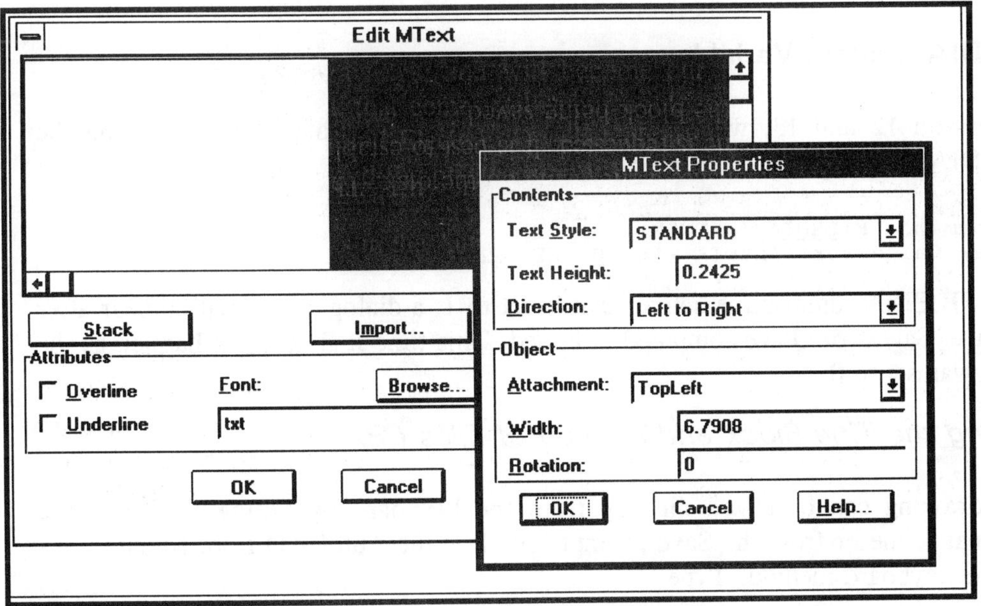

Figure 3-9d MTEXT and MTEXT Properties dialogue boxs of version 13.

In version 13 of AutoCAD, a complete text editor option was introduced that allows you to modify whole paragraphs or pages of text through the Mtext editor. To make these modifications, AutoCAD relies on an external text editor such as that found in DOS through the Shell Command. Figure 3-9d shows the Mtext editor dialogue box. When the selected text paragraph has been edited, clicking on OK will get you back to the AutoCAD Drawing Editor. *Note:* Your system must have a text editor as part of the operating system in order for this to work. AutoCAD does not supply the text editor.

To play with additional ways to input Text and select Fonts, complete Exercise 4A and 4B in Appendix E. Then refer to the "DTEXT" and "QTEXT" options in Chapter 9.

EXERCISE 3-4c. THE SAVE COMMAND

SAVE

Discussion

Heeding our warning to **save, save, save,** we'll now save the drawing of the title block to our work disk under the original name of "T-BLK".

TIP: Before saving a drawing, it is best to Zoom to All so the entire drawing can be viewed when it is restored back to the screen.

FILEDIA System Variable

In version 12 and 13, we will turn the Save dialogue boxes OFF by using the FILEDIA system command. Type:

```
Command: FILEDIA (RETURN)
New value for FILEDIA <1>: 0 (RETURN)
```

If the FILEDIA command is active, that is set to 1, a dialogue box will appear after the Save, Saveas or Qsave command is issued. This time it should not happen as the value was set to 0.

Saving the Title Block and Backup of Files (.BAK)

The drawing can be saved directly from the keyboard by typing SAVE at the Command line or from the Save prompt found on the AutoCAD Root Menu. Let's use the keyboard method. Type:

```
Command: SAVE (RETURN)
```

File Names And Backup Extensions

AutoCAD now requests a file name to save the drawing under. You are limited to *eight* alphanumeric characters as your file name. Windows 95 file names may have up to 256 alphanumeric characters. *Note;* A three-digit extension of .DWG is added automatically to the drawing file name by AutoCAD when saved. The drawing as it was at start-up is converted to a backup copy and receives the extension .BAK.

TIP: If you require the drawing to be saved to a diskette in drive A, then "A:" must by specified before the file name. If the drive is not specified, the drawing is saved to the "default" drive; that is, the drive established when you initialized AutoCAD.

We will save our drawing under the file name of T-BLK on the disk in drive A. Make sure your work disk is in drive A before proceeding any further.

```
Save current changes as <default>: A:T-BLK (RETURN)
```

Disk drive A now activates and our title block drawing is saved. Did you see the red LED come on?

SAVEAS

If at start-up, in versions 12 and 13, no file name was established, then use the Saveas command. This command requests a file name, saves the drawing and sets the current drawing name to the name just established. The Save Drawing As dialogue box (Fig.3-10) will appear when FILEDIA is set to 1. This time it should not happen as FILEDIA is set to 0.

QSAVE

Quick Save (Qsave) saves the current drawing without requesting a file name. If no file name has been established, it requests a name to be entered.

SAVETIME (AUTOMATIC TIMED SAVE)

In version 12 and 13, AutoCAD will automatically save your work at preset intervals by using the Savetime command. If set to 0, the command is disabled. The time entered is in minutes and becomes reset after a Save, Saveas, or Qsave. The file name established under the Savetime command is AUTO.SV$. This file can be renamed back to a standard AutoCAD.DWG file, using the DOS Rename command.

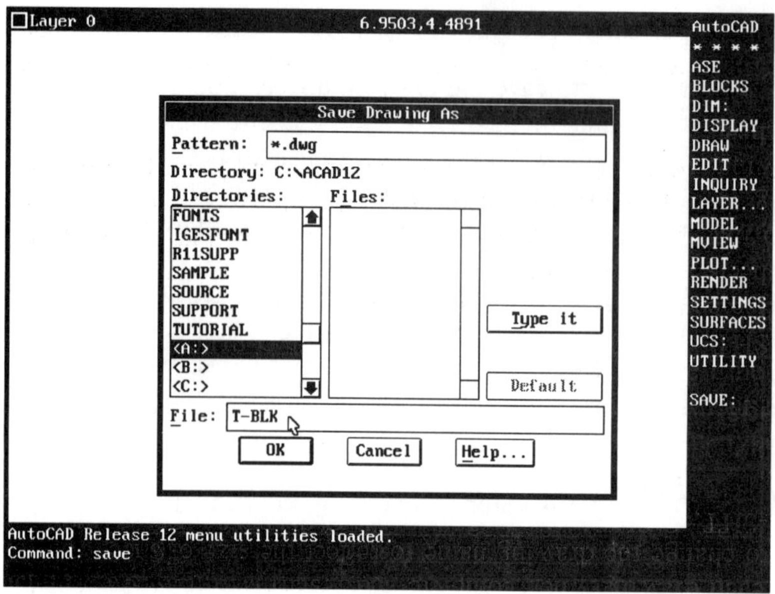

Figure 3-10 The Save dialogue box.

The Savetime command follows the format:

```
Command: SAVETIME (RETURN)
New value for SAVETIME <current>: 0 (RETURN)
```

The Savetime command is disabled by using 0 as the entry value.

EXERCISE 3-5. THE DRAWING SHEETS: A, B, C, & D

In this exercise, you will be introduced to the AutoCAD terms and commands of:

BLOCK	Insertion Points
CHANGE	WBLOCK
EXPLODE	INSERT

EXERCISE 3-5a. MAKING RUBBER STAMPS (BLOCKS, WBLOCKS AND ALL THAT JAZZ)

Discussion

We will use our title block drawing as a rubber stamp to create four new drawings. These drawings are the electronic sheets of paper that will be use in the future to develop our schematic, printed circuit board and assembly drawings on. But first we must capture our title block to disk as a rubber stamp.

To do this, we first capture our title block drawing as a Block called TITLBK1 and then save it to our work disk as a universal rubber stamp or Wblock under the name WTITLBK1.

TIP: To distinguish between regular drawings and universal Wblock drawings (rubber stamps), the letter W should be added as the first letter of the file name of all Wblocks, such as WTITLBLK. Using this convention will allow you to immediately see which drawing files are Wblocks and which are regular drawings when you do a directory listing of your disk.

BLOCKS AND WBLOCKS

Discussion - Symbol Library

As we progress with AutoCAD, we will use symbols such as diodes, resistors, and capacitors over and over again. It is not necessary to redraw these symbols each time we want to use them. We draw them once and then save them to our disk as a Wblock where they can be used many times over, like a rubber stamp. They are in effect a universal symbol. This collection of stored symbols is referred to as a symbol library and are usually found grouped together on a disk. The title block is a symbol we will use in several drawings, and will become our first entry in the symbol library. Likewise, the disk that comes with your book contains two symbol library files called Electemp.Dwg and Elecomp.Dwg which are used with the development of the schematic drawing and the printed circuit board.

Under the heading of Blocks in the AutoCAD Root Menu are two subheadings called Block and Wblock. In this discussion we will explore the use of both of these commands.

Block allows us to capture a portion of a drawing as a rubber stamp, or as an isolated entity, and Wblock saves the rubber stamp to disk as a file in the symbol library. The saved Wblock file can then be inserted into a second drawing at a later date.

To store a drawing object to disk, we must first isolate the object, using the Block command, by placing a window around it. Next, we establish an insertion point to be used when we insert the object into another drawing. The Block command stores the object as part of the present drawing file in a reserved area of memory dedicated to Blocks.

TIP: The insertion point should be chosen with some care. Using a physical location on the entity is best, rather than some point in space which has no reference point to the entity.

After the Block has been captured, we use the Wblock command to create a permanent record of our Blocked object on disk, along with all pertinent information (attributes) about it.

Attributes may be thought of as a data base that accompanies the drawing file. This data base may contain such things as parts lists, pricing, or any other useful data required for the drawing. As we are still at an introductory drawing stage, we will not make use of attributes at this time.

TIP: It should be noted that once a Block or Wblock is stored, it can no longer be edited unless special instructions are used. Through the Insert command, the entity can be broken back into its basic components using the asterisk (*) option while inserting the entity back into another drawing. When an entity has previously been inserted into a drawing, and not broken apart, the Explode command is used to create the original components.

A Block or Wblock acts as a single entity before it is exploded. Thus, if a line is erased, all lines become erased. Data may be added to the Wblock, such as the name of the drawing or date, but entity parts such as "HUMBER COLLEGE" cannot be changed. We will explore this process as we proceed.

Before we use the Block or Wblock commands, let's make sure the Files dialogue box is switched OFF, using the FILEDIA command. Remember how? If you are not sure, refer back to Exercise 3-4c.

Capturing a Block (Blocks and Block)

We will now capture the title block just created as a Block. At the command line type:

```
Command: BLOCK (RETURN)
```

AutoCAD will request a Block Name. We will call our title block TITLBK1.

```
Block name (or?): TITLBK1 (RETURN)
```

The next request is for an Insertion point. We'll choose the lower right corner of the title block by using the input device to position the cursor to this location (see Fig. 3-11).

```
Insertion point: SELECT LOWER RIGHT CORNER OF THE TITLE BLOCK.
```

Now, using the Window option, we will totally engulf the "outside" of our title block with the window, thus capturing it. Make sure you start outside the title block area, so the entire title block is captured.

```
Select objects or Window or Last: W (RETURN)
First point: - SELECT LOWER LEFT CORNER
Second point: - SELECT UPPER RIGHT CORNER

Press the RETURN key.
```

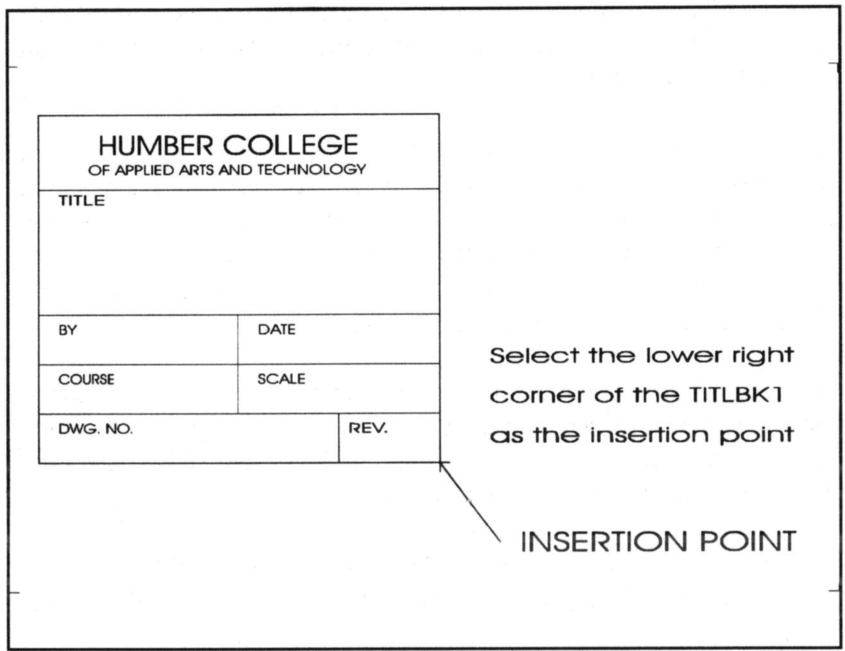

Figure 3-11. Title Block - Insertion point.

Oops

The title block disappears from the screen, captured by AutoCAD when the Return key was pressed. To recover it back to the screen, pick Oops from the Block menu.

```
Command: PICK OOPS FROM THE BLOCK MENU

or type OOPS at the Command Line
```

WHAT BLOCKS DO WE HAVE?

To see what blocks have been stored in the Block Storage Area of the drawing, we can use the "?" option of the Block command with the "*" option to list all blocks. To see if TITLBK1 is there, type:

```
Command: BLOCK (RETURN)
Block name (or?): ? (RETURN)
Block(s) to list <*>: (RETURN)
```

AutoCAD should respond with:

```
Defined blocks.

TITLBK1

User          External        Dependent        Unnamed
Blocks        References      Blocks           Blocks
  1               0               0                0
```

This indicates the title block, TITLBK1, has been isolated and is saved in the drawing area reserved for blocks.

WBLOCK

To save the Blocked entity called TITLBK1 to disk, we'll use the keyboard to enter the Wblock command.

TIP: Wblock first establishes a file name on the disk. There is nothing in this file until the Block name is requested. It is best to establish the Wblock file name so that it reflects the block being saved.

```
Command: WBLOCK (RETURN)
```

To store the Wblock on drive A, we must remember to specify the drive when typing the file name. We will use WTITLBK1 as our Wblock file name.

TIP: Note the W before the file name WTITLBK1 to signify that it is a Wblock. If we use this convention, we can easily pick out those files which are Wblocks when we do a Directory listing of our disk.

```
File name: A:WTITLBK1 (RETURN)
Block name: TITLBK1 (RETURN)
```

AutoCAD now writes our captured entity, TITLBK1, to disk under the file name of WTITLBK1 on drive A.

TRAP Any Block name can be used, as long as it is a legal Block. This way, you could store a Block called Resistor in a Wblock named Wdiode. This is something you would not want to happen because there is no correlation between the two names and the actual symbol used. A diode is not a resistor.

Ending Our First Drawing

With our title block (TITLBK1) saved to disk as a rubber stamp under the Wblock file name WTITLBK1, we can now end this drawing session.

Select END from the Utility menu, found under the AutoCAD Root Menu, or type END at the command line. This will save the present drawing and take you back at the Main Menu in version 10 and 11, or back to DOS in version 12 and 13.

```
Command: END (RETURN)
```

EXERCISE 3-5b. DESIGNING THE DRAWING PAPER

Discussion

Using our title block rubber stamp WTITLBK1, we'll now create four new drawings. These are the prototype sheets of electronic paper to be used to develop our schematic, printed circuit board and assembly drawings on. In industry, these sheets are known as A, B, C, and D size paper. Their equivalent metric sizes are called A4 through A0. Note Tables 3-1, 3-2 and 3-3.

To draw these sheets, we will use the Line and Insert commands. The file names assigned to these sheets are A-SHT, B-SHT, C-SHTA / C-SHTE, and D-SHT. These drawings will look similar to Fig. 3-1 when complete.

We start by creating a border around each sheet, using Tables 3-1, 3-2, and 3-3. The title block is then inserted into the lower right corner of each sheet. From Table 3-1, note that the suggested drawing area is less than the outside limits of the paper. This allows space for the border. For example, from Table 3-1, a D size sheet has an outside limit of 34 x 22 inches, but the suggested drawing area is only 32 x 20 inches, or approximately 800 mm x 500 mm. It is the suggested drawing area dimensions that are used when setting the Limits for our four pages. When the drawing is plotted, the plotter mechanism will leave about 3/4 to 1 inch (19 mm to 25 mm) of border area around all sides of the paper.

Table 3-1 (ANSI Y14.1 - 1980, IN.) Paper Sizes (Mechanical and Electrical)

Sheet Size	*Outside Limit (in.) X & Y*	*Suggested Drawing Area X & Y (in.)*	*Suggested Drawing Area X & Y (mm)*
A	11 x 8.5	9 x 7	230 x 170
B	17 x 11	15 x 10	380 x 250
*C	22 x 17	20 x 15	500 x 380
D	34 x 22	32 x 20	800 x 500
E	44 x 34	42 x 32	1060 x 800
F	40 x 28	38 x 26	950 x 650

Table 3-2 Paper Sizes (Civil and Architectural)

Sheet Size	*Outside Limit (in.) X & Y*	*Suggested Drawing Area X & Y (in.)*	*Suggested Drawing Area X & Y (mm.)*
*A	12 x 9	10 x 8	250 x 200
*B	18 x 12	16 x 11	400 x 275
*C	24 x 18	22 x 16	550 x 400
*D	36 x 24	34 x 22	850 x 550
E	48 x 36	46 x 34	1150 x 850

* Drawing sheets to be created in AutoCAD.

Table 3-3 International Paper Sizes (mm)

Sheet Size	*Outside Limit X & Y (mm)*	*Suggested Drawing Area X & Y (mm)*
A-4	297 x 210	230 x 170
A-3	420 x 297	380 x 250
A-2	594 x 420	540 x 380
A-1	841 x 594	800 x 540
A-0	1189 x 841	1130 x 800

Now we'll start a new drawing from the AutoCAD start up menu. The Grid, Snap, and Limits are set to allow us to create the drawings for the A, B, C, and D size architectural sheets. The dimensions for these are shown in Table 3-2. We use the Insert command to recall the title block drawing (WTITLBK1) from our disk and place it in the lower right corner of each drawing. Our finished drawing will look similar to Fig. 3-1.

From Table 3-2, you will note that the A size paper has a working Drawing Area of 250 mm x 200 mm. We'll use these units to establish the drawing limits.

Once the limits are set, we will draw the border around the outside of the sheet using the Line command. If you feel confident to do this on your own, go ahead and then proceed to the Insert command farther down the text. If not, let's walk through this exercise together.

Starting the A-SHT

If you are using version 12 and 13, restart the system using the method established by Exercise 3-2b and give the new drawing the file name of A:A-SHT.

For version 10 and 11 users, from the Main Menu select:

```
1 - Begin a NEW Drawing
```

Give the new drawing the name A-SHT. Don't forget to precede the name with the A drive letter to simplify the saving of this drawing in the future.

```
Drawing name <WTITLBK1>: A:A-SHT (RETURN)
```

AutoCAD proceeds to bring in the programs necessary to set up a new drawing. When ready, it displays the command line. For a review of the next commands, see Exercises 3-2a and 3-2b if needed.

Housekeeping for the A-SHT

Set up the following housekeeping for the A-SHT.

```
Command: GRID   - SET THE GRID ASPECT TO 10.
Command: SNAP   - SET THE SNAP ASPECT TO 10.
Command: LIMITS - SET LOWER LEFT CORNER TO 0.0000, 0.0000, AND
                  UPPER RIGHT TO 250,200.
Command: ZOOM   - ZOOM TO "ALL".
```

Your screen should now show a series of dots, 25 across the bottom by 20 high on a spacing of 10 units by 10 units. By turning the Coordinate toggle ON (Function Key F6) and placing the input device at the top right corner, you will see the coordinate locations of 250.0000, 200.0000. These are the outside drawing limits of the A size paper in mm.

ORTHO MODE

To help us draw straight lines, vertically and horizontally, we will turn the Ortho mode ON using the F8 function key.

```
Command: Press the F8 Function Key to toggle the ORTHO mode ON.
```

You will notice that an O (Ortho) appears in the Status line at the top left of the screen.

Using the Line command and the input device, we'll place a border around the outside perimeter of the drawing. This procedure is similar to making the title block.

Are the Coordinates turned ON (F6 Key)? Now to develop the border, type:

```
Command: LINE (RETURN)
From point: - PICK CURSOR LOCATION 0.0000, 0.0000
To point:   - MOVE CURSOR TO 200.0000 <90 and PICK
To point:   - MOVE CURSOR TO 250.0000 <0  and PICK
To point:   - MOVE CURSOR TO 200.0000 <270 and PICK
To point:   - PICK ''CLOSE'' IN LINE MENU, OR TYPE C (RETURN).
```

Your A-SHT drawing now has a border around the outside. Now to add the title block to our drawing, using the Insert command.

INSERT

Discussion

Stored on disk is the title block under the Wblock name WTITLBK1. We will use the Insert command to place the title block in the lower right corner of the A-SHT drawing.

Insert requests the block name to used and then a location on the drawing where the block should be placed. With the input device, we'll choose the lower right corner at location (250.0000, 0.0000). See Fig. 3-1.

AutoCAD next requests the X and Y scale factors and the Rotation angle for the block. For the "A" sheet, use a scale factor of three-quarter scale (.75), and a rotation angle of 0. For the "B", "C" and "D" sheets, use a scale factor of "1."

```
Command: INSERT (RETURN)
Block name (or?): A:WTITLBK1 (RETURN)
Insertion point: - PICK 250.0000, 0.0000
X scale factor <1>/Corner/XYZ: .75 (RETURN)
Y scale factor (default=x): .75 (RETURN)
Rotation angle <0>: (RETURN)
```

As you see, the title block is now in the lower right corner of the A-SHT drawing.

Using the Explode and Asterisk (*) Commands

Once an entity has been inserted into a drawing as a Block or Wblock, should it need editing, it can be broken into its original components by using the Explode command. This command is similar to using the Asterisk (*) option with Insert, except that it is used after the Block is inserted. We'll Explode the title block now. Type:

```
Command: EXPLODE (RETURN)
Select objects: PICK ANY LINE ON THE TITLE BLOCK. THIS WILL
                CAUSE IT TO BE REDRAWN AND EXPLODED IN THE
                PROCESS. (Watch closely for the screen to blink)
```

```
Select objects: (RETURN)
```

The Asterisk (*) option is used to break Blocks and Wblocks into their original components at the time of insertion. To do this, the Asterisk symbol (*) is placed before the Block or Wblock name during the Insert command. As an example:

```
Command: INSERT (RETURN)
Block name (or ?): A:*WTITLBK1 (RETURN)
```

Saving the A-SHT Drawing

Now that we have created our A-SHT, save your drawing to the work disk in drive A under the title A-SHT. This should be simply a matter of selecting Save and Return from the Utility or File menu, as we have already identified the drive and file name when the drawing was originally set up. You can also type:

```
Command: SAVE (RETURN)
File name <A:A-SHT>: (RETURN)
```

EXERCISE 3-5c. DEVELOPING THE B-SHT, C-SHTA, C-SHTE and D-SHT DRAWINGS

You're on your own again! Complete this exercise by creating the remaining four drawing sheets. They are a B size sheet that is 400 x 275 mm and *two* C size sheets. The first C size sheet is 550 x 400 mm (Architectural size paper), and the Second C size sheet is 500 x 380 mm (Mechanical/Electrical size paper). The last is a D size sheet that is 850 x 550 mm. The process of creating these four drawings is the same as creating the A-SHT, except the title block (WTITLBK1) is inserted at full scale (1:1) rather than one-half scale. Use Tables 3-1 and 3-2 as a reference for your border size.

Don't forget to End each drawing when complete and restart with the new Limits settings. Remember to change the drawing name to reflect the size of sheet you're developing, that is; B-SHT, C-SHTA for Architectural, C-SHTE for Electrical, and D-SHT.

Prototype Drawings

We will use the A-SHT, B-SHT, C-SHTA, C-SHTE and D-SHT as "prototype drawings." They are the correct size to meet the industry requirements for A, B, C and D size drawing sheets. As prototypes, they are available for us to use any time we need to work on that size of paper, thus saving us time, and your company money by not having to remake them each time. We will use them to draw the remainder of the drawings used in this tutorial. Now, go ahead and create the B, C and D size drawing sheets.

EXERCISE 3-6. PROBLEMS WITH THE LINE COMMAND AND PLAYING WITH THE BREAK, TRIM AND EXTEND COMMANDS

Discussion

Just before we close this chapter, let us look at a potential problem with the Line command and how to prevent from falling into its trap. We will also look at how to split a line in half and how to trim off an excess line length through the Break and Trim commands.

EXERCISE 3-6a. PROBLEMS WITH THE LINE COMMAND

TRAP Before we stray to far from the Line command, there is one trap we should be aware of when using it. Line allows you to draw lines on your drawing sheet. Often, we may misjudge the correct length of a line, so to correct this, we simply trace over the top of it again. WRONG! This is not what you want to do.

Let's look at what happens when we draw over a line for a second time. To AutoCAD, this looks like two lines, one on top of the other. This becomes a real problem when we try to hatch these multiple line areas. The second line appears as a barrier to the hatch routine and the hatch pattern starts to run outside the hatch boundaries, similar to paint running out of a can with a hole in it.

If you have two lines on top of each other, and you erase one, the top one, they will both disappear until you do a regeneration of the drawing. At this time, the original line will reappear. So the line you thought was gone, all of a sudden is back again, creating a real nightmare.

To demonstrate this, we will draw two lines, the first being 20 units long and the second 30 units long, on top of each other. Note Fig. 3-12.

In a clear area of your drawing, draw a line 20 units long starting at a reference point we'll call A. Now draw a second line 30 units long, again starting at the reference point A, and trace over top of the original line.

Now to erase the lines. Select the Erase command and pick the line to erase it. As expected, the line becomes dotted after picking it, and disappears from the screen when you press the Return key.

Now, do a quick regeneration of the drawing by toggling the Grid Function key, F7, OFF and ON.

What's this? Our first line appears again. This is because AutoCAD Erased only the topmost line and not all lines that are under one another.

As you can see, this could become a real problem if you are not aware this can happen.

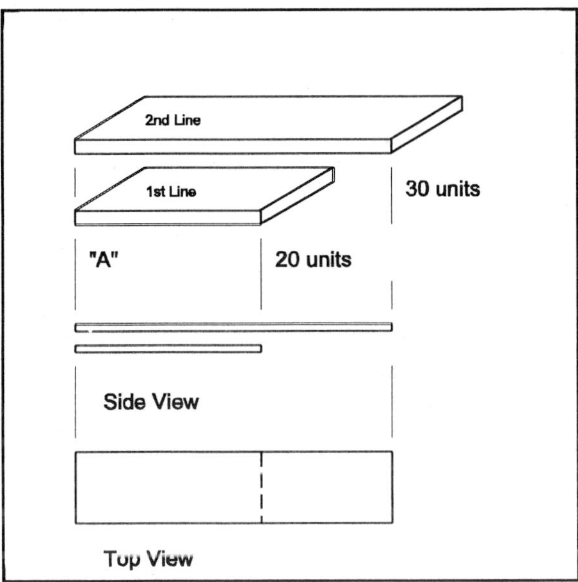

Figure 3-12 Lines on top of each other.

The solution is to extend the original line by grabbing the end, using the OSNAP End option, and draw it to the correct length, or to use the Extend command discussed in Exercise 3-6d. Also note the Stretch command in Appendix E, Exercise 7.

EXERCISE 3-6b. The BREAK COMMAND

The Break command is used to break and erase part of a line, circle, arc, trace, or polyline. It may also be used to trim the end of a line but with less accuracy than the Trim command offers.

In using the Break command, AutoCAD first requests that you pick a point on the object as the first break point and then a second point where the next break should take place.

<u>**TIP:**</u> To save time, AutoCAD assumes that the point you first selected during "object selection" is the first point of your break and immediately asks for the second point. You can reselect the first point by responding with "F" for First point at the command line.

To give us something to play with, draw a line on your drawing (Fig. 3-13A). Now to break it, type:

```
Command: BREAK (RETURN)
Select object: PICK A POINT NEAR THE CENTER OF THE LINE
```

This will become the first break point.

```
Enter second point (or F for first point): PICK THE SECOND
POINT JUST RIGHT OF THE FIRST LOCATION
```

Note that the line between the two selected points is removed. You now have two lines (Fig. 3-13B).

Now to trim off the right end of the line, again using Break. This is done by selecting the second point at a location past the end of the line. Use Fig. 3-13C as a reference.

```
Command: BREAK (RETURN)
Select object: PICK A POINT NEAR THE RIGHT END OF THE LINE
```

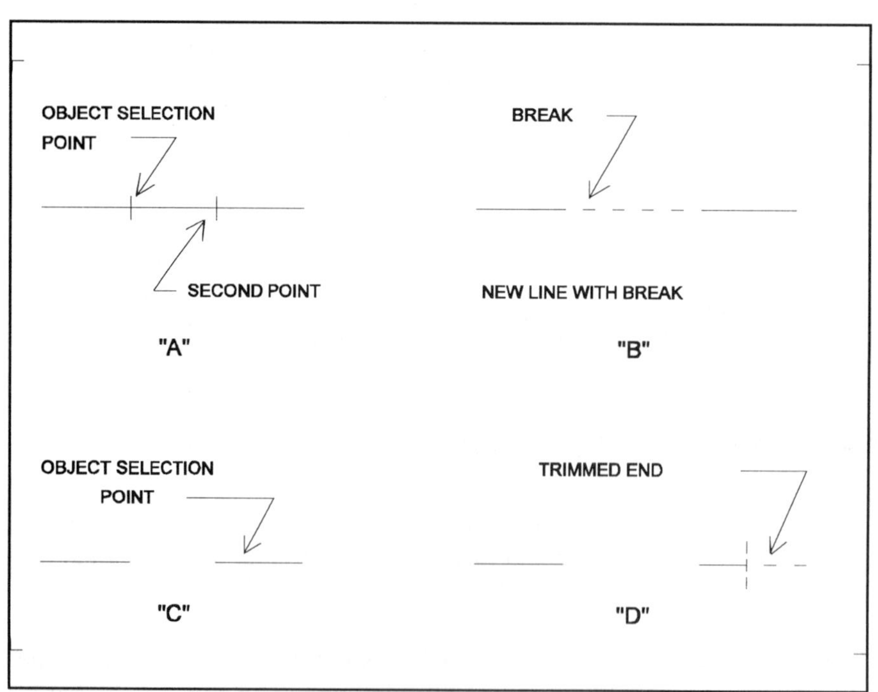

Figures 3-13 A, B, C, D. The Break command.

Here is where the line will be trimmed from (Fig. 3-13C).

```
Enter second point (or F for first point): PICK THE SECOND
POINT PAST THE RIGHT END OF THE LINE
```

Note that the section of line between the two points was trimmed off at the location of the first object selection (Fig. 3-13D).

Circles can be changed to arcs by removing the unwanted segment. The removed section of the circle is referenced counterclockwise from the object selection point.

EXERCISE 3-6c. THE TRIM COMMAND

Now to play with the Trim command.

The Trim command allows us to trim a line should it be too long. First we define a "cutting edge" as the first line in the selection process, and then point to the object to be trimmed. To understand this, look at Figure 3-14A. We will trim line A from the box, using line B as a cutting edge. Draw the figure so you can follow this.

```
Command: TRIM (RETURN)
Select cutting edge(s)...
Select objects: PICK LINE B(RETURN)
Select objects to trim: PICK LINE A(RETURN)
```

This caused the small line A to be removed from the box (Fig. 3-14B).

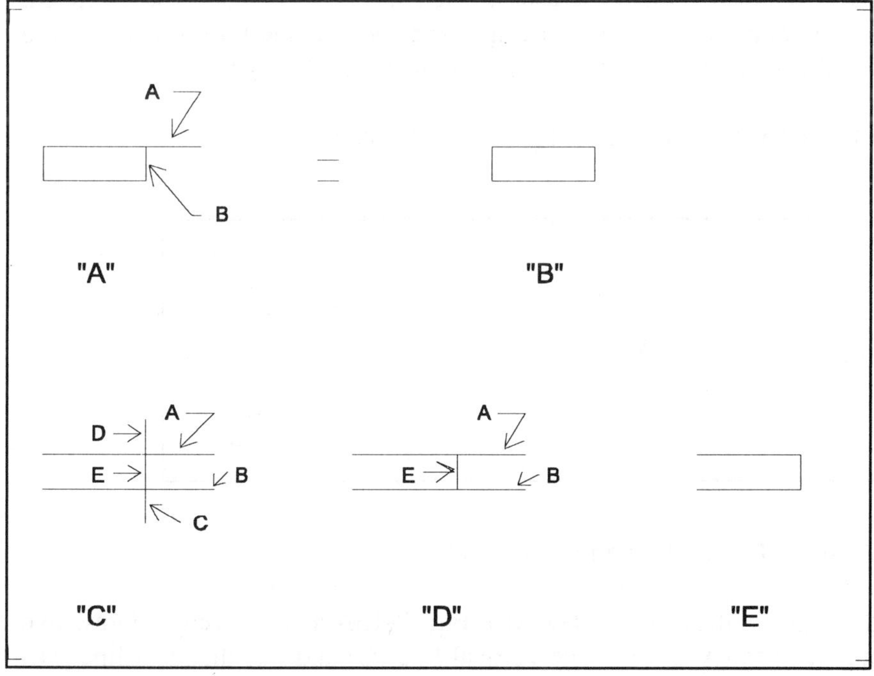

Figure 3-14 A, B, C, D, E Trim command.

Using Fig. 3-14C as a reference, draw it, and we will square up the end by first removing lines C and D.

```
Command: TRIM (RETURN)
Select cutting edge(s)...
Select objects: PICK LINE A AND B (RETURN)
Select object to trim: PICK LINE SECTION C
Select object to trim: PICK LINE SECTION D (RETURN)
```

The object now looks like Fig. 3-14D. Let's trim the remaining lines, A and B.

```
Command: TRIM (RETURN)
Select cutting edge(s)...

Select objects: PICK LINE E (RETURN)
Select objects to trim: PICK LINE A
Select objects to trim: PICK LINE B (RETURN)
```

This has removed all of the extended lines from Fig. 3-14C, and our object should now look like Fig. 3-14E.

EXERCISE 3-6d. THE EXTEND COMMAND

The Extend command is similar to using the Trim command in that you must first select (or create) a boundary line to extend to, then select the object to be extended. The extension is a continuation of an existing object or line and thus allowing us to never have to redraw one line on top of another, to make it longer.

Using Fig. 3-15 as a reference, we will play with Extend.

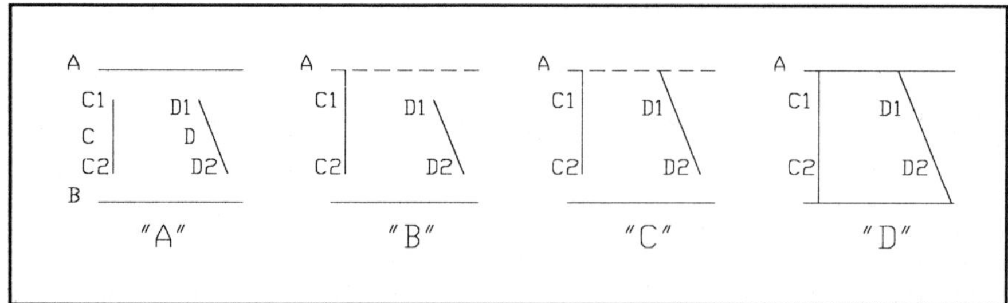

Figure 3-15 A, B, C, D The Extend command.

Create a small sketch to allow us to play. Use Fig. 3-15A as a reference. Draw two horizontal parallel lines (A and B), one vertical line (C) and one diagonal line (D).

Now type:

```
Command: EXTEND (RETURN)
```

```
Select boundary edge(s)...
Select objects: CLICK ON HORIZONTAL LINE A (RETURN)
```

Line A becomes dotted (Fig 3-15B).

```
Select object to extend: CLICK ON LINE C AT POINT C1 (RETURN)
```

Line C now extends to Line A. Let's try it again on line D, the diagonal line. Type:

```
Command: EXTEND (RETURN)
Select boundary edge(s)...
Select objects: CLICK ON HORIZONTAL LINE A (RETURN)
```

Line A, again becomes dotted (Fig 3-15C)

```
Select object to extend: CLICK ON LINE D AT POINT D1 (RETURN)
```

Note how line D was extended, at the same angle, straight to line A.

Play with Extend on your own using Line B and points C2 and D2 as a reference. When you are done, your drawing should look similar to Fig. 3-15D.

Get Rid Of The Garbage

We have added a number of unwanted lines to our drawing. Remove these before you end this session, as we don't want them as part of our drawing. This ends our title block exercise.

REVIEW

Before we proceed any further with our drawing, let's review what we have done so far.

1. We have examined a number of AutoCAD commands, gotten our feet wet with a play session called *getting the itch out*, and explored the differences and similarities between various versions of AutoCAD during start-up.

2. We have given much thought to the layout of a simple AutoCAD drawing, our title block.

3. We have chosen to work in decimal (metric) units, played with Limits and the Status of our drawing, and discovered one of AutoCAD's gremlins, the orphan.

4. We have explored the Snap command and found how to lock on to grid points, as well as how to move freely around the drawing or at only right angles with

the Ortho command.

5. We have zoomed to All to view our complete drawing, or by a Window which allowed us to observe the tiny details through a close up view.

6. We have discovered a number of ways to use and modify text through the Text and Dtext command, as well as change the text style.

7. We have saved an image of the title block to disk. In doing so, we used the Block and Wblock command.

8. We have created four sizes of paper (A, B, C, & D) and used Insert to add a title block to each.

9. We have looked at the Line command and examined some problems we could run into if lines are drawn on top of each other, as well as explored the Break Trim and Extend commands.

10. Most of all, you probably had *FUN*, which is what learning is all about.

This is quite an accomplishment for a newcomer to AutoCAD! You should feel proud of your accomplishments! Now; are we ready to continue?

Make a Backup Disk

Before we go any further, it is highly recommended that you make a backup copy of the work disk. This will protect these many hours of hard work you have done so far, should a mishap occur. Keep your backup disk and work disk in different locations, again as added insurance against mishaps. To make your backup, refer to the DOS Copy command in Chapter 2.

USEFUL HINTS

It should be noted that when a drawing is saved, the Grid, Snap, Limits, and Zoom reference, as well as many more attributes of the drawing are also saved. These attributes are reinstated when the drawing is recalled to the screen and do not require being re-established.

Before saving a drawing, it is best to zoom to All. This will allow all of the drawing to be seen when it is recalled, thus preventing some confusion.

QUIZ: GETTING AutoCAD RUNNING

Name ____

Circle the most correct answer or fill in the blanks.

1. Which of the following commands will start AutoCAD running?
   a. ACAD or ACADR12 or ACADR13.
   b. RUN AutoCAD.
   c. START.
   d. None of the above.

2. After most keyboard commands, you should press:
   a. The Return key.
   b. The Enter key.
   c. The carriage Return key.
   d. All of the above.

3. "Housekeeping" in AutoCAD refers to:
   a. Doing a "chkdsk".
   b. Making surc your disk is clcancd oncc a wcck.
   c. The process of setting up an AutoCAD drawing before any work is performed on it.
   d. Formatting a disk.

4. To Cancel a command, which of the following keys are pressed together?
   a. SHIFT + C
   b. CTRL + C
   c. ALT + C
   d. CTRL + ALT + DELete

5. What are four things that are usually set up at the beginning of most drawings, before any lines are drawn?

   a.____

   b.____

   c.____

   d.____

6. What are the two steps that must be done in order to view a portion of your drawing?

   a.____

   b.____

7. What command is used to bring the cursor directly to a grid line or mark?

   Answer:____________________________________

8. What command restores an object when it disappears from the screen during a Block or Erase command?

   Answer:____________________________________

9. What two *simple* methods can be used to restore or repeat an AutoCAD command?

   a.____

   b.____

10. What two commands must be issued in order to save an object from your drawing to disk?

    a.____

    b.____

chapter 4

The Schematic

DISCUSSION

In this chapter, we will develop the drawing for the electronic schematic for our project, the dual tracking 0 to +/- 15 volt power supply and its side kick, the fixed +5 volt supply. In doing so, you will:

a) Convert the hand-drawn sketch of Fig. 4.2 into a full-blown industrial quality schematic drawing.

b) Be introduced to the Electronic Symbol Template (ELECTEMP.DWG) drawing found on the disk supplied with this tutorial.

c) Create a Symbol Library.

d) Walk through the procedure required to lay out the SCHEMA drawing.

e) Become familiar with drawing notations, and

f) Be introduced to a host of new AutoCAD commands.

EXERCISE 4-1. GETTING OUR DRAWING PAPER

In this exercise, you will be introduced to the AutoCAD terms and commands of:

ARC	CHANGE	CIRCLE
COLOR	DDLMODES	DOS Copy
DOS Shell	Dotline	Hidden
Linetype	LAYER	OSNAP Options

EXERCISE 4-1a. DUPLICATING THE C-SHTE INTO THE SCHEMA DRAWING

Discussion

Our schematic (SCHEMA) is drawn on the C size prototype sheet already developed and stored on our work disk under the file name C-SHTE (500 x 380mm) for Mechanical and Electrical drawings. To make use of this sheet and not destroy it for future use, we will duplicate it. This can be done using two methods; In versions 10 and 11 of AutoCAD, by loading the file C-SHTE from the Main Menu and saving it under the name SCHEMA, and then ENDing. In version 12, by using a method called the "Prototype Drawing", which you'll find is quick and easy to use. We'll first look at version 12.

For those of you with version 10 or 11, go to Exercise 4-1b.

Using Versions 12 & 13, The "Prototype Drawing" Method

In version 12, there again are two methods that can be used to duplicate the drawing. These depend upon the status of the FILEDIA command, which controls whether the dialogue boxes are active or not.

To check the status of the FILEDIA (File Dialogue), type:

```
Command: FILEDIA (RETURN)
```

If the status returned is a 1, then the dialogue boxes are active which is its default setting. If the status is a 2, not active, the dialogue boxes can only be requested via an internal condition set by the file ~(Tilde). When they are not active, you can type your request at the Command line. To activate the dialogue boxes from the FILEDIA Command, reply with a "1" to the FILEDIA Command requester.

Dialogue Boxes Turned OFF

If the dialogue boxes are turned OFF, the duplication of the C-SHTE is done through the Command line. With your work disk in drive A, type:

```
Command: NEW (RETURN)
```

When AutoCAD asks for the drawing name, type:

```
Command: A:SCHEMA=A:C-SHTE (RETURN)
```

We now have two identical files, one on our work disk (C-SHTE), and one displayed at the drawing editor called SCHEMA which has all the attributes of the C-SHTE.

Now, save this new copy of the C-SHTE as SCHEMA.

```
Command: SAVE (RETURN)
File name: A:SCHEMA (RETURN)
```

Dialogue Boxes Turned ON

If the dialogue boxes are active, (turned ON), we can use this next method. After typing NEW at the Command line, the "Create New Drawing" dialogue box (Fig. 4-1) appears.

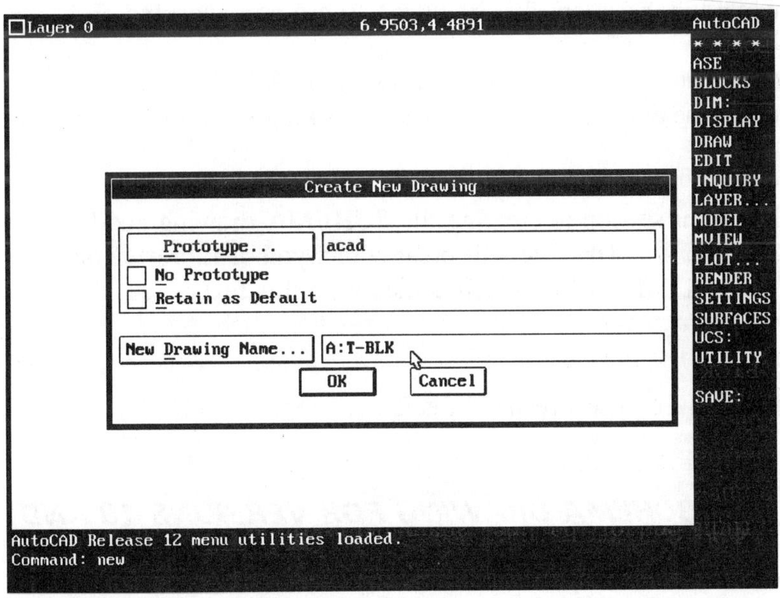

Figure 4-1 Create New Drawing dialogue box.

```
Command: NEW (RETURN)
```

The Create New Drawing dialogue box appears.

Using the Create New Drawing dialogue box, we'll double click on the Prototype

Edit Box. This is the box to the right of the "Prototype.." button. The drawing name "ACAD" should be the default name in this box.

TRAP Be careful not to pick the Prototype button as this will take you to another dialogue box, the Prototype Drawing File dialogue box. If, by accident you did, pick OK in the Prototype Drawing File dialogue box. This will get you back to the Create New Drawing dialogue box.

```
Command: DOUBLE CLICK THE "PROTOTYPE EDIT BOX,"
         AND THEN TYPE A:C-SHTE (RETURN)
```

The text in the Prototype Edit Box now contains C-SHTE, replacing ACAD.

Now pick the New Drawing Edit Box, the box to the right of the "New Drawing Name..." Button.

```
Command: PICK THE NEW DRAWING EDIT BOX AND TYPE A:SCHEMA (RETURN)

Command: CLICK ON THE OK BUTTON.
```

When you pick the OK button, AutoCAD imports the C-SHTE drawing with all its attributes and calls it SCHEMA. This is the name of the new drawing now being displayed by the drawing editor screen, the drawing you are now working on.

TIP: Note that you do not have to save this drawing as SCHEMA and restart AutoCAD because the drawing you are already working on is the SCHEMA drawing.

At this point we will END our session of creating the SCHEMA drawing and briefly look at the Hand Sketch, Fig. 4-2. Then we will create our Symbol Library from the Electronic Template ELECTEMP.DWG which comes with this text.

```
Command: END (RETURN)
```

Now move on to Exercise 4-1c, The Freehand Sketch.

EXERCISE 4-1b. THE SCHEMA DRAWING FOR VERSIONS 10 AND 11 USERS.

For versions 10 and 11 users, boot AutoCAD. The Main Menu appears:

```
Main Menu:

  0.  Exit AutoCAD
  1.  Begin a NEW drawing
  2.  Edit an EXISTING drawing
  3.  Plot a drawing
  4.  Printer Plot a drawing
  5.  Configure AutoCAD
```

```
6.  File utilities
7.  Compile shape/font description file
8.  Convert old drawing file
9.  Recover damaged drawing

Enter selection:
```

Insert your work disk in drive A, and select 2 (Edit an EXISTING drawing).

```
Enter selection: 2 (RETURN)
Enter Name of drawing: A:C-SHTE (RETURN)
```

AutoCAD now brings into the computer's memory the C-SHTE drawing and sets up the screen for use. When this process is complete, type or select Save at the command prompt.

```
Command: SAVE (RETURN)
File name: A:SCHEMA (RETURN)
```

We now have two identical drawings, C-SHTE and SCHEMA on disk. Now type:

```
Command: END (RETURN)
```

This brings you back to the Main Menu selection.

The Limits, Grid, and Snap attributes of the SCHEMA drawing are the same as those set for the C-SHTE drawing.

EXERCISE 4-1c. THE HAND SKETCH

Discussion

Before starting any work on our SCHEMA drawing, we will briefly look at the hand sketch of Fig. 4-2 and then copy to our work disk the Electronic Symbol Template (ELECTEMP.DWG), found on the disk that comes with this text.

The hand sketch of Fig. 4-2 is similar to that which you may receive as a reference drawing from a company's engineering office. As a drafter, you will be responsible to translate engineering sketches into a full set of production working drawings. These drawings include:

a) The Electronic Schematic Drawing.
b) The Printed Circuit Board Drawing Artwork.
c) The Printed Circuit Assembly Drawing, and
d) The Mechanical Assembly Drawing.

The production of these drawings, in essence, is what this tutorial is all about; that is, developing a set of working drawings which can be used by a company's production shop to produce a product.

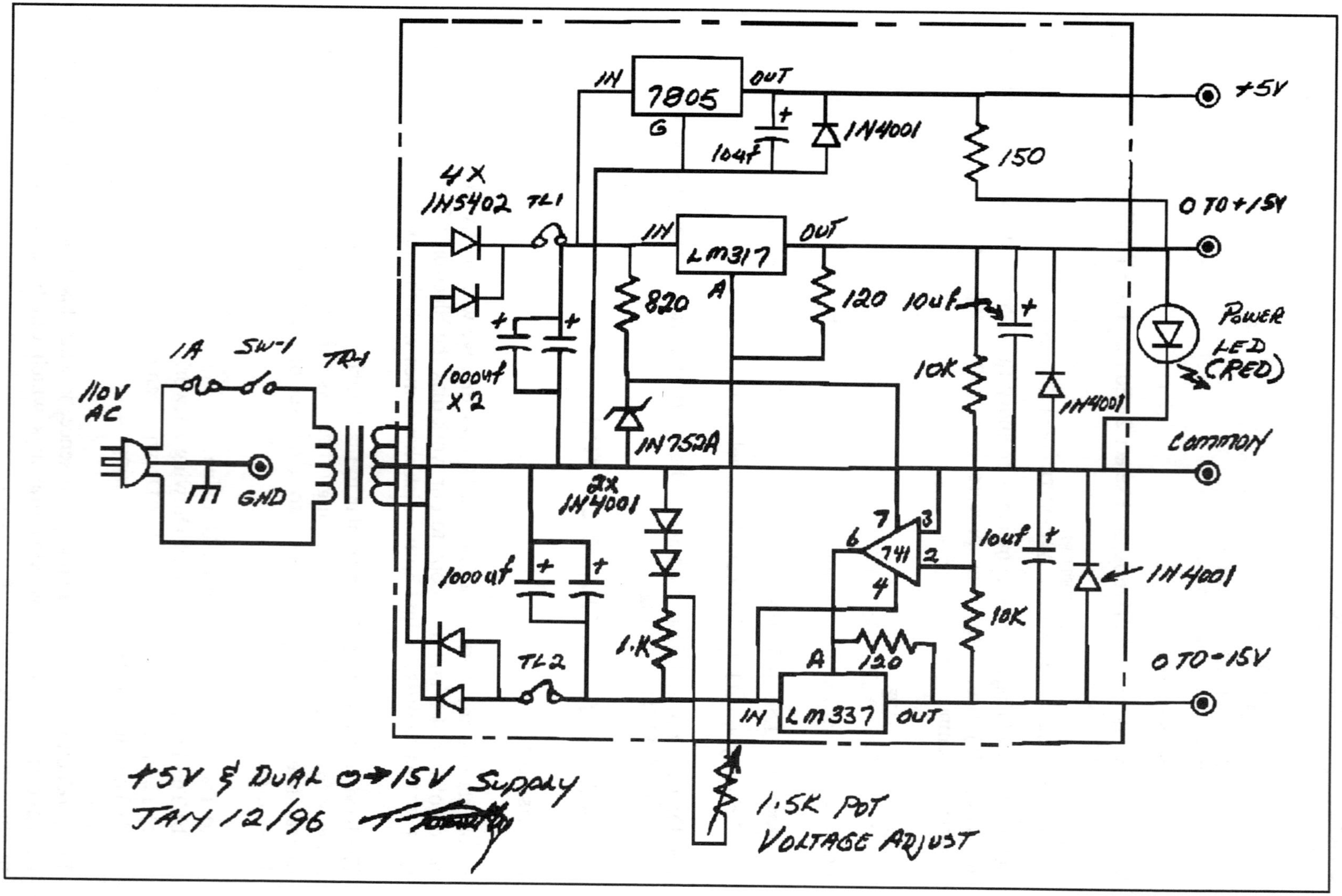

Figure 4-2 Hand Sketch of Power Supply Schematic.

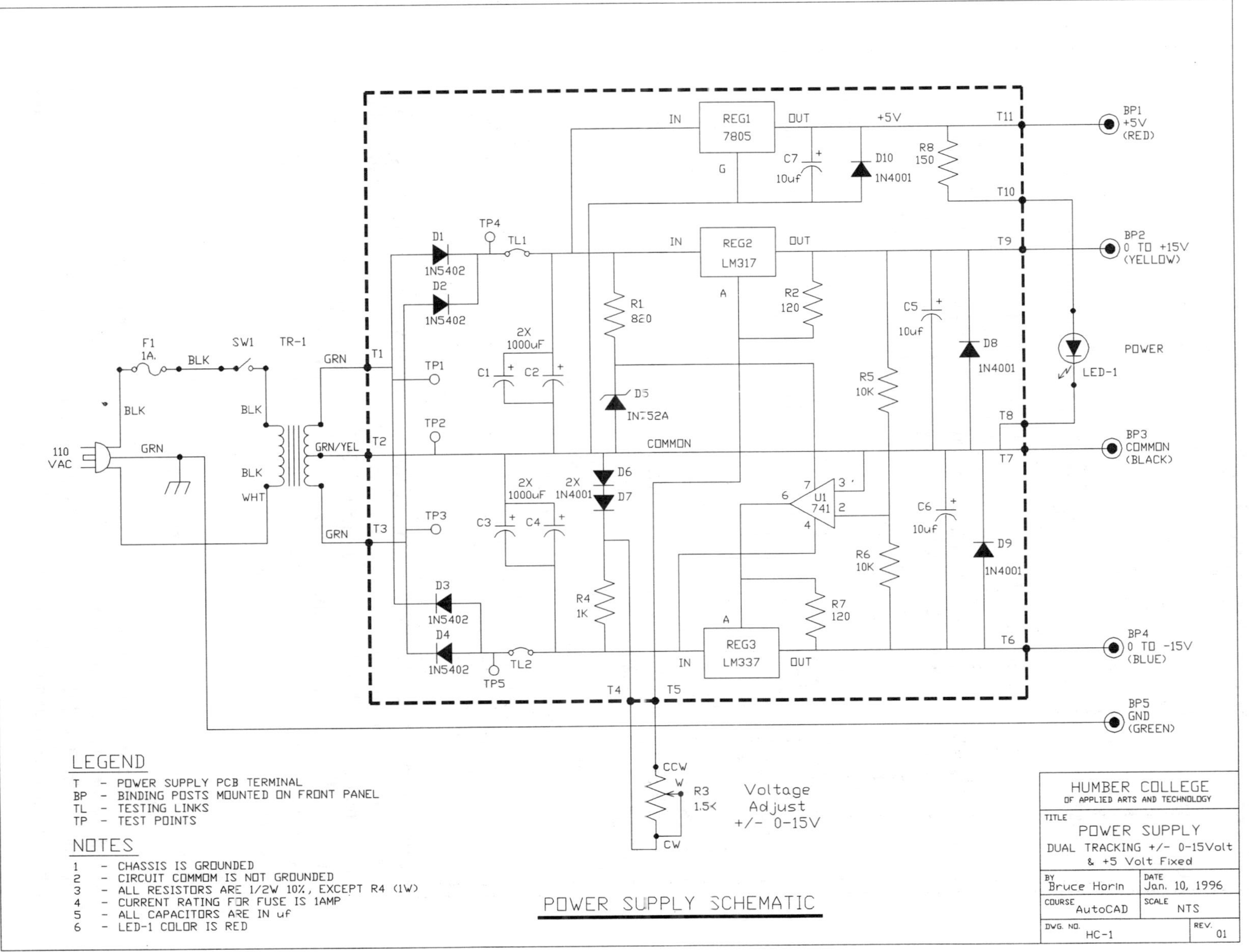

Figure 4-3 Schematic Drawing.

ANALYZING THE CIRCUIT AND THE SCHEMATIC

Let's spend a few moments to look at the schematic and analyze the circuit. Use Figure 4-3 as a reference.

The AC Power, Transformer Circuit, Fuse and Power Switch

The circuit is designed to meet Underwriter Labs or local Hydro standards. The AC line cord must be a three conductor cable. The green (GRN) wire of the cord is connected to the chassis (Chassis ground), the white (WHT) wire is the power "return" line and the black (BLK) wire is the power or hot wire. The first component in series with the power wire must be the fuse to protect the circuit. Power is switched On or Off through switch SW1.

Transformer TR-1 reduces the applied 110V AC power to a center-tapped 35 volt AC level. This is used to supply the +5 volt supply and the dual +/- 0 to 15 volt supplies.

Diodes D1 and D2 convert half of the 35V AC from transformer TR-1, to produce the DC source voltage for the LM7805 (+5Volt regulator) and the LM317 regulator (0 to +15 volts), and diodes D3 and D4 create the DC source voltage for the LM337 negative voltage regulator.

Capacitor C1 and C2 filter the positive voltage for regulators REG1 and REG2. Capacitor C3 and C4 does the same for the negative supply.

+5 Volt Circuit

Regulator REG1 is a standard +5 Volt, three terminal voltage regulator which is completely fool proof. That is, it is internally short circuit protected and shuts down on thermal overload. Remove the short and everything returns to normal, thus it is almost indestructible. Just don't apply 110V AC across it. The output lead of the regulator is connected to capacitor C7 which removes high frequency spikes and diode D10 is used to reduce negative spikes.

The +/- 0 to 15 Volt Dual Tracking Circuit

The dual tracking supplies will track each other to within .5 percent. The heart of the circuit is the Operational Amplifier U1 (LM741 or LF356) that controls this circuit by monitoring the differential voltage at the two output leads of REG2 and REG3 and then adjusts the output of REG3 to track REG2. It does this by comparing the voltage produced at the node or center point of resistors R5 and R6 which should be 0 volts. If it is not 0, REG3 is automatically adjusted until it becomes so.

R1 and the Zener Diode D5 (1N752A) are used to produce a +5 volt supply for the 741 OpAmp.

Diode D6 and D7 are used to provide a -1.2 volt reference voltage so the positive 0-15 volt supply (REG2) can be adjusted down to zero volts.

Output Voltage Adjustment

Output voltage adjustment is controlled by R2 and the variable resistor R3 which varies the output of the positive supply. When this voltage changes, the voltage at the center tap of resistors R5 and R6 also changes, and the OpAmp automatically adjusts the negative supply to track the initial voltage change.

Remaining Components

R7 (120 ohms) provides a load current for the LM337 to assure minimum load current is present at all times on the negative supply. R2 (120 ohms) does the same for the positive supply.

R8 (150 ohms) is a current limiting resistor to hold the LED current to a maximum of 20mA, and also provides a minimum current for the LM7805.

LED1 is used as a "Power ON" lamp and to indicate the +5 volt supply is operating.

Capacitor components C5 - C7 and Diodes D8 - D10 provide transient protection for the outputs of their associated regulators.

A Few Words Of Caution

It should be noted that the input, output, and common connections on regulators REG-1, REG-2 and REG-3 are quite different, even though they look the same. This must be remembered when the regulators are placed on the printed circuit board and wired into the circuit.

All electrolytic capacitors have the potential to explode if wired into the circuit backwards. Therefore, care must be taken when labeling the capacitor and showing its orientation on the schematic drawing.

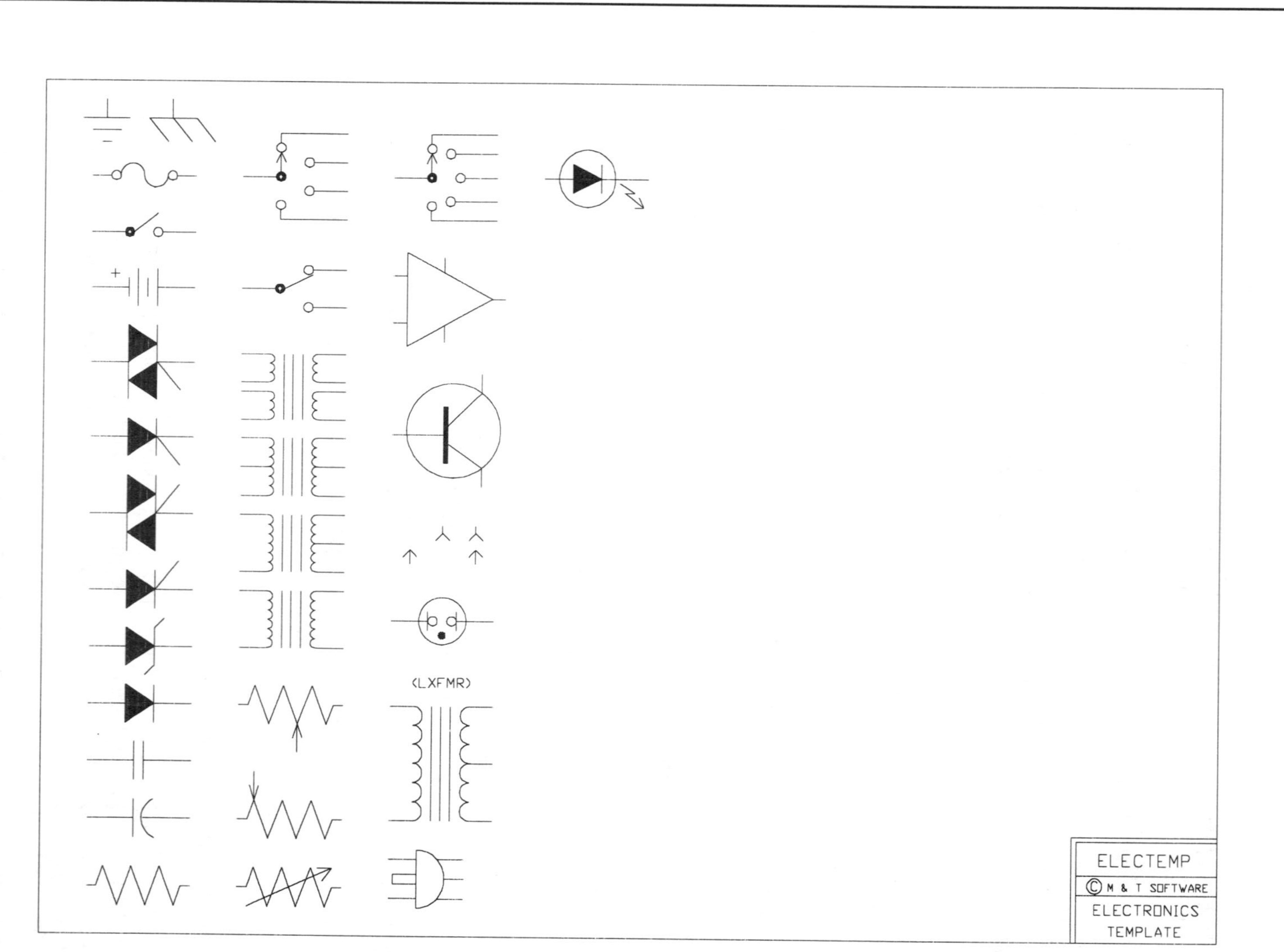

Figure 4-4 Electronic symbol template.

EXERCISE 4-2. ELECTRONIC SYMBOL TEMPLATE

Discussion

The Electronic Symbol Template drawing on the disk that comes with this text (ELECTEMP.DWG), Fig. 4-4, contains all the necessary symbols required for the layout of the schematic drawing. Each symbol is drawn on a 1 mm grid and is approximately 10 mm x 25 mm in size, making them the correct scale for the C size drawing.

The Electronic Symbol Template (ELECTEMP.DWG) and the companion drawing, Electronic Component Template (ELECOMP.DWG) will speed up the drawing process as they already contain many electronic symbols and components needed for drawing AutoCAD electronic projects, thus saving you time to create them. When you have played with AutoCAD and developed symbols on your own, you may want to store them on this disk.

As each symbol is a stand-alone entity, it will be extracted from the template and placed on your work disk as a Wblock (rubber stamp) to create a Symbol Library. By doing the selection, using the AutoCAD commands of Block, Wblock, and Insert, you have the option to select which point you want as an insertion point. We will walk through this process as we proceed.

Our next task will be to transfer the Electronic Symbol Template, Fig. 4-4, to your work disk. This may be accomplished by using the DOS Copy command or through AutoCAD. In this discussion, we will use AutoCAD, as it is up and running in the computer. As your computer setup may have a different hardware configuration than mine, I have included several methods to copy the Electemp.dwg file to your work disk.

EXERCISE 4-2a. COPYING FILES TO YOUR DISK ON A TWO-DRIVE SYSTEM FOR VERSIONS 10 AND 11.

Note: This procedure is for a two-disk-drive floppy system with similar capacity. If you have only a single drive and a hard disk, skip down to the next section, Exercise 4-2b. If you are using version 12, skip to Exercise 4-2c.

Discussion

There are several ways to copy a file on a two-drive system from within AutoCAD. The first method uses the Copy File routine found under File Utility Menu, while the second uses the Shell command. An alternative method is directly from DOS using the Copy command. See Copy in Chapter 2 for a detailed explanation.

Method 1. Using the File Utility "Copy"

In this exercise, the B drive will contain your "write protected" source disk, the one with the ELECTEMP.DWG file, and the A drive will contain your work disk.

For versions 10 and 11 users, you gain access to Copy from the Main Menu under item 6 (File utilities). We'll walk through this process for releases 10 and 11 here. Version 12 users should advance to Exercise 4-2c. At the command prompt type:

```
Command: FILES (RETURN)
```

After the File Utility menu is displayed, place the Electronic Symbol Template disk in drive B and your work disk in drive A.

```
File Utility Menu

   0. Exit File Utility Menu
   1. List Drawing files
   2. List user specified files
   3. Delete files
   4. Rename files
   5. Copy files

Enter selection: 5 (RETURN)
Enter name of source file: B:ELECTEMP.DWG
Enter name of destination file: A:ELECTEMP.DWG (RETURN)
```

TIP: The .DWG extension must be used in the file names, because AutoCAD recognizes only files with this extension. After completion of the above, the ELECTEMP.DWG file is now on your work disk.

Method 2. Using the Shell Command to Copy Files

Note: To use the AutoCAD Shell command, you must be in the Drawing Editor, not the Main Menu.

In this example, we'll copy the file ELECTEMP.DWG, found on the Electronic Symbol disk, from drive B to our AutoCAD work disk in drive A. The names of these two drawings will be kept the same: ELECTEMP.DWG. This procedure makes use of the Shell command found within the standard AutoCAD command set, which in turn invokes a DOS operation. After you have used the desired DOS command, return back to AutoCAD by using the Exit command.

Insert the "write protected" Electronic Symbol disk into drive B, and make sure that your work disk is in drive A. Then type the following at the command prompt.

```
Command: SHELL (RETURN)
Dos Command: Copy B:ELECTEMP.DWG  A:ELECTEMP.DWG (RETURN)
```

TIP: The .DWG extension must be used with both file names, as AutoCAD recognizes only files with this extension.

The A and B drive lights become active during this operation as an indication that all is well, and the drawing is transferred to your work disk.

EXERCISE 4-2b. COPYING FILES USING A FLOPPY AND A HARD DISK VIA DOS

On a single-drive system with a hard disk, you can use the hard drive as a temporary storage while you swap the source disk for the destination disk.

Method 1. Single Floppy and Hard Disk Using DOS

The first method uses the hard disk as an intermediate step to receive the file being transferred from the source drive (A). This file is stored in a temporary file on the hard disk. The new destination disk is now placed in drive A, and the temporary file on the hard disk is transferred back to drive A. When you have completed this operation, you may want to clean up your hard disk by deleting the old file. The following command sequence accomplishes this.

Insert your "write protected" source disk in drive A.

```
COPY A:ELECTEMP.DWG  C:ELECTEMP.DWG (RETURN)
```

Insert the destination disk in drive A.

```
COPY C:ELECTEMP.DWG  A:ELECTEMP.DWG (RETURN)
```

Now delete the temporary file from the C drive.

```
ERASE C:ELECTEMP.DWG (RETURN)
```

Method 2. Using a Single Floppy "A" and a Pseudo Floppy "B"

In Chapter 2, under the Copy command, we discussed using a Pseudo B drive when we really only had a single A drive.

First write-protect your source disk and insert it into drive A, then from the A:\ prompt type:

```
A:\ COPY A:ELECTEMP.DWG  B:ELECTEMP.DWG
```

DOS activates and responds with:

```
Insert diskette for drive B:
Strike any key when ready.
```

Now remove the source disk and insert your work disk in drive A, even though DOS thinks it's drive B.

Now press any key. DOS eventually responds with:

```
1 File(s) copied
```

Your work disk now contains the file ELECTEMP.DWG.

EXERCISE 4-2c. USING THE DIALOGUE BOXES IN VERSION 12 TO COPY A FILE FROM DRIVE B TO DRIVE A

To assure the dialogue boxes are active for version 12 users, we will turn the FILEDIA program ON. Type:

```
Command: FILEDIA (RETURN)
New value for FILEDIA <1>:1 (RETURN)
```

We are now ready to start our Copy process. Place your work disk in drive A and the disk with the file ELECTEMP.DWG in drive B.

We will initiate the File Utilities dialogue box (Fig. 4-4a) which contains six option buttons. They are: List files, Copy file, Rename file, Delete file, Unlock file and Exit.

```
Command: SELECT FILE, AND THEN DOUBLE CLICK ON UTILITY.
```

This displays the File Utility dialogue box, Fig. 4-4a.

Figure 4-4a File Utilities dialogue box.

Now,

```
Command: CLICK ON THE COPY FILE BUTTON.
```

This activates the Source File dialogue box. See Fig. 4-4b. This dialogue box contains two windows marked "Directories" and "Files". We will select drive B in the Directories window. This will show us what files exist on drive B.

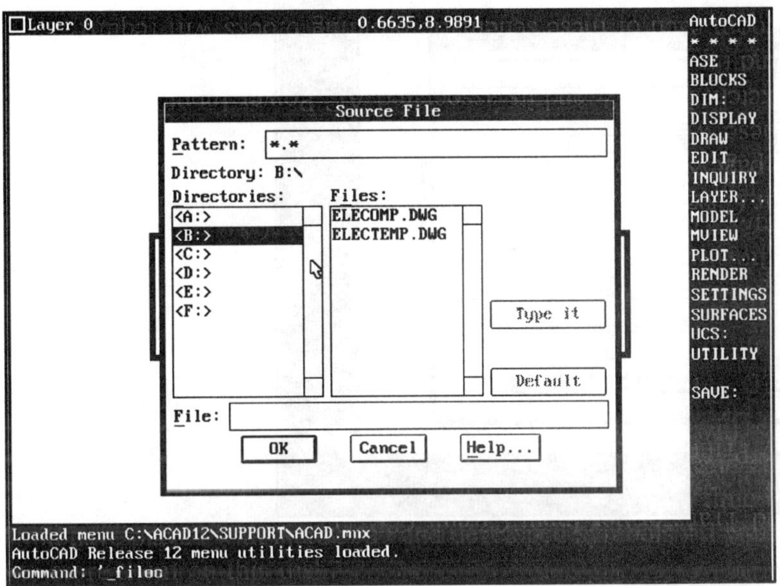

Figure 4-4b Source Files dialogue box.

In the Directories window:

Command: **MOVE THE DIRECTORIES SCROLL BAR TO REVEAL THE B DRIVE.**

Command: **DOUBLE CLICK ON THE B**

This causes a directory listing of the B drive to appear in the Files window.

Command: **DOUBLE CLICK ON THE "ELECTEMP.DWG" FILE.**

The Destination Files dialogue box appears. See Fig. 4-4c.

Command: **MOVE THE DIRECTORIES SCROLL BAR DOWN TO REVEAL THE A DRIVE, AND DOUBLE CLICK ON "A".**

This caused a directory listing of the files on the A drive to appear in the Files section of the dialogue box. Now to copy the ELECTEMP.DWG file from the B drive to the A drive.

Command: **CLICK ON THE OK BUTTON.**

This causes the ELECTEMP.DWG file to be copied to the disk in the A drive. We are now ready to start our Symbol Library.

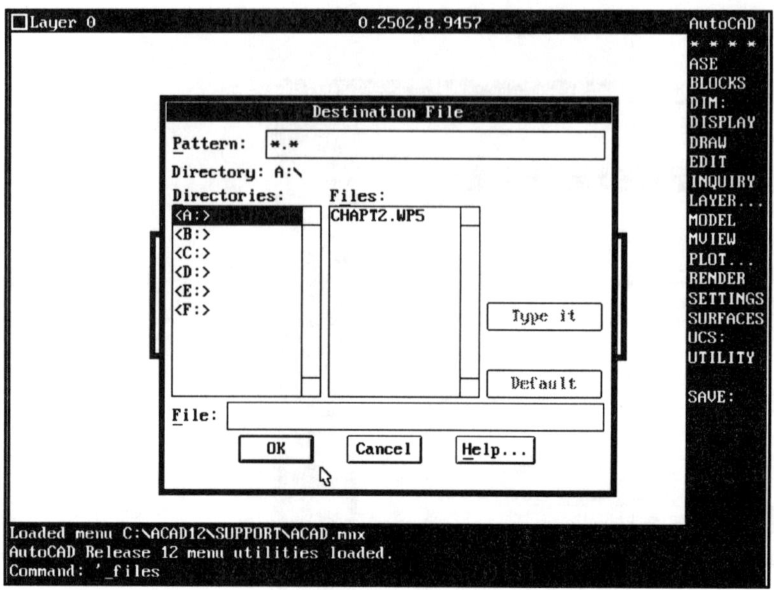

Figure 4-4c Destination File dialogue box.

EXERCISE 4-3. CREATING THE ELECTRONIC SYMBOL LIBRARY

Discussion

We now have the Electronic Symbol Template (ELECTEMP.DWG Fig. 4-4) transferred to our work disk. It's time to extract the symbols we will use in our schematic and save them to our work disk as Wblocks in a Symbol Library. The process is identical to that used to save our Title Block in Chapter 3. If you have doubts concerning how to proceed, review Exercise 3-5a again.

Capturing and Saving the Symbols as Blocks and Wblocks

What Symbols Will Be Used? To get started, load the drawing ELECTEMP into AutoCAD. Our schematic drawing will require the use of 15 symbols from the ELECTEMP, as shown in Fig. 4-5. They are listed in Table 4-1.

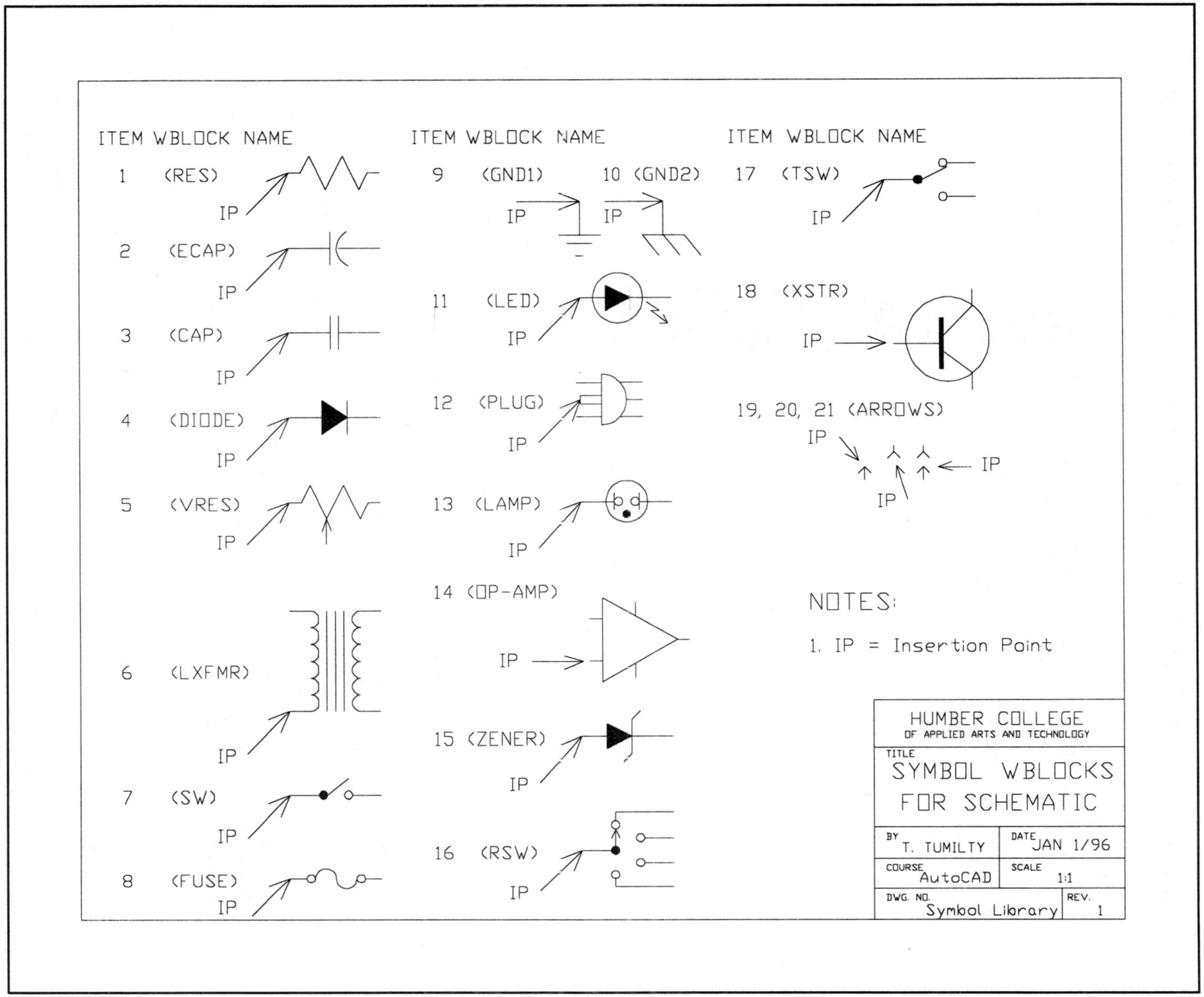

ITEM	WBLOCK NAME	ITEM	WBLOCK NAME	ITEM	WBLOCK NAME
1	(RES)	9	(GND1)	17	(TSW)
2	(ECAP)	10	(GND2)	18	(XSTR)
3	(CAP)	11	(LED)	19, 20, 21	(ARROWS)
4	(DIODE)	12	(PLUG)		
5	(VRES)	13	(LAMP)		
6	(LXFMR)	14	(OP-AMP)		
7	(SW)	15	(ZENER)		
8	(FUSE)	16	(RSW)		

NOTES:

1. IP = Insertion Point

HUMBER COLLEGE
OF APPLIED ARTS AND TECHNOLOGY

TITLE SYMBOL WBLOCKS FOR SCHEMATIC

BY T. TUMILTY	DATE JAN 1/96
COURSE AutoCAD	SCALE 1:1
DWG. NO. Symbol Library	REV. 1

Figure 4-5 Wblocks required for the schematic.

EXERCISE 4-3a. CAPTURING THE SYMBOLS AND USING BLOCKS

<u>**TIP:**</u> Zoom into the Electronic Symbol template so that you get a clear view of the symbols you want to extract. Figure 4-5 suggests typical insertion points. I highly recommend you use these, but you may choose any point you desire.

TABLE 4-1
Electronic Symbol Library (ELECTEMP)

Item	*Symbol Name*	*BLOCK*	*WBLOCK*
1	Normal resistor	(RES)	(WRES)
2	Electrolytic capacitor	(ECAP)	(WECAP)
3	Normal capacitor	(CAP)	(WCAP)
4	Diode	(DIODE)	(WDIODE)
5	Variable resistor	(VRES)	(WVRES)
6	Large transformer	(LXFMR)	(WLXFMR)
7	Switch SPST	(SW)	(WSW)
8	Fuse	(FUSE)	(WFUSE)
9	Ground earth/power line	(GND1)	(WGND1)
10	Ground chassis	(GND2)	(WGND2)
11	Light Emitting Diode	(LED)	(WLED)
12	Power plug	(PLUG)	(WPLUG)
13	Pilot lamp	(LAMP)	(WLAMP)
14	Operational amplifier	(OPAMP)	(WOPAMP)
15	Zener diode	(ZENER)	(WZENER)

We are now ready to start selecting symbols. If you feel confident to select these on your own, do so and then continue to the next discussion section. Good luck.

We will first select the symbols as Blocks and then save them as Wblocks.

If you are still getting your feet wet, let's walk through the selection of item 1, the resistor (RES) together. Use Fig. 4-5 as a reference. We will capture it using a window.

```
Command: PICK "BLOCKS" FROM THE ROOT MENU

Command: PICK "BLOCK" FROM THE BLOCKS MENU
Block name (or?): RES (RETURN)
Insertion base point: - PICK YOUR INSERTION POINT
Select objects: W (RETURN)
First point: - PICK 1ST POINT OF WINDOW
Second point: - DRAG THE WINDOW AROUND THE RESISTOR
Press the RETURN key.
```

When you press the Return key, the object selected will disappear from the screen (captured by AutoCAD). To bring the object back, pick OOPs from the screen menu, or simply type OOPS at the keyboard.

```
Command: PICK OOPS FROM SCREEN MENU
```

You have just created a Block called RES.

TIP: To save time, repeat this procedure with the remaining symbols, giving each its appropriate name. When all symbols are captured, we'll save them to our work disk, using the Wblock command as shown in the next exercise.

TRAP Should you make a mistake in selecting a symbol, don't worry. It's easy to correct. Just select it again and add "1" after it (that is, RES1 instead of RES). The mistake can be deleted the next time you enter AutoCAD using the Purge command. (*Note:* To purge an entity, Purge must be the first command you use in AutoCAD.)

EXERCISE 4-3b. CREATING THE SYMBOL LIBRARY USING WBLOCK

Saving Our Symbols as Wblocks: We'll use the keyboard to create our Wblock and assign the associated Block. The Wblock name is identical to the Block name, except it is prefixed by the letter "W" for easy identification later. Each Wblock is like a rubber stamp, a unique entity of our Block symbol, but is stored on the work disk. We have, in effect, a Symbol Library. We'll use the resistor (RES) just captured as the first Block to save as a Wblock in our Symbol Library.

```
Command: WBLOCK (RETURN)
Command: Wblock file name: A:WRES (RETURN)
```

This creates the Wblock file name WRES on the A drive.

```
Block name: RES (RETURN)
```

This places the Block RES in the drawing file, named WRES.

Repeat the above procedure for each of the symbols captured in Exercise 4-3a.

TIP: Don't forget to precede the Wblock file name with the drive letter A: so that it is written to your work disk and not the hard disk. Likewise, remember the "W" in front of the Wblock file name.

EXERCISE 4-4. STARTING OUR SCHEMATIC DRAWING

Discussion

Some Food For Thought Before We Start - Drawing Scale

In the normal process of doing manual drafting, much time and thought is given to the spacing and scaling of the building blocks of the drawing before we put pen to paper. We must consider what size the lettering should be, what dimensions the components should be and where to start the placement of these components in the first place. If we choose blocks that are scaled too large, we will run out of paper by the time we're done. If we choose too small a scale, our drawing will look lopsided and unbalanced on the sheet. Here is where AutoCAD excels.

In creating a drawing in AutoCAD, we can use any scale we desire. Most often, the drawing is developed at full scale. That is, if you were drawing the layout for a city, you would draw it at full scale where 1 foot on the screen is equal to 1 foot of the city street. Of course, zooming allows you to see the complete drawing at any time. But the scale is still one to one. After the drawing is complete you have the option, at the plotting stage, to scale the drawing to fit the size of paper available. This allows us to take something many miles in actual size and plot it on a C size sheet.

This does not mean that we start to draw anywhere on the sheet. We must still give some thought as to where we should begin.

TRAP You must remember to take into account, when working with full scale drawings that will eventually be reduced in scale to fit a standard paper size, that any text on the drawing will also be reduced by the same scale factor. This might mean that you may not be able to see the text, let alone read it because it is too small.

TABLE 4-2a Scale Conversion Factors

Scale Factors for Engineering Drawing Scales								
1" = n	10′	20′	30′	40′	50′	60′	100′	200′
Scale factor	120	240	360	480	600	720	1200	2400

Scale Factor for Architectural Drawing Scales								
n = 1′.0"	1/16"	1/8"	1/4"	1/2"	3/4"	1"	1 1/2"	3"
Scale factor	192	96	48	24	16	12	8	4

To solve this problem, you must know the scale factor you will use when plotting, and then enlarge the text by this factor when you work at full scale. AutoCAD includes tables for these scale factors, as shown in Tables 4-2a and 4-2b.

TABLE 4-2b AutoCAD Scale Factors

	Plotted Text Size	Scale	Drawing Text Size
Architectural Drawing	3/16"	1/4" = 1'0"	8.0"
		1/2" = 1'0"	4.5"
		1" = 1'0"	2.25
Mechanical	.25	1:2	.5"
		1:10	25.0"
		2:1	.125"
	.125	1:2	25.0"
		1:10	12.50"
		2:106	25.0"
	.1875"	1:2	0.3875"
		1:10	18.75"
		2:1	0.09375"
Metric	3mm	1:10	30mm
		1:100	300mm
		1:500	1500mm
		1:1000	3000mm
	2mm	1:10	20mm
		1:100	200mm
		1:500	1000mm
		1:1000	2000mm

Using Table 4-2a. Setting Up Outside Drawing Limits

Using the Architectural Drawing Scale conversion factors. If you intend to plot a drawing on an Architectural C size sheet that is 36" x 24" (34" x 22" suggested drawing area from Table 3-2 on page 90), what are the limits you must choose for a drawing to fit this size of paper if you want to reduce it using a scale of 1/4" equals 1 foot? From the Architectural Drawing Scale, we see that 1/4" uses a scale factor of 48. That is, there are 48 quarters in a foot (12 inches). Thus the outside drawing limits should not exceed 1632" x 1056" (48 x 34 = 1632 and 48 x 22 = 1056).

Using Table 4-2b. Setting Up Sizes For Text

If we want the text size on the final plotted drawing to be 2mm in height, and we plan to reduce the drawing by a 100 to 1 reduction factor when plotted, (i.e., scale = 1:100) then the text on the original drawing must be 100 times larger. Thus the text must be 2mm x 100 = 200mm (20cm) high on the original drawing in order for it to reduce to 2mm when plotted.

An alternate method to drawing at full scale is to choose a paper size, say a C size, and scale the drawing portion down to fit the paper being used. When text is inserted into the drawing, it is scaled to a size that is readable for the paper size chosen.

The best method of course, occurs when the object is small enough to be drawn at full scale and still fits on the paper at one to one scale.

TIP: No matter which scale method is chosen, AutoCAD is accurate to sixteen decimal places. Thus, you could be drawing the Universe, at full scale, and still be accurate when zoomed down to the molecule level, as accuracy is maintained at any zoom level.

Schematic Building Blocks

The development of the schematic is broken into three building blocks or areas of the circuit diagram. These blocks include:

1 The AC Power and Transformer Input Circuit (TR-1)

2 The Power Supply Printed Circuit Board Schematic (PCB-1)

3 The associated LED, Voltage Control Pot, and Output Circuit.

Preliminary Layout

The working area of our C size drawing, (C-SHTE) from Table 3-1 (Mechanical and Electrical Drawings on page 89) is 500 mm x 380 mm, with the midway points being 250 mm x 190 mm. These will become the reference points for the next discussion. Looking at the Hand Sketch, Fig. 4-2, and its associated AC Power Input circuit, the Printed Circuit Board and the LED, Switches and Output Circuit, we can get a rough idea as to how much room each of these schematic building blocks will require.

Notice that the Hand Sketch is mostly in a horizontal plane. If we follow this horizontal format, we can arbitrarily block the schematic into the three blocks of 1) A.C. Input Circuit, 2) Power Supply (PCB-1), and 3) LED and Binding Posts.

If we allocate about 30% of the drawing space to block 1 and another 10% to block 3, this will leave us 60% of the drawing area for the Printed Circuit Board. The result of this trial-and-error hand layout is Fig. 4-6.

Using the technique of blocks gives us an appreciation for how much room can be allocated for each section. We will know if there is a problem should we find that PCB-1 is spilling over into the space allocated for the Output Circuit. If this occurs, we'll have to squeeze the layout for the PCB-1 block down in size.

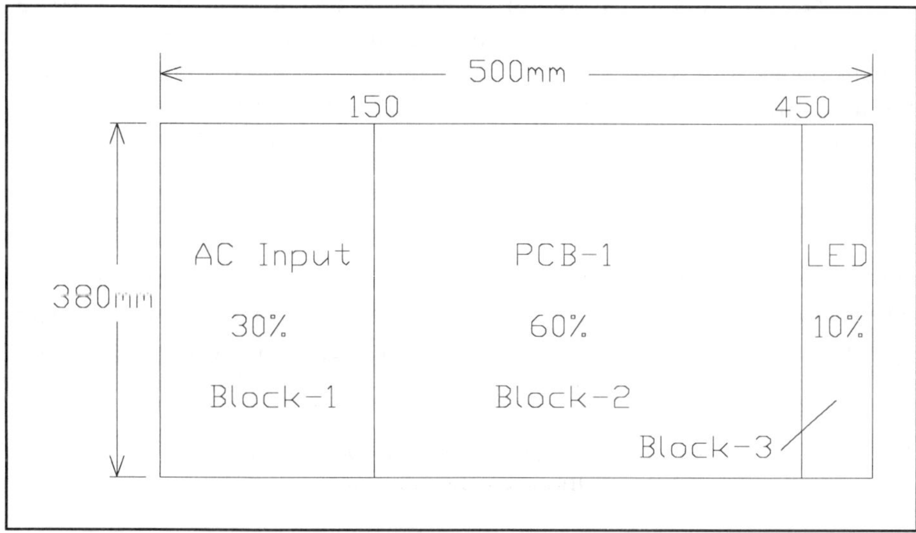

Figure 4-6 Rough layout of C size sheet.

DRAFTING STANDARDS FOR SCHEMATIC DRAWINGS

Electronic Schematics and Interconnect Drawings

On an Electronic Schematic drawing, all lines that interconnect components are drawn with either horizontal or vertical straight lines. No diagonal lines are permitted. On "Wiring Interconnect" drawings, diagonal lines are permitted. Our drawing is a combination of an electronic schematic and wiring interconnect. The lines outside the Printed Circuit Board are considered to be interconnect wiring, while the lines falling within the board area are electronic schematic wiring.

A Wiring Interconnect diagram is usually representative of where the connection is made on the physical part, and is sometimes highlighted with a dot. For an example of this, look at the AC circuit area of the schematic, Figure 4-3.

You will note that the fuse and power switch have a black dot on the outside of the component. The dot is representative of the connection point between the wire and the component. This point may be part of the component or may be a terminal lug mounted to the chassis.

Connections and Crossovers (Dots and No-Dots)

Sooner or later when you are drawing the schematic, a line will cross over another. The question arises, "Is this a connection or not?" A recent standard adopted to prevent confusion of this type, between junction points (circuit connections) and cross overs (no circuit connection) is the "No-Dot" method of designation. In this standard, connections are indicated by lines that terminate at other lines, forming a "T" intersection. Lines that cross over other lines have no connection with the line crossed over, as shown in Fig. 4-7.

The older and more time consuming, outdated standard, was to place a dot at each junction point where lines were connected. This method was quickly dropped, since errors could easily be introduced should a dot be missed. Likewise, in an age where productivity is essential, spending time to dot these hundreds of junction points is too expensive.

The standard we will use for our schematic is the No-Dot method.

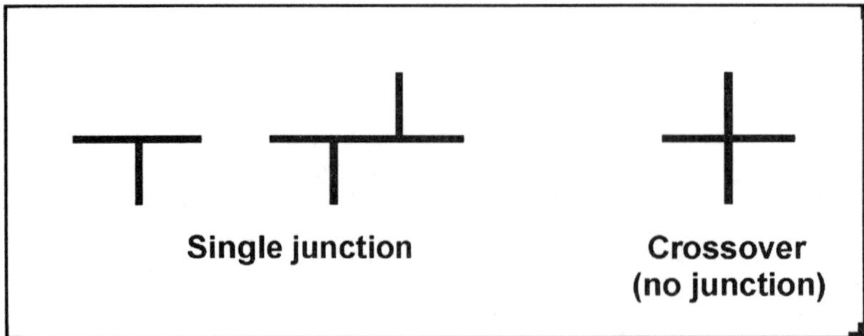

Figure 4-7 No-Dot connections and crossovers.

EXERCISE 4-4a. STARTING OUR SCHEMATIC DRAWING

Discussion

In Exercise 4-1a and b, we copied the C size sheet (C-SHTE) and called it SCHEMA. We are now ready to draw our schematic on this sheet. When complete, our schematic will appear similar to Fig. 4-3. As we progress, we will increase the revision number of our SCHEMA drawing. For example, our bare-bones drawing is presently called "SCHEMA." Just before we finish the first drawing session, we will use the Save command to save a copy of SCHEMA as "SCHEMA1."

In our second drawing session, we use SCHEMA1 as our working drawing. This keeps SCHEMA as a permanent record on the disk as something to fall back on, if our drawing becomes corrupted.

Likewise, as we proceed with SCHEMA1 and save interim copies, the reference "1" should be increased to successive numbers. For example, the second version would be called SCHEMA2. We use this technique to protect ourselves should the current working drawing accidentally become corrupted with garbage or even destroyed due to a virus. Today, viruses are found in almost every high school, college, university and in the work place.

TIP: You should, after each drawing session, save a complete copy of your work disk on a backup disk. This gives you a duplicate disk to fall back on should your work disk accidentally become damaged or lost. Having this backup means that all the time and effort just spent in making the drawing is not lost.

A second backup, kept at a friend's home, is double insurance. I, for example, keep 3 backups of the text for this book: one in my home, one at a neighbor's home, and one at my parent's home. Thus, if a burglar breaks into my home and my copy becomes stolen, I still have a copy next door. If my home and my neighbor's home burn down, then I still have a final backup at my parent's place. You can't be too safe.

If you are conscientious and save very frequently, you will use up disk space. That's okay, because the more backups you have the less updating time is required if a catastrophe occurs. You can save approximately 14 SCHEMA-type drawings on a 360K disk, or about 50 on a 1.44M disk since each SCHEMA drawing contains about 25,000 bytes. Don't forget you can delete those extremely old SCHEMA generations and any AutoCAD.BAK (back up drawing) files that are not needed, to give you that extra disk space.

TIP: When AutoCAD asks for the file name, remember to precede the file name with the drive letter in order to activate the correct drive to retrieve your SCHEMA drawing. For example, A:SCHEMA.

It is not necessary to set the limits or units for the SCHEMA drawing, as this was done on the C-SHTE, and these attributes were transferred across when we created the SCHEMA drawing. The grid and snap aspect ratio should be set at 10.

If you have forgotten how to do the above procedure, refer to Chapter 3, "Getting AutoCAD Up and Running." The following is an extract.

```
MAIN MENU

0.
1.  Begin a NEW drawing
2.  Edit an EXISTING drawing
-
-
8.

Enter selection: 2 (RETURN)
Enter name of drawing <A:TITLE>: A:SCHEMA (RETURN)
```

Housekeeping

The units and limits (500 mm x 380 mm) were previously set as part of the C-SHTE drawing. All that is necessary for us to do now is to set:

GRID: = 10
SNAP: = 10
ZOOM: = ALL
COORDINATES: = ON (Function Key F6)

After zooming to All, use the input device to check the top right corner position of the drawing to see if the extents of the sheet is 500 mm by 380 mm.

Schematic

Discussion. We will first draw the AC circuit, Fig. 4-8, using Insert to position the plug, fuse, switch, and transformer. You should try to keep this circuit within the confines of our preliminary layout, Fig. 4-6, where we restricted ourselves to a width of 150 mm for the AC input circuit. Our symbols will be placed on the left side of the drawing, and slightly above the midway point of the sheet.

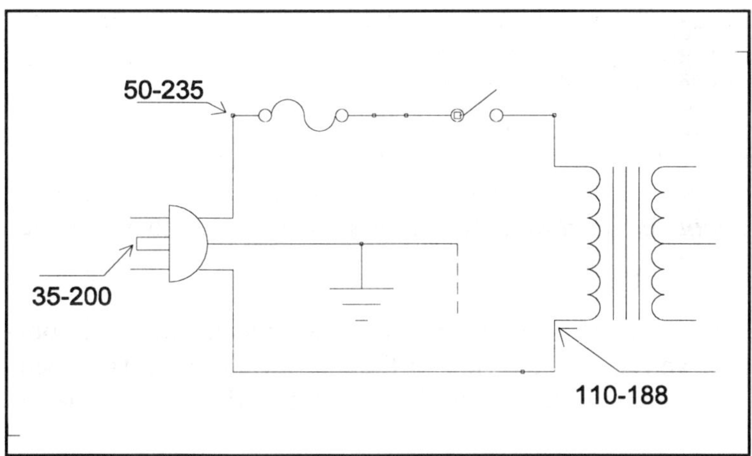

Figure 4-8 AC Circuit

The first symbol, the plug, will be positioned about 35 mm in from the left edge at coordinates 35.0000, 200.0000. The remaining symbols will then be added and interconnected by using the Line command. Let's do it.

Inserting the Plug.

```
Command: INSERT (RETURN)
Block name (or?): A:WPLUG (RETURN)
Insertion point: PICK LOCATION 35,200
X Scale factor <1>/Corner/XYZ: 1 (RETURN)
Y Scale factor (default=X): 1 (RETURN)
Rotation Angle <0>:  (RETURN)
```

The use of the above procedure will insert the plug into our drawing at the specified location. Did you precede the block name with the drive letter "A"? If not, AutoCAD may try to retrieve the plug drawing from the hard disk, where it won't be found.

Now to continue with the insertion of the remaining blocks of Fig. 4-8.

Inserting the Fuse, Switch, and Transformer

```
Command: INSERT (RETURN)
Block name (or?): A:WFUSE (RETURN)
Insertion point: PICK 50,235
X Scale factor <1>/Corner/XYZ: 1  (RETURN)
Y Scale factor (default=X): (RETURN)
Rotation Angle <0>: (RETURN)

Command:  INSERT (RETURN)
Block name (or?): A:WSW (RETURN)
Insertion point: PICK THE RIGHT END OF THE FUSE
X Scale factor <1>/Corner/XYZ: (RETURN)
Y Scale factor (default=X): (RETURN)
Rotation Angle <0>: (RETURN)
```

```
Command: INSERT (RETURN)
Block name (or?): A:WLXFMR (RETURN)
Insertion point: PICK 110, 188
X Scale factor <1>/Corner/XYZ:  (RETURN)
Y Scale factor (default=X):  (RETURN)
Rotation Angle <0>:  (RETURN)
```

You are on your own again! By now you should be skilled enough to insert the "Ground" symbol on the drawing, on your own.

With these four blocks in place, interconnect them, using the Line command. Zoom in as best you can, to allow you to pick the ends of the lines or components. When these components are all interconnected, save the drawing under the file name A:SCHEMA1.

You have just seen the technique of using the Electronic Symbol Library to create this portion of the drawing. The remaining part of the drawing follows the same procedure.

Before we continue with the drawing, we'll look at a few more AutoCAD commands that will make life much easier, especially when you are trying to connect lines to the ends of components and other lines.

EXERCISE 4-4b. OSNAP, YOUR BEST FRIEND

Discussion

You might have noted how difficult it was to join some of the lines to components and have them meet with no space between them. To solve this problem, especially when the drawing gets large, use the "object snap" or "OSNAP" mode.

OSNAP places a box or target area at the cross hairs of the drawing cursor. To use OSNAP, place the target area over the object you want to snap to and press the Pick button on the input device. Once you become familiar with this command, it will probably become your most useful tool for the remaining part of this exercise.

Because OSNAP is so popular, it has its own special symbol which has appeared in every screen menu area of AutoCAD since version 2.5. The selection of the four stars "* * * *" will take you directly to the OSNAP menu.

OSNAP can be toggled ON or OFF (NONE) at the command line. To switch OSNAP OFF, type OFF or NONE:

```
Command: OSNAP (RETURN)
Object snap modes: OFF (RETURN)
```

When OSNAP is toggled ON, all subsequent points selected will be under OSNAP Mode selection.

OSNAP is used in conjunction with the basic drawing commands such as Line, and allows you to capture objects in eleven different modes. These are:

1. CENter captures the exact center of a circle, or arc.

2. ENDpoint snaps to the nearest endpoint of a line, polyline segment, or arc.

3. INSert captures the "origin" or insert point of a block, or next entity.

4. INTersect snaps to the intersection of two objects that fall within the target area. Arcs and circles that are part of a block are not captured, although line intersections within blocks are.

5. MIDpoint snaps to the midpoint of a line, polyline segment, or arc.

6. NEArest snaps to the nearest point of an object. Arcs and circles that are part of a block are not captured.

7. NODe snaps to a point.

8. PERpendicular snaps to a point on a line, polyline segment, arc, or circle that forms a perpendicular from the last point. Arcs and circles must not be part of a block.

9. TANgent snaps to a point on an arc or circle that will become tangent to the last point entered.

10. QUAdrant snaps to the nearest quadrant point, that is, 0, 90, 180, 270 degree points, on an arc or circle. It will only capture the visible points on an arc.

11. QUIck is an option used before each OSNAP function to quickly select a location within the target area. "QUINEA" will QUIckly select the NEArest point on a line, arc, or circle that it comes to in the evaluation of the point, rather than looking at all points and selecting the best. The main function is to save search time.

Target Area Control (Aperture Command)

The size of the target area is controlled by the Aperture command. The size chosen should be such that it is large enough to capture wanted objects, yet does not include any unwanted items. See Chapter 8, section OSNAP, for the setup of the aperture size.

Let's Play with OSNAP

To give us something to work with, in a clear area of the drawing we'll draw two straight lines. Now with Snap OFF, we will connect to the end of these lines, using the OSNAP "END" option.

Note: Don't become too alarmed with drawing extra lines and circles on your drawing. We'll erase them from our SCHEMA1 when we're done.

```
Command: TOGGLE THE SNAP "OFF" WITH F9 FUNCTION KEY
Command: DRAW TWO STRAIGHT LINES ON YOUR DRAWING
```

Now to OSNAP to the end of these lines.

```
Command: LINE (RETURN)
```

at the From point type,

```
From point: END (RETURN)
```

Note that a square aperture appears around the cross hair. Place the aperture area over the end of a line.

```
Of: CHOOSE THE END OF A LINE AND PRESS THE PICK BUTTON.
```

The crosshair snaps to the end of the first line selected when Picked. Now we'll OSNAP it to the end of the second line.

```
To point: END (RETURN)
Of: PICK ANY PLACE NEAR THE END OF THE SECOND LINE. (RETURN)
```

Note: The two lines are connected with no gaps or spaces at the connecting points.

Now, let's connect the open end of the first line to the "MIDdle" of the second line.

```
Command: LINE (RETURN)
From point: END (RETURN)
Of: PICK THE REMAINING OPEN END OF THE FIRST LINE
To point: MID (RETURN)
Of: PICK ANY POINT ON THE SECOND LINE (RETURN)
```

The first line is now connected to the middle of the second line.

Easy to use? You bet! Before we continue, erase the above lines from your drawing.

OSNAP Dialogue Box (DDOSNAP)

Selection of the OSNAP modes is also a function of a Dialogue Box, (Fig. 4-8a), used with the DDOSNAP command. You can easily click on one or more object snap modes to allow all subsequent point selections to be under OSNAP control. These become your "Running Object Snap" modes and remain part of the drawing attributes until modified or turned OFF.

Aperture Size

The Dialogue Box uses a slider to govern the size of the Target Box or Aperture that appears around the outside of the cross-hairs. The maximum Target Box size is limited to 20 pixels or 20 dots across the screen.

You can gain access to the OSNAP Dialogue Box through two means, first by using the command DDOSNAP or secondly by selecting it from the Settings pull-down menu at the top of the screen and clicking on "Object Snap...:"

```
Command: DDOSNAP (RETURN)
```

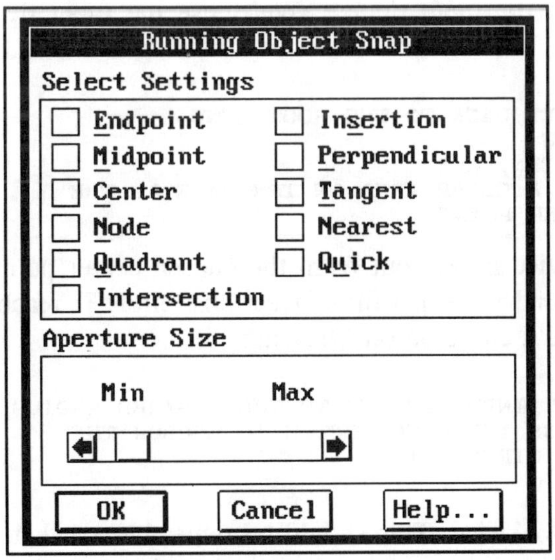

Running Object Snap

Select Settings

- [ ] Endpoint
- [ ] Midpoint
- [ ] Center
- [ ] Node
- [ ] Quadrant
- [ ] Intersection
- [ ] Insertion
- [ ] Perpendicular
- [ ] Tangent
- [ ] Nearest
- [ ] Quick

Aperture Size

Min Max

OK Cancel Help...

Figure 4-8a "Running Object Snap" dialogue box.

Let's look at Circles and Arcs before we end this exercise.

CIRCLES AND ARCS

In the process of completing the schematic, you will need to draw a few circles and arcs. These will become test point locations and binding post positions.

Circle Command

The Circle command allows us to specify the circle, using six methods:

1. The center point and radius	CEN,RAD:
2. The center point and diameter	CEN,DIA:
3. Two points on the circle	2-POINT:
4. Three points on the circle	3-POINT:
5. Two tangent points and radius	TTR:
6. Tangent circles.	TC:

Let us draw a circle, using the center point and radius method.

We gain access to the Circle command from Draw, through the AutoCAD Root Menu. The center point and radius may be selected by using either the keyboard and specifying the coordinates or from the input device. We'll use the input device because it's quick and easy.

```
Command: PICK AutoCAD TO GET BACK TO THE ROOT MENU
Command: PICK THE DRAW MENU
Command: PICK CIRCLE AND "CEN,RAD"
3P/2P/<Center Point>: PICK A CLEAR AREA ON THE PAGE AS THE
                      CENTER POINT
```

You will note that as the input device moves out from the chosen center point, a circle is dragged into position around it. When the correct size circle is reached, press the Pick button. We now have a circle on our drawing.

```
Diameter/<Radius>: DRAG THE INPUT DEVICE TO THE DESIRED RADIUS
                   AND PRESS THE PICK BUTTON TO ANCHOR THE
                   DESIRED CIRCLE SIZE
```

TIP: Circles are drawn counterclockwise. This should be remembered when the two- or three-point drawing method is used, as some thought is required as to where the first point should be located.

To erase the circle, select Erase from the menu and then Last. This will remove the circle from our drawing.

Arc

There are ten methods to construct an arc, as shown below and in Chapter 8, and none of them are attributed to our friend, Noah. Arcs, like circles, are drawn counterclockwise. Remember this in the selection of the reference points for the arc.

3-point ARC choices:

1. S,C,E (Start, Center, End)
2. S,C,A (Start, Center, included Angle)
3. S,C,L (Start, Center, Length of chord)
4. S,E,A (Start, End, included Angle)
5. S,E,R (Start, End, Radius)
6. S,E,D (Start, End, starting Direction)
7. C,S,E (Center, Start, End)
8. C,S,A (Center, Start, included Angle)
9. C,S,L (Center, Start, Length of chord)
10. CONTIN (Connect to previous arc)

Playing with the Arc Command

In this short tutorial on the Arc Command, we will draw an Arc using the Startpoint, Center, and Endpoint (S,C,E:) options. This was the method chosen to draw the semicircles found in the transformer symbol of our symbol library. You will find the Arc command under the Draw menu.

Using the Input Device, do the following:

```
Command: PICK "DRAW" (from ROOT MENU)
Command: PICK "ARC"
Command: PICK "S,C,E:"
Command:ArcCenter/<Start point>: PICK THE STARTING POINT
Center/End/<Second point>: C Center: PICK THE CENTER POINT
Angle/Length of chord/<End point>: DRAG - SELECT END POINT
```

TIP: You will notice that the arc was drawn counterclockwise from the starting point, using the center point as a pivot, and finished at the chosen end point.

This is the method that should be used to draw the jumper wires or test links, TL-1 and TL-2, on the schematic.

Now that we have finished playing, erase this arc from your drawing before proceeding any further.

COMPLETE THE EXERCISE

You now have the necessary knowledge to complete the placement of the remaining electronic symbols. Do so at your own speed, and be careful not to venture beyond the limits of your drawing.

When inserting some of the symbols, such as diodes, electrolytic capacitors (ECAPs), and LEDs, you may have to change the rotation angle to fit the desired position on the drawing. Otherwise, no other commands are needed to complete the schematic drawing.

After the schematic symbols have been inserted and connected, we will proceed with Exercise 4-5, working with layers, and see how to place text on the drawing. *Have you **"saved"** your drawing lately?*

EXERCISE 4-5. LAYERS

Discussion

Drawings can become confusing if too much information is displayed. In mechanical drafting, the clearest way to show a customer how the product will finally look, is to display only the text and main assembly drawing, leaving the hidden features, special notes and mechanical overlays hidden from view. In an architectural drawing of a house, the customer may become confused with the floor layout if the heating ducts, air conditioning, electrical wiring and water pipes are all shown on one view. In the layout of an eight-up multilayer printed circuit board, seeing all the plated through holes on all layers becomes an impossible task if all layers are displayed at once.

The solution to this confusion is to show only that which is necessary. The Layers feature in AutoCAD is a means of doing this. It allows us to reduce the clutter on a drawing, by displaying only that information relative to the view being shown.

We can think of a layer as a sheet of clear cellulose on top of the drawing sheet. Layers can be placed one on top of another to create multiple layers. The first or default layer is always termed layer 0. As with single sheets of transparent cellulose, we can write on the layer that is uppermost or active at the time. This is called the current or active layer. To see data on other layers, they must be turned on or made available. When a layer is turned off, it disappears from view as if it did not exist. In this state, it will not be seen, printed or plotted.

Drawing on different layers has many special benefits, such as:

- Each layer may be assigned a different color to help clarify the drawing.

- Each layer may be plotted separately with the main layout. Thus, in the installation of air ducts of a building, it would not be necessary to clutter the drawing with the underground parking requirements.

- The drawing may be plotted with different layers active, thus protecting the security of the project. An example of this would be to exclude the part numbers or identification numbers on a printed circuit board so cloning of the board would become more difficult.

- Turning Layers OFF will prevent them from being seen and Freezing them will speed up the regeneration process when developing the drawing.

For the lettering on our schematic drawing, we will use two layers, one to type the text identifying the electronic symbol codes, such as IC1, C1, and R5, and the other to type the text identifying the value of these components, such as 741, 25uf, 1000 ohms.

The Layer name of the first layer containing COMPonent symbol CODes is "COMPCOD," and will use the layer color of red. The second layer, to identify the component values, is called COMPonent VALue "COMPVAL," and will use the color of green.

Setting up a Layer

There are two methods to establish a new layer on a drawing. The first is to type LAYER from the keyboard. The second is through the Modify Layer/Layer Control dialogue boxes found in the pull-down menu under Settings. We will first explore the manual method to become familiar with the various options available and then go on to discuss the dialogue box.

When you type LAYER, AutoCAD requests, through the Command line, the following data:

Command: Layer
?/Make/Set/New/ON/OFF/Color/Ltype/Freeze/Thaw/LOck/Unlock:

?	- Do we want information about the layer status?
(M)ake	- One step to create a new layer and make it current.
(S)et	- Sets any layer to become the "current" layer, no matter what its status is, i.e., On or Off.
(N)ew	- Establish a New layer without affecting status of current layer.
(ON)	- Turn a layer On.
(OFF)	- Turn a layer Off.
(C)olor	- Changes layer Color being used. See Trap Note*
(L)type	- Changes layer Line Type, i.e., continuous or dotted?
(F)reeze	- Freeze or inhibit a particular layer/s from regenerating, (saves time).
(T)haw	- Undo the effects of the Freeze command.
(LO)ck	- Locks a layer from being edited, but are visible.
(Unlock)	- Unlock locked layers

TRAP You may change the color being used on a layer at any time by using the Color Command option. This will make the current Layer Color different from that which you originally chose as the Layer Color during Layer setup. This may become confusing if the new color is similar to colors used by other layers. To switch back to the color of the layer, type BYLAYER at the Color command, such as:

```
Command: COLOR (RETURN)
Layer name(s) for color n <current>: BYLAYER (RETURN)
```

To understand the operation of a few of these options, we will set up several layers using the keyboard. Then we'll modify the layer using the dialogue box to show you the simplicity of this latter method:

(N) A New Layer. We will now establish three "New" layers called COMPCOD, COMPVAL, and DOTLINE.

TIP: Layer names can be up to 31 alphanumeric characters long. When specifying the name, it is best if you keep it short, under eight letters, so the full name will be displayed in the Mode Area of the video screen (See Fig. 1-2A).

```
Command: LAYER (RETURN)
?/Make/Set/New/ON/OFF/Color/Ltype.../Unlock: NEW (RETURN)
New layer name(s): COMPCOD,COMPVAL,DOTLINE (RETURN)
```

TIP: Note that there is only a comma separating the layer names (no spaces).

When we press Return, the Layer Prompt Line returns instead of the Command line. AutoCAD is now ready to receive another Layer command. If we were to press Return for a second time, (don't), it would bring us back to the command prompt again.

(?) Layer Status. Let's examine the layer status of all layers so far created by replying to the prompt with the question mark (?) symbol. Type:

```
?/Make/Set/New/On/Off/Color/Ltype..../Unlock: ? (RETURN)
```

We indicate "all layers" through the use of a "wild card" or the star symbol (*) for the name.

```
Layer name(s) for listing <*>: * (RETURN)
```

The following listing will appear:

```
   Layer name       State     color        Linetype
---------------------------------------------------
   0                On      7 (white)   continuous
   COMPCOD          On      7 (white)   continuous
```

```
COMPVAL          On      7 (white)   continuous
DOTLINE          On      7 (white)   continuous

Current layer:0

?/Make/Set/New/ON/Off/Color/Ltype... /Unlock:
```

Note: 1. COMPCOD, COMPVAL, and DOTLINE were turned on automatically when their names were established and the color was set to white.

2. The current layer is 0, the layer AutoCAD always starts with AutoCAD'S default.

3. We will change the linetype of Layer DOTLINE, now shown as "continuous," to a dotted line before this exercise is finished.

(C) Setting the Layer "Color." We will change the color of the COMPCOD Layer to red through the keyboard. The COMPVAL color will be changed to green using the dialogue box.

The color choices are:

1. Red, 2. Yellow, 3. Green, 4. Cyan, 5. Blue, 6. Magenta, 7. White.

The Layer Color may be selected either by number or by the first letter of the name.

```
?/Set/New/On/Off/Color/Ltype... /Unlock: COLOR (RETURN)
Color: RED (RETURN)
Layer name(s) for color 1 <Red>: COMPCOD (RETURN)
```

(S) "Setting" a Layer to Current.

Now we'll set the current layer to COMPCOD and then check the layer status again.

```
?/Set/New/On/Off/Color/Ltype... /Unlock: SET (RETURN)
New current Layer <default>: COMPCOD (RETURN)
?/Set/New/On/Off/Color/Ltype... /Unlock: ? (RETURN)
Layer name(s) for listing <*>: * (RETURN)

Layer name    State    Color       Linetype
-------------------------------------------
0             On       7 (white)   continuous
COMPCOD       On       1 (red)     continuous
COMPVAL       On       3 (white)   continuous
DOTLINE       On       7 (white)   continuous

Current layer: COMPCOD

?/Set/New/On/Off/Color/Ltype... /Unlock: (RETURN)
```

Note the Layer status is now COMPCOD!

LINETYPE COMMAND

There are 24 basic linetypes available for AutoCAD to use. You can also customize your own linetypes if you so desire. When a new drawing is first created, there is only one linetype available, the default linetype "Continuous". The remaining linetypes are stored in the Linetype Library File. This saves space in memory. When you are ready to use one of the remaining linetypes on a layer, it must be loaded from the Library into your drawing file. The Linetype command allows us to do this. (Note Appendix A-2 for the complete list of Linetypes.)

TIP: Remember, the linetype must be loaded from the Linetype Library File into your drawing before it can be assigned to a layer.

Linetypes are loaded from the linetype file "acad.lin" by either using the keyboard or a dialogue box. We will look at what linetypes are available using the "?" option and then load the Linetype "HIDDEN" from the Library file. Type:

```
Command: LINETYPE (RETURN)
?/Create/Load/Set: ? (RETURN)
```

If FILEDIA is active, the "Select Linetype File" dialogue box appears. (Note Exercise 7-5 in Chapter 7.) Otherwise use the following prompt:

```
File to list <ACAD>: ACAD (RETURN)
```

A list of 24 linetypes available in the file Acad.lin now appears. See Appendix A2. The following is a summary of these linetypes:

```
Linetypes defined in file C:\AutoCAD12(AutoCAD13)\ACAD.LIN

 Name                        Description
-----------   ----------------------------------
 Dashed       __ __ __ __ __ __ __ __ __ __ _
 Hidden       _ _ _ _ _ _ _ _ _ _ _ _ _ _ _ _
 Center       ____ _ ____ _ ____ ____ _ __
 Phantom      ____ _ _ ____ _ _ ____ _ _ ____ ____

 Dot          ..............................
 Dashdot      __ . __ . __ . __ . __ . __ . _
 Border       __ __ . __ __ . __ __ . __ . __
 Divide       __ .. __ .. __ .. __ .. __ .. _
```

Now to load the hidden linetype file, type:

```
?/Create/Load/Set: LOAD (RETURN)

Linetype(s) to load: HIDDEN (RETURN)
File to search <ACAD>: (RETURN)

Linetype HIDDEN loaded.

?/Create/Load/Set: (PRESS RETURN TO TERMINATE THE COMMAND)
```

Now press the F1 Function key to get back to the drawing editor screen.

The Hidden linetype file is now transferred from the Linetype Library File to your drawing file.

You may find that the dashes are not at the desired spacing for your drawing scale. In this case, use the LTSCALE Command to modify them. It is recommended that you use a scale factor of 25 for the dotted lines on the schematic drawing.

```
Command: LTSCALE (RETURN)
New scale factor <1.000>: 25 (RETURN)
```

We are now ready to use the *component code layer* (COMPCOD) as it should now be the current layer. On this layer, we'll type the text codes used for each component on the schematic, such as, "R12" or "C4."

THE "MODIFY LAYER/LAYER CONTROL" DIALOGUE BOXEs

Discussion

In version 10, the Layers dialogue box (DDLMODES) was called the Modify Layer dialogue box. In versions 11, 12 and 13 the *name* was changed to Layer Control dialogue box. Versions 10 and 11 Layers dialogue boxes look very similar, see Fig. 4-9a, and in versions 12 and 13 have a totally new look. See Fig. 4-9b page 144.

EXERCISE 4-6. MODIFY LAYER DIALOGUE BOX

Playing With Versions 10/11 Modify Layer/Layer Control Dialogue Box

Now to have some fun with the Modify Layer/Layer Control dialogue boxes. A dialogue box is one of the time-saving features AutoCAD introduced with release 9. Having just gone through the process of establishing additional layers, setting the color and linetype, as well as making a layer current, you can see it is a lengthy operation from the keyboard. With the Modify Layer/Layer Control dialogue boxes, all of these operations can be performed by picking the item from one menu.

The Modify Layer/Layer Control menus (Fig. 4-9a & b) can be accessed either by typing DDLMODES at the keyboard, or by selecting the Settings pull-down menu at the top of the screen and clicking on Modify Layer/Layer Control.

Let's look at these two dialogue boxes. If you are using version 10, type:

```
Command: DDLMODES (RETURN)
```

The Modify Layer dialogue box menu has six panels, which are used to indicate and change the status of each layer. These are:

The Current panel, which highlights the layer that is presently current.

The Layer Name panel, which indicates the layers available.

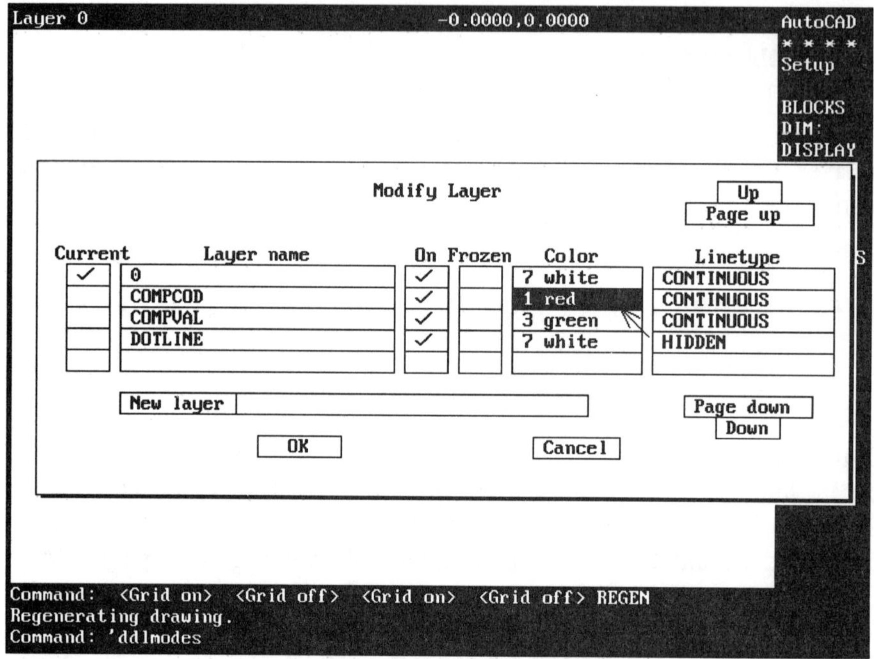

Figure 4-9a Modify Layer dialogue box.

The ON panel shows which layer is turned ON or OFF through the use of a check mark beside those turned on.

The Frozen panel, again through a check mark, highlights those layers that will not be displayed or regenerated.

The Color panel shows the selected layer color.

The Linetype panel displays the type of line that will be active when that layer is made current.

Changing the Layer Status. To set a layer current, simply point to the layer you want under the Current panel and press the input device button. The check mark in the panel window will be transferred to the new location, indicating that it is now current.

To turn layers ON or Off, or to Freeze them, point to the layer and press the input device button. They toggle with each click of the button, and their status is shown by a check mark.

When the color panel is selected, a Color menu is displayed that highlights the seven colors to choose from. When you have selected your color, choose OK, and the Modify Layer menu reappears.

Let's modify the color of the COMPVAL Layer

We will do this by Picking the color "White" under the COMPVAL Layer name which will bring up the Color Panel. From the palette shown, we will pick the color GREEN and then pick OK. Let's do it:

```
Command: PICK THE COLOR WHITE AT THE COMPVAL LINE
```

The Color Panel dialogue box is displayed

```
Command: PICK THE COLOR GREEN
Command: PICK OK
```

We now see the color setting for the COMPVAL layer has changed to Green.

In addition to the six status panels, there is a Page Up and Page Down panel to allow you to scroll up and down the layer names. To add another layer, simply point and click on the New Layer panel to activate it and then type the layer name.

When you are satisfied with your changes, click on the OK panel, and the Modify Layer Dialogue box disappears. As you can see, it is quick and simple.

EXERCISE 4-6a. LAYER CONTROL DIALOGUE BOX

Playing With Versions 12 & 13 Layer Control Dialogue Box

In AutoCAD versions 12 and 13, the Modify Layer dialogue box is replaced by the Layer Control dialogue box which is evoked by either typing DDLMODES at the command prompt, or by selecting "Layer Control" from the pull-down Settings menu, or by picking LAYER from the screen menu. We'll use the keyboard. Type:

```
Command: DDLMODES (RETURN)
```

The Layer Control dialogue box appears, Fig. 4-9b.

Within the Layer Control dialogue box is the Layer Name, State, Color and Linetype box (Layer Name List Box). It also displays the status of the layers, arranged alphabetically. By dragging the input device across them, you can scroll through the list of layers. Pressing the pick button will select and highlight that layer. It is now possible to highlight several layers so multiple operations such as Freeze or Thaw can be performed at the same time on the selected layers.

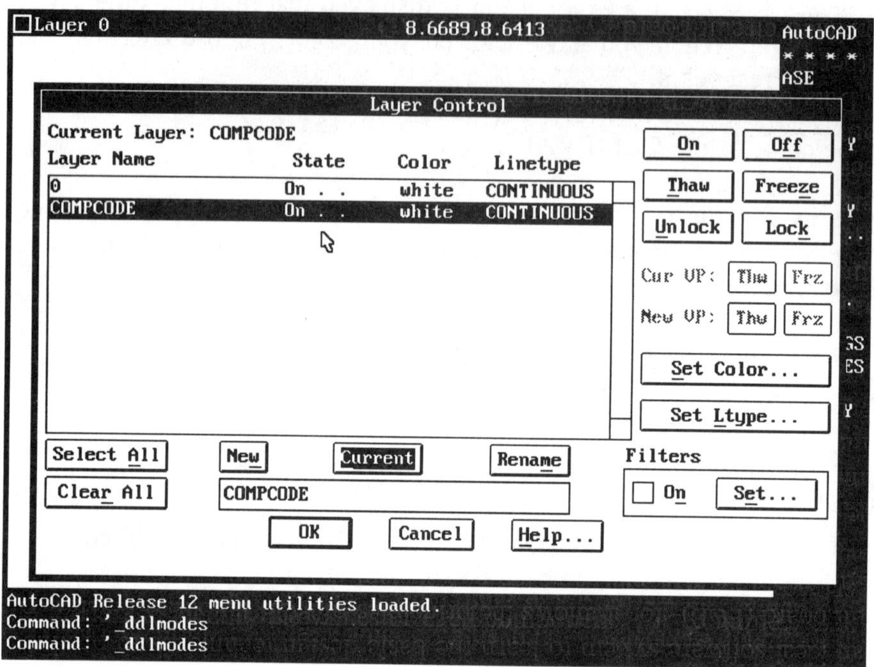

Figure 4-9b Versions 12 and 13 Layer Control dialogue boxes.

Through the Layer Control Dialogue box, there are buttons that will allow you to turn layers On and Off, Freeze and Thaw them, Lock and Unlock them, and Set their Color and Linetype. All of these functions are normally available to you through the Layer Keyboard controls.

The "Select All" And "Clear All" Buttons.

In the bottom left corner of the Layer Control dialogue Box are the Select All and Clear All buttons. Picking Select All automatically selects all layers in the list and picking the Clear All button clears all selections picked.

The Layer Name Edit Box and the New, Current, Rename Buttons

At the bottom of the dialogue box is the Layer Name Edit Box and the New, Current and Rename buttons. The currently selected layer's name appears in the Edit box. See Fig 4-9c.

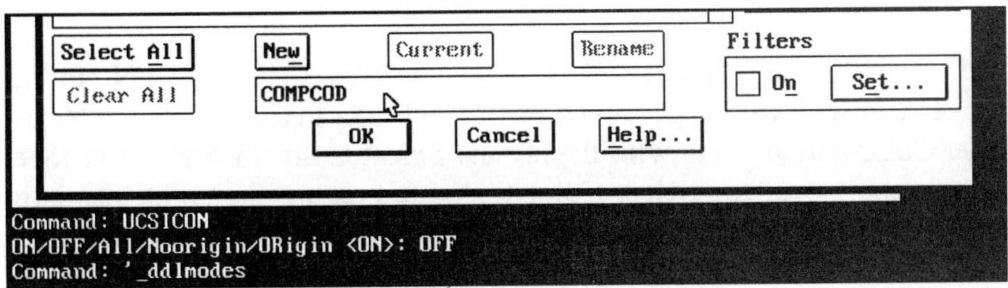

Fig. 4-9c The Layer Name Edit Box.

Adding a New Layer

To add a new layer to your drawing, Click on the Layer Name Edit Box to activate it and then type in the new layer name. Next, pick the NEW Button and the new layer is added to the list of layer names.

Renaming an Existing Layer

You cannot give a new layer an existing layer's name, but you can rename an existing layer. To do this, pick the layer name to be changed from the Layer Name List Box. The Layer Name now appears in the Edit Box. Now edit the name using the cursor control keys. When satisfied, pick the Rename Button to confirm your decision. The newly named layer now appears in the Layer Name list box.

Making a Layer Current

Using the input device, pick the layer you want to become current. It becomes highlighted in the Layer Name Edit Box. Now pick the Current Button, and it's current.

TRAP Make sure only one layer is active before you pick the Current Button, as only one layer can be current at any one time. If more than one layer is selected, the Current Button becomes inactive. Select Clear All and try again.

You can also use the CLAYER system variable to make a layer current by typing at the Command line:

```
Command: CLAYER (RETURN)
New current layer <0>: TYPE LAYER NAME TO MAKE CURRENT
```

Changing a Layer's Color

In this exercise, we will change the color of Layer COMPVAL to GREEN. First, we'll pick the layer you want to change the color of, and then make it current. Next, we'll pick the Set Color button. This will display the Select Color dialogue box (See Fig. 4-9d). Across the top of the Select Color dialogue box are the eight Standard Colors of the color palette. If your graphics card supports 256 colors, the Full Color Palette and Gray Shade Palette area will display the remaining colors.

```
Command: PICK THE COMPVAL LAYER
Command: PICK THE SET COLOR BUTTON
Command: PICK COLOR GREEN FROM THE STANDARD COLOR PALETTE
Command: PICK THE OK BUTTON
```

You have just changed the color of the COMPVAL Layer to Green and should be back at the Layer Control dialogue Box. If you're still at the Select Color dialogue box, pick the OK button again.

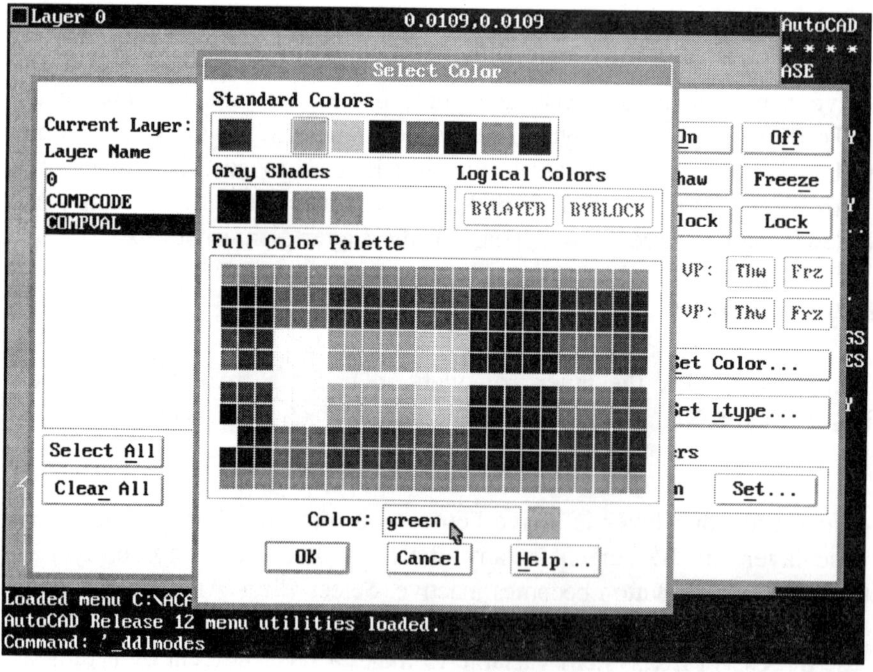

Figure 4-9d Select Color dialogue box.

Filter Buttons

Some AutoCAD drawings may contain hundreds of layers. This means scrolling through the Layer Name List Box becomes quite time consuming.

The Filters option allows you to "filter" out only those layers that are relative to the current drawing session, by selectively showing the Layer Names. The Filter can be based on options such as Layer Names, Color, Linetypes, On or OFF conditions, Frozen or Thawed conditions, and Locked or Unlocked status. When Filters is used, the "Set Layer Filters" dialogue box (Fig. 4-9e) activates to allow you to choose the conditions.

Let us say we want to use only those layers that are green. In the Colors option of the Set Layer Filters, you would type GREEN and then pick OK. The only layers that would appear would be the green layers. To disable the filter, pick the On button under Filters which toggles between On and Off.

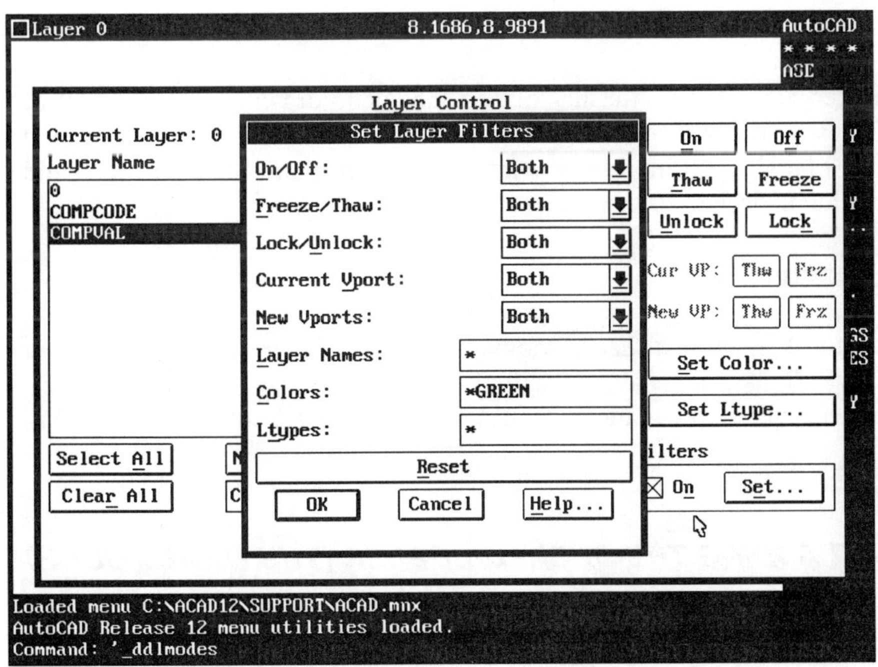

Figure 4-9e Set Layer Filters.

The Layer Control dialogue box supports Viewport operation which is a feature we will explore in another chapter.

EXERCISE 4-6b. TEXT

Lettering the Component Codes

You are on your own again. Using Figs. 4-2 and 4-3 (pages 108 and 109) as a reference, letter each of the components with their respective symbol codes; that is, the R1 and C1 type codes. Check the mode area (upper left corner) of the screen; it should say COMPCOD. If it does not, set the current layer to COMPCOD.

Is your lettering coming out in red? If not, are you on Layer COMPCOD and was its color set to RED in Exercise 4-5?

Lettering the Component Values

To letter the component's values, switch the layer to COMPVAL where the color was set to Green. Again using Figs. 4-2, and 4-3 (pages 108 and 109) as a reference, add the text to identify the values of each component, for example, "22k" or "10uf."

Text Size

The recommended size of the lettering for our schematic drawing is 3 units (mm) high for the General (small) letters, 5 units (mm) high for the Header Subtitles, and 7 units (mm) for Main Titles. Subtitles should be underlined and Main Titles should have a 1 mm Pline underline to make them stand out.

Notes and Legend Text

Add the Notes and Legend text that is shown in Fig. 4-3 (page 109) to the bottom left corner of layer 0, as well as any remaining text that you have not added as yet. This will complete this drawing and it should appear similar to the Schematic, Fig 4-3.

EXERCISE 4-7 ADDING THE DASHED BORDER (THE FINAL TOUCH)

Discussion

In Exercise 4-5 on Layers, we established the Linetype HIDDEN, a series of dashed lines. We will now use HIDDEN to add a 1 mm Pline border around the area which represents the outline of the printed circuit board schematic information, giving our drawing that *final touch*. See Fig. 4-3 on page 109.

TIP: When you are drawing on a layer, the default linetype of the layer becomes the line style used by that layer. For example: We have been drawing on

Layer 0 where the default linetype is Continuous, and the lines drawn are continuous type lines (not dashed).

In this exercise we will change our layer to DOTLINE, select the Linetype of HIDDEN and draw the border lines around the schematic. This will produce the dashed lines shown in Fig. 4-3 (page 109). To select our Layers Control dialogue box, select Settings and Layer Control from the pull-down menu, or type:

```
Command: DDLMODES (RETURN)
```

At the previously established layer DOTLINE (Exercise 4-5), click on the DOTLINE Layer name to highlight it, and then select Linetypes/Ltype... .

```
Command: CLICK ON DOTLINE AND SELECT LINETYPE(LTYPE...).
```

This brings up the Select Linetype dialogue box. Now select the HIDDEN linetype.

```
Command: CLICK ON HIDDEN (LINETYPE) AND OK TO EXIT.
```

You will now see the Layer DOTLINE linetype is HIDDEN, and any lines drawn on the DOTLINE layer will be of the HIDDEN line style (dashed). See Linetype Styles on page 140 or in Appendix A-2.

Now, select DOTLINE as the current layer and draw the printed circuit board outline using a Pline 1 mm wide.

```
Command: Pline (RETURN)
SET THE WIDTH TO 1 mm AND DRAW THE PRINTED CIRCUIT BOARD SCHEMATIC
BORDER LINES USING FIG. 4-3 (page 109) AS A REFERENCE.
```

You should be able to see the dashed lines, but if you can't, it's probably because the line type scale LTSCALE, which sets the distance between dashes is defaulted to 1. To widen these out, set the LTSCALE to 25.

```
Command: LTSCALE (RETURN)
New scale factor <1.000>: 25 (RETURN)
```

REVIEW - BLOCKS AND LAYERS

Discussion

A word of caution. We have seen how to use blocks, but what about the layer on which they should be developed?

TIP: In creating an object that will later be saved as a Block, be careful which layer is used for the development. ***It is imperative to use only layer 0 to create your entities for Blocking and Wblocking.***

TRAP Should an object, such as a resistor, be developed on layer "1" and then saved as a Block, it can not be inserted into any other layer. This is because the Blocked object retains the original layer (1) definition and attributes. When this object is Wblocked, it retains the Layer 1 attributes. When the Wblock entity is inserted into a new drawing, the new drawing will have a "Layer 1" created, with all the attributes of the original Block.

If you try to insert the blocked object developed on layer 1 into another layer, it will appear as if it was on the specified layer, but in reality it is still on Layer 1, the wrong layer. You may not realize this until it is too late. On the other hand, if the resistor was developed on layer 0, the universal layer, and then saved as a Block, it can be inserted into any active layer.

Changing the Layer of an Inserted Block

Should you accidentally insert a blocked object into the wrong layer, it can be changed by using the Change command. *Note:* This correction is used as the result of a wrong Insert, not as the result of using the wrong layer for development.

The command sequence for a change is

```
Command: CHANGE (RETURN)
Select objects: PICK THE OBJECTS (RETURN NEEDED)
Properties/<Change point>: P (RETURN)
Change what property(Color/Elev/LAyer/LType/Thickness): LA (RETURN)
New layer<current>: TYPE IN NEW LAYER NAME.
```

This removes the block from its initial layer and reinserts it into the changed layer.

MAKE YOUR BACKUP

When you have completed this drawing, your Schematic should appear similar to Fig. 4-3. You should **save** this drawing on a backup disk and then place it in a different location from your regular work disk for safekeeping.

QUIZ: THE SCHEMATIC
Name ____

Circle the most correct answer or fill in blanks.

1. To save destroying the prototype drawing called C-Sht, you should:

   a. Duplicate it under a new name such as SCHEMA.
   b. Draw on it, but use a layer called SCHEMA.
   c. Redraw it from scratch.
   d. None of the above.

2. Which of the following commands will copy a single file from one disk to another, so it can later be used by AutoCAD?

   a. COPY A:*.* B:*.*
   b. DISKCOPY A: B:
   c. COPY A:ELECTEMP B:ELECTEMP
   d. COPY A:ELECTEMP.DWG B:ELECTEMP.DWG

3. To capture an "entity" from your drawing, you should use the command sequence:

   a. COPY A:*.* and B:*.*
   b. BLOCK and WBLOCK
   c. SAVE and BLOCK
   d. WBLOCK and BLOCK

4. Before starting a drawing, you should:

   a. Examine the hand sketches to establish sheet size.
   b. Perform a "rough" block layout of the drawing.
   c. Give some thought as to where to begin.
   d. All of the above.

5. "DOTS" and "NO-DOTS" refer to:

   a. The small marks plotter pens make.
   b. Reside on the drawing paper.
   c. A drawing standard.
   d. None of the above.

6. What command is used to place a BLOCK or WBLOCK into your drawing?

ANSWER:____

7. To latch to the end of a line previously drawn, you should use the command:

a. OSNAP.
b. SNAP.
c. TARGET.
d. All of the above.

8. Briefly describe the process in Question 7.

____

____

____

9. In AutoCAD, an ARC may be drawn through:

a. All degrees up to 359.99.
b. 360 Degrees.
c. Only to 180 Degrees.
d. None of the above.

10. Where is a quick reference place on the screen to see if a LAYER is "current"?

ANSWER:____

11. Layers in AutoCAD may be thought of as:

a. New drawings.
b. Colored areas.
c. Sheets of paper.
d. Sheets of clear cellulose .

chapter 5

The Printed Circuit Board

DISCUSSION

AutoCAD is an ideal software system to learn the basic design techniques to produce a printed circuit board (PCB), for it allows the principles of PCB layout to be shown. It is in effect a simple "electronic hand taping" system.

Once you have mastered the design concepts to manually lay out a PCB, you will be ready to examine the more complex computer-assisted PCB layout systems. These systems fall into two categories: data-based and nondata-based. Data-based systems generate the printed board directly from the schematic. AutoCAD is not a data-based system. You must do the translation from schematic to board, yourself.

In the data-based system, the computer extracts the information needed to produce the complete drawing package. This includes all the drawings you are developing in this tutorial, using AutoCAD, and then some. The data-based drawing package includes a netlist which shows the interconnect path between components, drilling drawings, assembly drawings as well as provides test data, inventory requirements, ordering information and then, if you're lucky, will tell you how much overtime you will be paid. Data-based PCB capture systems, as you can see, are in themselves a topic for another book. In this tutorial, we will take a more simplified approach to laying out a printed circuit board by using AutoCAD.

In this chapter, you will discover the technique of using AutoCAD to produce a basic PCB assembly drawing and its associated PCB artwork. See Fig. 5-1. We will use the schematic drawing developed in Chapter 4, the A size sheet developed in Chapter 3, and a number of new AutoCAD commands to create our PCB drawing.

The development process involved in laying out a PCB with AutoCAD is similar to that when a manual system is used. It involves: 1) The analysis of the schematic. 2) Relating the schematic to the equivalent hardware components. 3) Physically arranging the hardware on a grid layout so it fits the area of the board, and 4) Interconnecting the components with lines (traces) to finally create the circuit described by the schematic.

Component size, orientation, and spacing must all be taken into account in the layout of a printed circuit board. The drawing can be made at one-to-one scale because of the accuracy of AutoCAD. When manually laying out a printed circuit board, the PCB drawings are usually made at twice and even three times their normal scale and then photographically reduced to their correct scale at the manufacturing stage. This will shrink or reduce any drawing errors that may occur in the manual development and layout of the board.

PRINTED CIRCUIT BOARD TYPES

Discussion

Printed circuit boards fall into three major types determined by the number of sides or layers that have conductors. These types are:

1. Single-sided board
2. Double-sided board
3. Multilayer board

Single-sided boards have copper foil on only one surface, called the "trace" side. The components are mounted on the opposite side, which is referred to as the "component" side.

A double-sided board has copper foil on both surfaces. The two sides are interconnected by what is known as a "via." This may be either a "Plated-Through-Hole," an "Eyelet," or a "Jumper-Wire," and allows for higher component densities. A via also provides for greater flexibility in track layout.

A multilayer board is comprised of several very thin printed circuit boards compressed together to form the final product. As many as eight to twelve boards can be sandwiched together to produce the multilayer boards that are used to satisfy the demand of today's high-density computer systems.

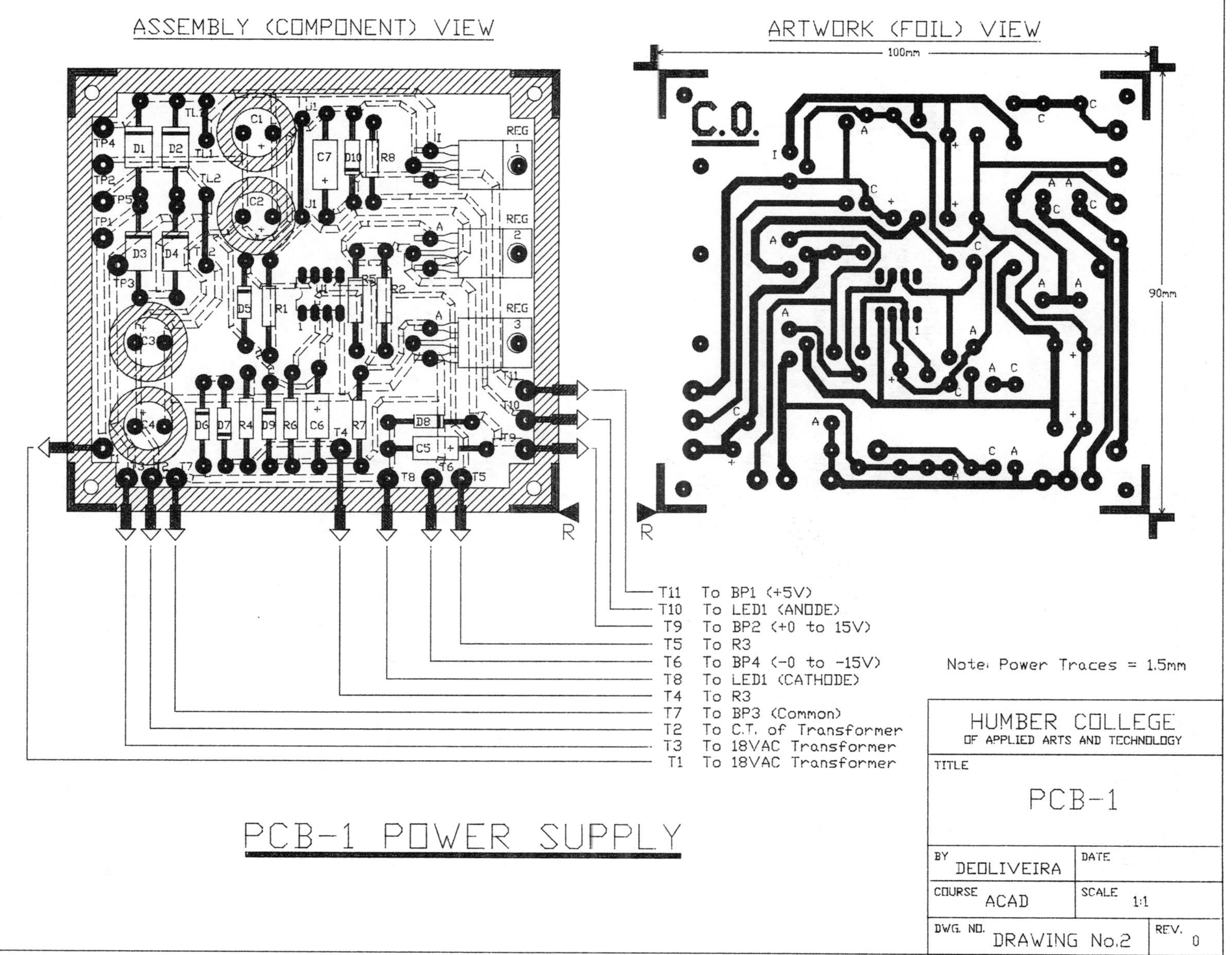

Figure 5-1 The printed circuit board Assembly View and Artwork View.

Our project will use a single-sided board having copper traces on only one side. The board is mounted permanently to the chassis through a heat sink, and hard-wired into the circuit. It is 100 mm long by 90 mm wide.

POWER SUPPLY BOARD

Discussion

In the development of the Power Supply Board, two drawing views are generated: an Assembly (Component) view and an Artwork (Tape) view. These views are developed on a single A size Architectural sheet. The drawing title is called PCB-1 (Fig. 5-1). The drawing will almost completely fill your work disk, so start with a new formatted disk, and don't forget to back it up periodically.

The steps to create the Power Supply Board are:

For the Assembly (Component) side,

Exercise 5-1. Starting PCB-1.

Exercise 5-2. Hatch PCB-1.

Discussion: Understanding Components, Their Polarity and Orientation.

Exercise 5-3. Transferring The Component Template.

Exercise 5-4. Positioning Components On The Layout Sketch.

Exercise 5-5. Wide Llines And their Use.

Exercise 5-6. Adding The Traces.

For the Artwork (Tape) side:

Exercise 5-7. Creating The Artwork.

Exercise 5-8. Adding The Remaining Text.

Exercise 5-9. Datum Lines and Targets.

PCB Layout Design Criteria and Specifications

Before any work can be done on the layout of the printed circuit board, we must first set down the "Design Criteria and Specifications" for the board. This requires answering some of the following questions:

* How big is the board to be?
* What are the trace size and conductor spacing?
* What are the number and size of the components?
* What are the heat and temperature requirements?
* Are there certain components, such as op-amps, that require special layout procedures due to environmental noise and temperature problems?
* How many PCBs are to be made?
* What voltages and current will be used?
* Are there any mechanical limitations?
* What should the cost of the board be?

PCB-1 Specifications

We will choose the following as our design criteria, which are specified at 1:1 scale.

Electrical:

Board size = 90 mm by 100 mm (+/- 0.5 mm)
Drawing scale = 1:1
Temperature range = 0°C to +70°C (32°F to 158°F)
Voltages and current used on the board:
35 V AC (Power in @ 2A max)
+5 V Fixed at 500 mA nominal (1 A max)
0 to +15 V at 500 mA nominal (1 A max) Dual Tracking
0 to -15 V at 500 mA nominal (1 A max) Dual Tracking

Mechanical:

Conductor width	= 1.5 mm (min) for common and power traces
	= 1 mm (min) for control traces
Inter-conductor width:	
track spacing	= 1 mm (min)
pad spacing	= 1 mm (min)
I.C. pad spacing	= .5 mm (min)
Pad size: large (for wire)	= 4 mm (OD), 1.5 mm (ID),
small (components)	= 3 mm (OD), 1 mm (ID)
round I.C. pad	= 2 mm (OD), 1 mm (ID)
rectangular I.C. pad	= 3 mm x 1.5 mm (OD), .5 mm (ID)
Component spacing	= 1.0 mm (min)

Setting Up the Drawing

In Chapter 3, we developed various size drawing sheets, A, B, C, and D. For this drawing, we will copy the A size prototype sheet to our work disk as we did in Exercise 4-1, "Getting Our Drawing Paper" (page 104) and call it PCB-1.

In these exercises, you will be introduced to the AutoCAD terms and commands of:

ARRAY	FREEZE	POLYLINE (PLINE)
BLOCK	HATCH	PLINE (Arc)
Component (Polarity)	HATCH (Ignore)	SHELL
COPY	HATCH (Normal)	Targets
DASHED (Linetype)	HATCH (Outermost)	THAW
Datum Lines	INSERT	TRACE
Dimension Lines	LAYER	WBLOCK
DONUT	MIRROR	
FILL	MOVE	

Housekeeping:

Set up the new drawing, using the following parameters:

Drawing Title: PCB-1
Limits = A size Architectural paper: 250,200
Grid = 5 units
Snap = 5 units
Text = 2 mm normal text, 3mm titles

Now, establish the following layers:

Layer	Comment	Color	Linetype	Pattern Use
0	Main Dwg	White	Continuous	Line
HATCH	Hatched area	Yellow	Continuous	ANSI31
COMP	Components	White	Continuous	Line
DONUTS	All Donuts	Red		
TRACES	Traces	Magenta	Continuous	Pline/Trace
TEXT	Text info	Green	Continuous	Text
DIM	Dimensions	White	Continuous	Line/Dim
TAPE	New Trace layer for plotting	White		

TRICK In Chapter 7, you will see how we switch the left side of the Traces layer to a new layer called Tape. This allows the Polylines on the Traces layer (the right side) to be plotted solid and the Polylines on the Tape layer (left side) to be plotted as a dashed line.

EXERCISE 5-1. STARTING PCB-1

On the left side of the drawing we will develop the Assembly view of the board and on the right, through the magic of AutoCAD, the Artwork view will automatically be created. Each view is 100 units long by 90 units high. See Figures 5-1 and 5-2.

Where Should We Start to Draw?

Exactly where should we start drawing on the sheet? Let's look at Fig. 5-2. Horizontally, our A size drawing area is 250 units long. Subtracting 200 units for the two PCB layouts leaves us with 50 units left over. This will divide nicely into 15 units of space at each end of the drawing and 20 units between the two views (15+15+20 = 50).

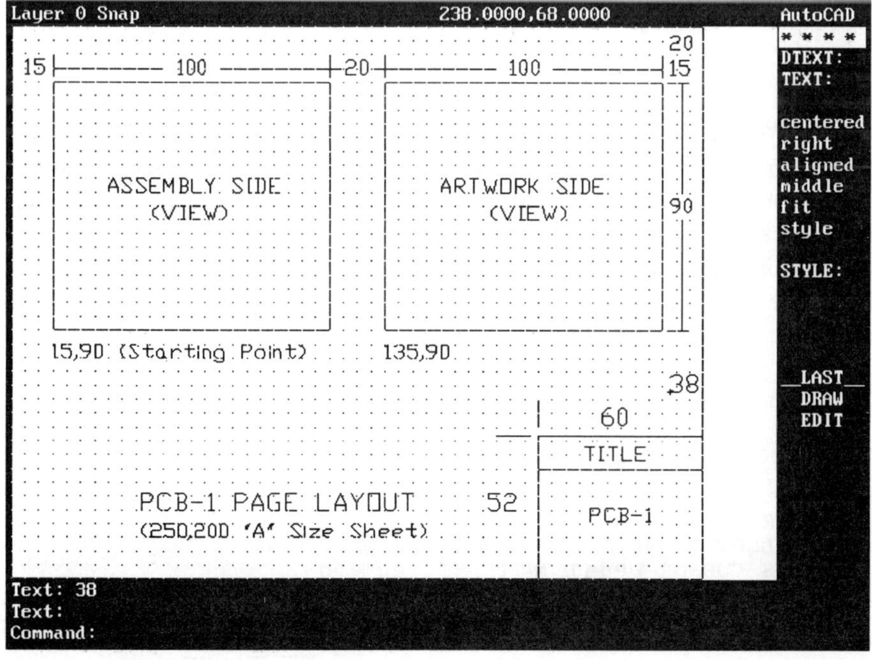

Figure 5-2 PCB-1 page layout

In the vertical positioning, our A size sheet is 200 units high. We must consider the Title Block which is 52 units high at 3/4 (.75) scale factor. The board vertical height is 90 units, which leaves 58 units of free vertical space. This can be divided such that 20 units of space is above the board and 38 units is below the board. This will allow us more room at the bottom for the lettering, notes and Title Block.

From the above calculations, we can position the lower left corner of the Assembly View at location 15,90.

To provide better resolution, zoom into that area of the drawing where the Assembly View will take shape. A good window size for the zoom would be to place the lower left corner of the window at 5, 80 and the upper right corner at 125, 190, thus giving us a 10 mm area around the desired view. See Fig. 5-2.

TIP1: Remember, when drawing the lines to use the OSNAP mode to grab (SNAP) onto the end of a line with the End option.

TIP2: Don't draw one line on top of another if you have to adjust the line length, just snap to its end and extend it. Refer to Exercise 3-6a on page 94.

DEVELOPING THE BOARD OUTLINE

Discussion

On layer 0, we will develop the board layout using Fig. 5-3 as a reference. Draw the outside of the Assembly View first, using the Line command. This is followed by the inner area of the board, the corners, and the mounting holes. When you have finished, it should appear the same as Fig. 5-3 without the dimension lines and text. Remember, the dimensions shown are the actual dimensions of the board. When this is complete, save your drawing as A:PCB-1 and then continue with the exercise on hatching.

```
Command:- USE THE DDLMODES DIALOGUE BOX TO SELECT LAYER "0"
```

Now to draw the outside of the board.

```
Command: LINE (RETURN)
From point:- SELECT 15,90
To point: - MOVE CURSOR UP 90 UNITS (90.00 <90>)
To point: - MOVE CURSOR RIGHT 100 UNITS (100.00 <0>)
To point: - MOVE CURSOR DOWN 90 UNITS (90.00 <270>)
To point: - MOVE CURSOR LEFT 100 UNITS (100.00 <180>)
```

At this point you are on your own to complete the remainder of the board. Use Fig. 5-3 as a reference.

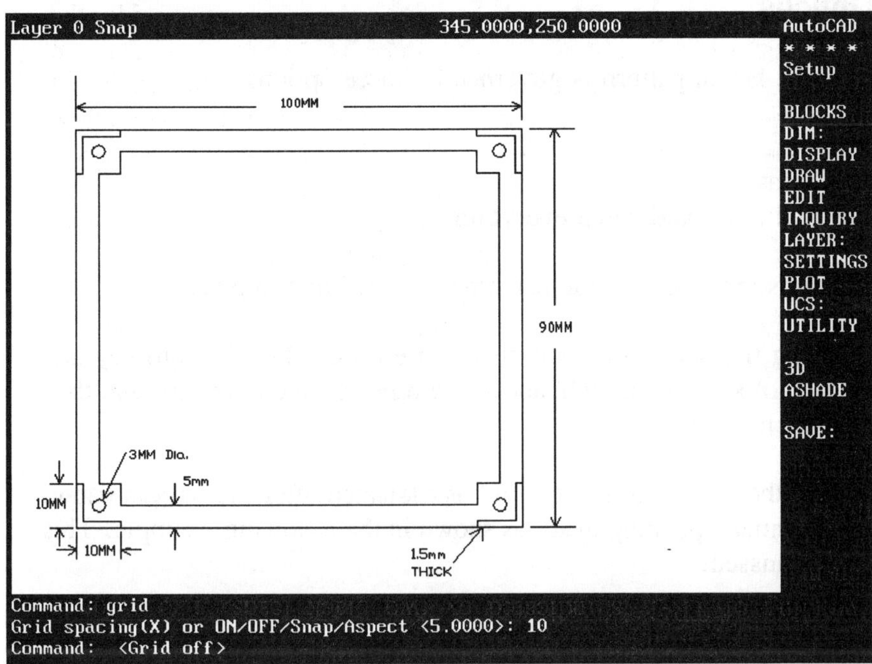

Figure 5-3 PCB-1 board dimensions.

```
Command: SAVE (RETURN)
File name <default>: A:PCB-1 (RETURN)
```

HATCH COMMAND

Discussion

Hatching is the process of filling an area with a series of dotted or dashed lines to highlight it from the rest of the drawing. AutoCAD has 53 built-in hatch patterns, 11 PostScript Fill patterns and also allows us to construct our own if these are not sufficient.

TIP: The hatched item is considered as one entity; therefore, editing may be undertaken only on the complete pattern.

In our drawing, all hatched patterns will be placed on the layer called "HATCH." The following is an in-depth look at the Hatch command. In the next discussion, I will show you some of the problems you can get yourself into if you are not careful.

Hatch Area Options

The area to receive the Hatch pattern is governed by three options:

1. (N)ormal
2. (O)utermost areas only
3. (I)gnore internal areas and hatch everything

If none of these options are selected, the hatching area defaults to Normal.

TIP: When selecting the Outermost option, be sure to use the "O" (oh) key and not the "0" (zero) key; one is right above the other on the keyboard and they often get interchanged.

You should further note that the "option" is indicated after the pattern name, with only a comma separating them, as shown in the following example. This can easily be missed!

```
Command: HATCH
Pattern(? or name/U,style)<?>:"PATTERN NAME","OPTION"
```

"Normal" Hatched Area Option. Using the Normal (default) option will hatch alternate areas of the drawing. Using Fig. 5-4A as a reference, the Normal option will cause the drawing to appear as Fig. 5-4B. Note the alternate areas of hatching as well as that area around the text which has a small open window to protect the text from being overwritten by the hatch pattern.

TRAP Caution should be exercised when using the Normal Hatch command, as the pattern will fill in alternate areas of the drawing, including areas you may not want filled. One way of working around this problem is to draw only that portion of the drawing you want to be hatched, then hatch it, and then continue with the rest of the drawing. Another solution to use is the next hatch option, Outermost.

TIP: You should be aware of leaks that may occur when hatching, the result being an area you do not want filled, being filled. This is usually the outcome of several lines being placed on top of each other rather than being connected at their ends, or because lines are broken, allowing the hatch line to "leak" through.

"Outermost" Hatched Area Option. Using the Outermost hatch option and the "window" select function allows you to zero in on the exact area to be hatched. For example, to hatch "Area two" of our test Fig. 5-4A without affecting the other areas, you would perform the following steps:

1. Use the Outermost (O) option when the pattern and style is requested.

2. Select the object with the "Window" option and place the window just outside the "Area Two" line.

The result is shown in Fig. 5-4C.

Figure 5-4A, B, C, D Hatch reference block.

"Ignore" Hatch Area Option. If you choose the "Ignore" option, the complete hatch pattern appears within the area selected, with no regard to boundaries or text. Your drawing will appear as Fig. 5-4D.

As you will see in the next exercise, AutoCAD has a few idiosyncrasies or a mind of its own. Be on the lookout for these!

EXERCISE 5-2. HATCHING PCB-1

Discussion

Having drawn the framework of the power supply board and examined the Hatch command, we are now ready to hatch the noncomponent area (outside perimeter) of our board; see Fig. 5-3.

Note: I'm going to lead you into a "trap" and then show you two solutions to avoid this type of problem in the future.

Before we start to do any hatching, we must be sure we are in the right layer. That is; the current Layer must be switched to the Hatch Layer. We'll also use the Hatch pattern of ANSI31. (Note, this is 31 not 3i.) The hatch scale will also be set to 10.

Selecting the Layer and Hatching

```
Command: LAYER (RETURN)
?/Set/New/On/OFF/Color/Ltype... /Unlock: SET (RETURN)
New Current Layer <Default>: HATCH (RETURN)
?/Set/New/On/OFF/Color/Ltype... /Unlock: (RETURN)
```

The layer prompt line returns again; pressing Return a second time will get the command line back.

Discussion: Now to hatch. We'll use the input device to select the outside perimeter of the board by placing a window around the complete board. AutoCAD highlights the area chosen by dotting it. If you are satisfied with your selection, press the Return key, and the hatch pattern will slowly fill the area selected. Should you change your mind, after seeing the results, use the Erase Last command to remove the pattern. During the hatching process, the keyboard and input device are disabled.

```
Command: HATCH (RETURN)
Pattern(? or name/U,Style)<Default>: ANSI31 (RETURN)
Scale for pattern <1.0000>: 10 (RETURN)
Angle for pattern <0>: (RETURN)
Select objects or Window or Last: W (RETURN)
First corner: PICK THE FIRST WINDOW CORNER
Other corner: WRAP THE WINDOW AROUND THE BOARD
Select objects: (RETURN)
```

Is All Well?

You expected to see a hatched area similar to Fig. 5-5, but what you probably got was something that looked like Fig. 5-5a, where the corner areas did not become hatched. Welcome to the world of hatching. You have just discovered one of those glitches or idiosyncrasies of AutoCAD.

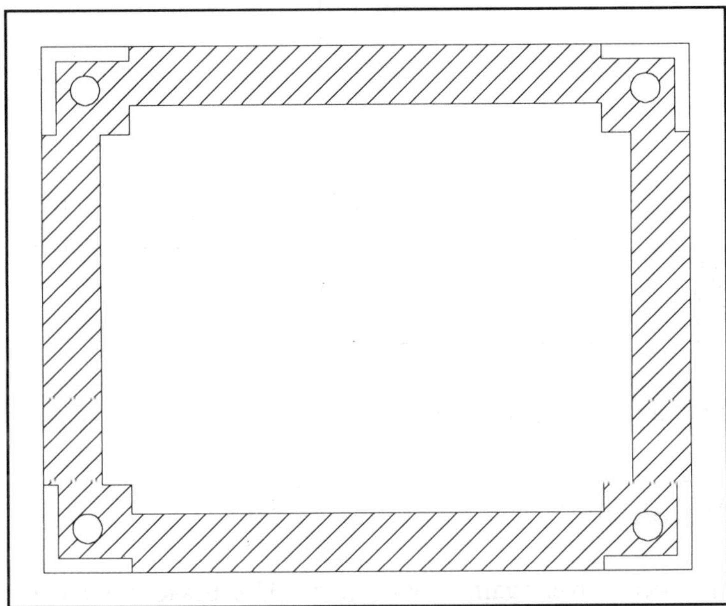

Figure 5-5 PCB-1 hatch area

The "Trap:" What Went Wrong?

Because of the shape of the corner areas which includes the outside corner bracket, the mounting holes, and the inner right angle; the hatched area was blocked from becoming completely hatched. The corner bracket acted as a roadblock to the hatch software routine. In this case, it's not your fault. *Let's see how we can overcome this problem.*

Solving the Hatch Problem

I will now show you two methods to solve this Hatch problem. The first avoids the problem by eliminating the offending obstacles, while the second tackles it head on.

Figure 5-5a PCB-1 hatched area with problems.

Hatching Solution, Method 1

We will begin the Hatching process over again, from scratch. Use Erase to remove all the hatching lines from the Hatch Layer.

Have you done this? Good, now let's begin again.

On the Hatch layer redraw only that area to be hatched. See Fig. 5-6a.

TIP: Having changed to the HATCH layer which uses the color yellow, you will be able to see the outline of the hatch area much more easily. Include the mounting holes in your reworked hatched drawing.

When you are finished, your drawing should look similar to Fig. 5-6a. You are now ready to rehatch the assembly view of your PCB-1 board, but before we do so, turn OFF layer 0.

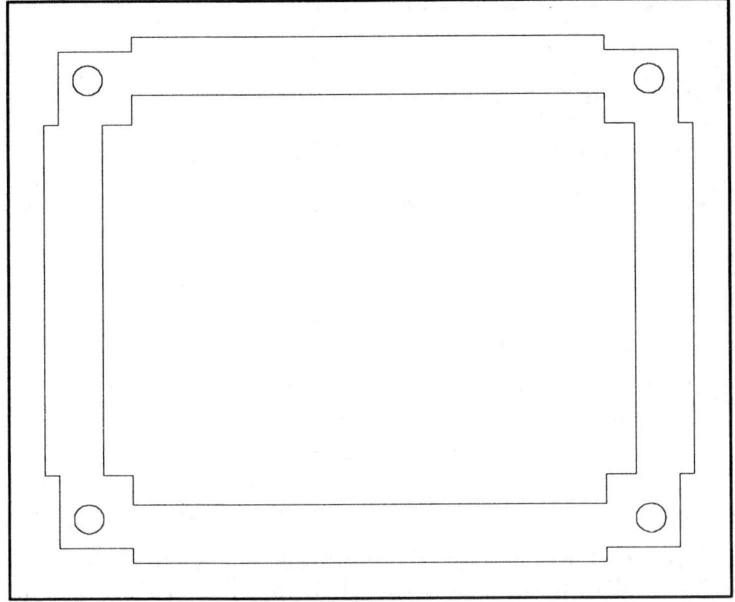

Figure 5-6a Rehatched outline.

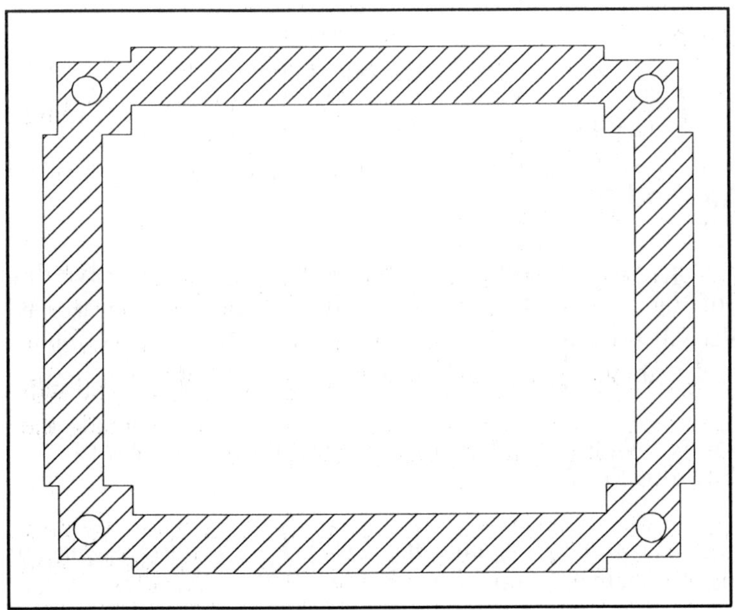

Figure 5-6b New hatched area.

Rehatching

```
Command: LAYER (RETURN)
?/Set/New/On/OFF/Color/Ltype... /Unlock: OFF (RETURN)
Layer name(s) to turn OFF: 0 (RETURN)
?/Set/New/On/OFF/Color/Ltype... /Unlock (RETURN)
```

TRAP: ***Are you in the Hatch layer?*** *You should be!*

```
Command: HATCH (RETURN)
Pattern(? or name/U,style)<Default>: ANSI31 (RETURN)
Scale for pattern <1.0000>: 10 (RETURN)
Angle for pattern <0>: (RETURN)
Select objects or Window or Last: W (RETURN)
```

Your hatched area should now appear similar to Fig. 5-6b.

TIP: If you have hatching lines in the middle of your board area, check all the lines for breaks at the corners. The slightest break will allow the hatching to leak out into the remaining part of the drawing. Likewise, when drawing lines, do not retrace one line on top of another. This buildup of lines acts as a barrier when the entity is hatched, creating void areas in the hatching. See Exercise 3-6a on page 94.

Turn your Layer 0 back ON again.

```
Command: LAYER (RETURN)
?/Set/New/On/OFF/Color/Ltype... /Unlock: ON (RETURN)
Layer name(s) to turn ON: 0 (RETURN)
?/Set/New/On/OFF/Color/Ltype... /Unlock (RETURN)
```

Hatching Solution, Method 2

There is a unique series of steps you can follow, in this second method, to solve the unhatched area problem of Fig. 5-5a or Fig. 5-7B. To do this, each corner area is worked on separately. We will use the top left corner of Fig. 5-7B as an example. Zoom to an area similar to that shown in Fig. 5-7C and perform the following steps.

Step 1. Isolate the area in question by inserting two lines into it to close off the unhatched area, as shown in Fig.5-7D.

Step 2. Implement the hatch command, again using ANSI31. A window is placed around the remaining "unhatched" area as in Fig. 5-7E. When the hatch command is complete, the corner area appears as Fig. 5-7F within the window area.

Step 3. (Fig. 5-7G) Now erase the two lines added at step 1 above. The top left corner now looks like Fig. 5-7H, completely hatched in.

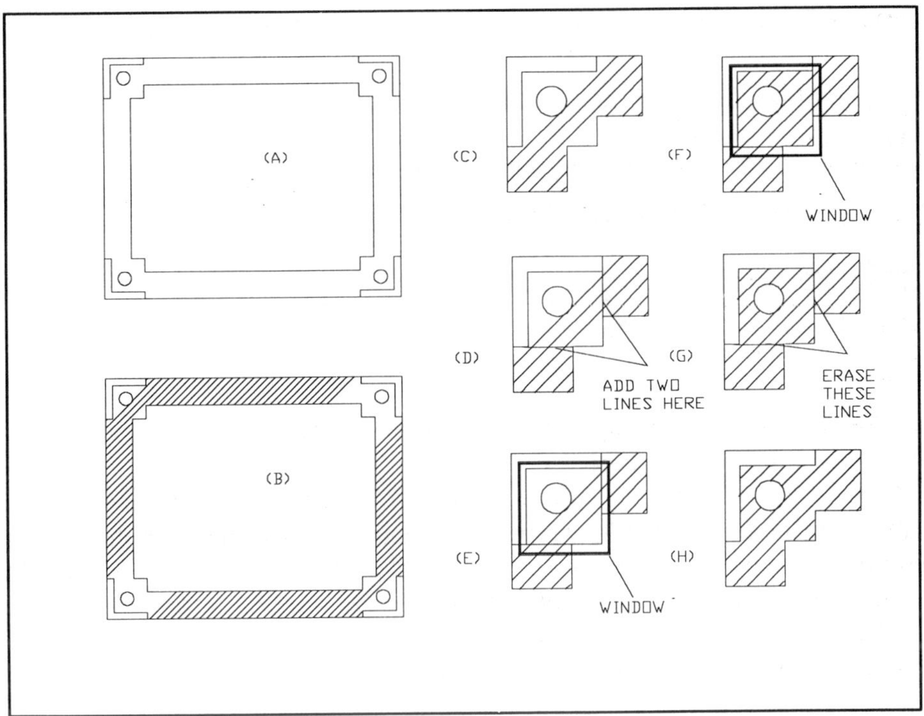

Figure 5-7A, B, C, D, E, F, G, H Hatch solution, Method 2.

These three basic steps can now be applied to the remaining unhatched areas. When complete, your drawing will look like Fig. 5-5.

Turn Off the HATCH Layer

TIP: It is preferable to Freeze the Hatch layer when the hatching is complete. This is because the hatching process consumes much time when the drawing is redrawn or regenerated. If you are using a slow computer go for a coffee break after you initiate the Hatch command and hope the drawing will be regenerated when you come back.

Hatching should be used sparingly because it consumes memory space by the bucket full. A drawing can quickly go from 30,000 bytes to 180,000 bytes or more just by incorporating a few hatched areas. Remember this in Chapter 6 when you design the Assembly Drawing and Front Panel view.

How do we Freeze a layer? Easy: a) Through the Layer command or b) through the Layer Control dialogue box.

```
Command: LAYER (RETURN)
?/Set/New/On/OFF/Color/Ltype/Freeze/Thaw: FREEZE (RETURN)
Layer name(s) to Freeze: HATCH (RETURN)
```

Upon redrawing the screen, the hatched layer will disappear.

At this point, we'll **save** our drawing with a new revision name and then End, but before doing so, we had better Thaw the Hatch Layer.

```
Command: LAYER (RETURN)
?/Set/New/On/Off/Color/Ltype/Freeze/Thaw: THAW (RETURN)
Layer name(s) to Thaw: HATCH (RETURN)
```

Now to save it,

```
Command: SAVE (RETURN)
File name <A:PCB1>: A:PCB1A (RETURN)
Command: END (RETURN)
```

Before we go on to Exercise 5-3, we will examine a few important points about components.

UNDERSTANDING COMPONENTS, THEIR POLARITY, AND ORIENTATION

Discussion

In placing components on a PCB board, one should understand that some components are polarized and must be placed in the circuit using a specific orientation. Other considerations to be taken into account are lead length, lead diameter, and component spacing off the board and between components.

In all cases, the manufacturer's data sheets should be consulted before designing any component into the circuit. Appendix D contains the data sheets for the components used in this project, and the following highlights much of the design considerations and requirements you should take into account when selecting components.

Diodes and Light Emitting Diodes (LEDs)

A diode is a polarized device that has an "anode" and a "cathode" lead. See Fig. 5-8. To produce the correct voltage polarity, positive or negative, it must have the anode/cathode correctly oriented in the circuit, as indicated by the schematic. Should the diode be mounted backwards, it may short-circuit and burn out. Note the end with the "bar," which represents the cathode end, when mounting the device.

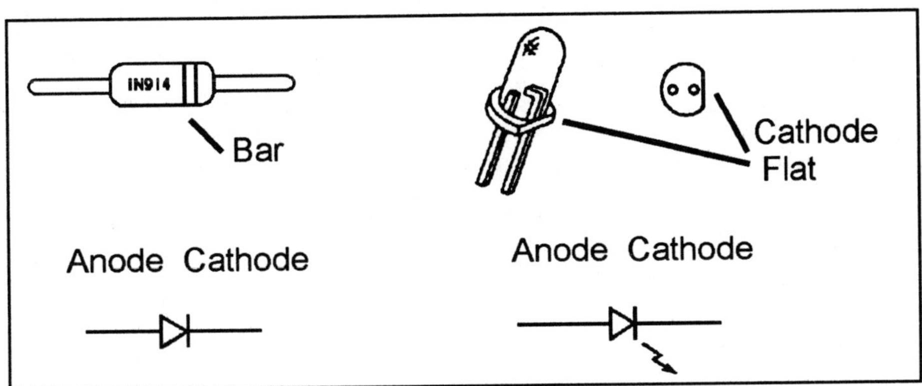

Figure 5-8 Diode

Figure 5-9 LED

Light Emitting Diodes (LEDs), like diodes, are polarized. See Fig. 5-9. The cathode is the lead which is closest to the flat edge on the side of the device. LEDs require a series resistor to limit the current to around 20 mA when the device is powered ON. If an LED is mounted backwards no harm will come to it; it just won't light up.

Voltage Regulators

There are three types of voltage regulators used in the power supply circuit, a 7805 (+5V) fixed output regulator, an LM317 positive voltage variable regulator, and an LM337 negative voltage variable regulator. See Fig. 5-10. You cannot tell from the schematic, but the pinouts of these three devices are different. Before placing any semiconductor device on the board, their pinout connections must be checked against the manufacturer's data sheet. Note Appendix D.

TRAP *Heat Sinking Of Regulators Can Be A Problem.* The heat sink tab on the 7805 series is connected to the ground terminal of the device. On the LM317 series, the heat sink is connected to the output lead, and on the LM337 regulator, the heat sink is connected to the input lead. If you try to place a common metal heat sink between these three devices, without insulating them, you will short the complete power supply out. To use a common heat sink, the heat sink tabs must be insulated from each other by using a thin mica washer between the regulator and the metal heat sink, and then fastened together with a Teflon screw. This fastening method will isolate each of the voltage supplies from each other.

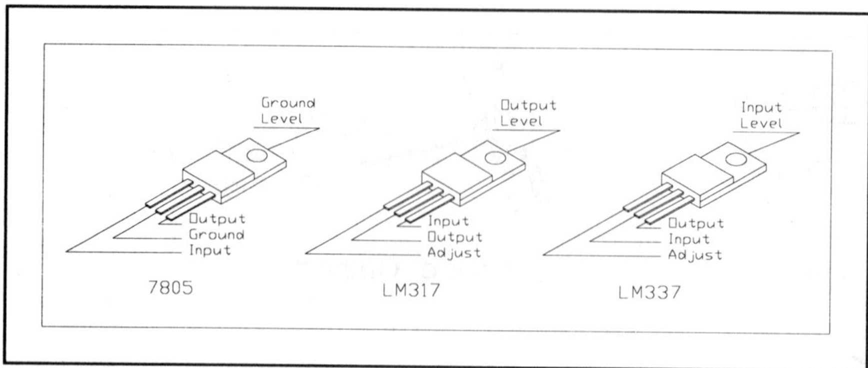

Figure 5-10 Power regulators.

Transformer

The transformer (TR-1) is placed on the chassis so that the leads will reach the required components. The primary winding is connected to the 115 V AC power, and the center tapped secondary winding (35 V AC) to the power supply board at points T1, T2, and T3. These points are shown in the schematic drawing, Fig. 4-3 (page 109), and in the schematic and PCB reference figures, Fig. 5-14 and Fig. 5-15, on page 176.

The primary winding of TR-1 is indicated by two black leads. The secondary winding, as seen in Fig. 5-15, has three leads. In most cases, the two outside ends of this winding (T1 & T3) are indicated by two solid green wires, while the center tap (T2) usually is a green wire with a yellow stripe. When wiring, make sure the center tap is connected to the "Common Lead" of the board.

Capacitor

Capacitors (Fig. 5-11) may or may not be polarized. If they are, such as electrolytic capacitors, there is usually a plus (+) sign at one end. The other end may use a bar (-) or stripe to indicate the negative side.

TIP: When inserting electrolytic capacitors into the circuit in the PCB Assembly drawing, double-check the polarity because they may explode if powered up incorrectly. Nonpolarized capacitors may be inserted in any direction without fear of explosion.

Note: Capacitors have been around for many years and many standards have evolved to indicate the component value. To determine the real value of a capacitor, always refer to the manufacturer's data sheet.

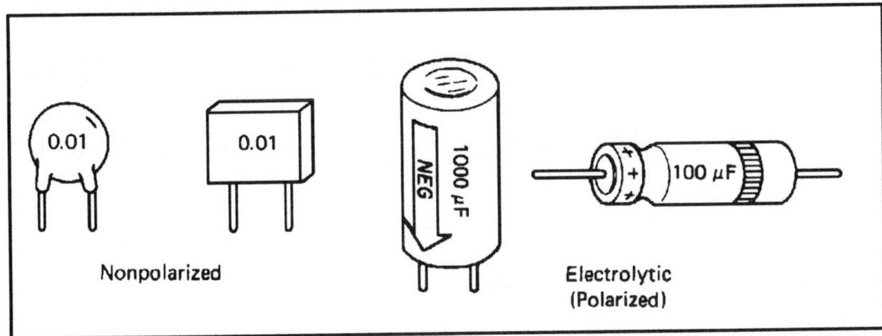

Figure 5-11 Capacitors.

Resistor

Resistors are nonpolarized and may be placed in the circuit in either direction.

TIP: From a "quality" point of view, it looks best if all resistor color codes are positioned in the same direction. Some quality control departments will reject an assembled PCB on this point alone.

CHART - 1 RESISTOR COLOR CODE CHART

COLOR	DIGIT	ZEROS	(MULTIPLIER)
BLACK	0	none	(X 1)
BROWN	1	0	(X 10)
RED	2	00	(X 100)
ORANGE	3	000	(X 1,000)
YELLOW	4	0000	(X 10,000)
GREEN	5	00000	(X 100,000)
BLUE	6	000000	(X 1,000,000)
VIOLET	7	-	
GRAY	8	-	
WHITE	9	-	
GOLD	-	(X .10)	

TOLERANCE	COLOR	TOLERANCE	COLOR
+/- 20 %	NONE	+/- 5 %	GOLD
+/- 10 %	SILVER	+/- 2 %	RED

Transistor

Transistors (Fig. 5-12) usually have three leads, an emitter, a base, and a collector. Often the designation "E B C" is indicated on the bottom edge of the transistor along with its JEDIC number. When the designation is not shown, refer to the manufacturer's data sheet.

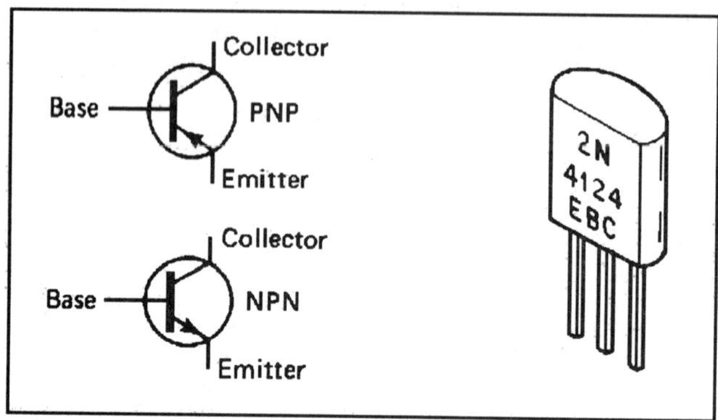

Figure 5-12 Transistors.

Integrated Circuits (ICs)

Dual-in-line integrated circuit pinouts are universally defined. The device has a notch at one end for orientation purposes. See Fig. 5-13 for pin one identification. Some manufacturers place a small indent over pin one to identify it. All other pins are numbered counterclockwise. Many surface-mount ICs also follow this convention.

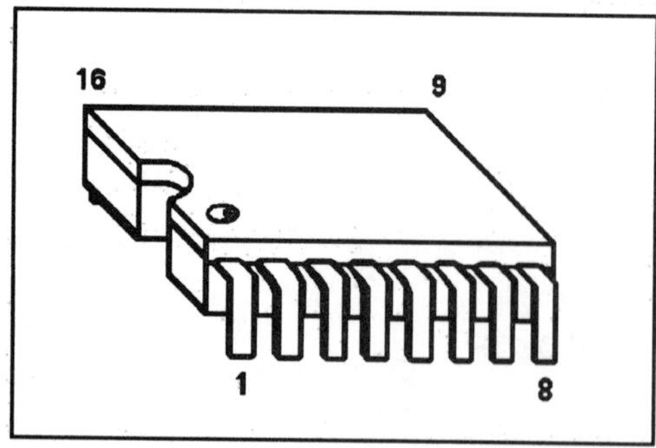

Figure 5-13 Integrated Circuit (IC).

HOW DO WE LAY OUT THE BASIC PRINTED CIRCUIT BOARD?

Discussion

The basic method to "manually" lay out a PCB is as simple as 1, 2, 3. The steps are:

1. Observe the schematic from the logical flow of the circuit.

2. Select a component, noting its orientation and polarity, and position it on the drawing, allowing sufficient space around it for minor adjustments and easy connection.

3. After an number of components have been positioned, join the component leads with a line to represent the schematic connections.

This procedure is repeated for each component until the complete circuit has been developed. Often the components will require some slight shifting in order to place the next component in the space available. With patience and perseverance, you will eventually complete the circuit. AutoCAD is ideal for this process, as you can easily move or shift things around without having to erase every component every time. To edit the line, we will explore the Move, Extend, Stretch and Grips commands as we proceed with this tutorial.

Let us walk through this process, using the Power Supply Board as our demonstration.

Step 1 (Observe the Schematic)

Referring to the schematic (Fig. 4-3 or Fig. 5-14), note how terminal points T1, T2, T3, diodes D1 and D2, test-link TL-1, capacitors C1 and C2, and regulator REG-1 are connected. This circuit has been redrawn as Fig. 5-14 and includes the secondary of the transformer TR-1.

By observing the logical flow of the schematic starting at terminals T1, T2, and T3, it can be seen that T1 connects to the anode of D1, and T3 to the anode of D2. Therefore, our two diodes D1 and D2 should be placed close to the edge of the board and oriented so the anodes point toward T1 and T3.

You should note that the cathodes of D1 and D2 are connected together along with one side of test-link jumper wire TL-1. The other side of TL-1 is connected to the positive lead (+) of capacitors C1 and C2, as well as to the "input" of regulator REG-1. Terminal T2 (Common) connects to the negative side of capacitors C1 and C2, as well as to the Gnd (ground) lead of regulator REG-1.

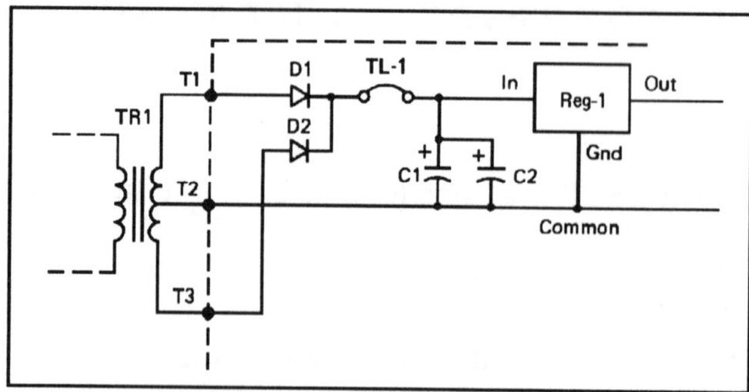

Figure 5-14 Schematic reference.

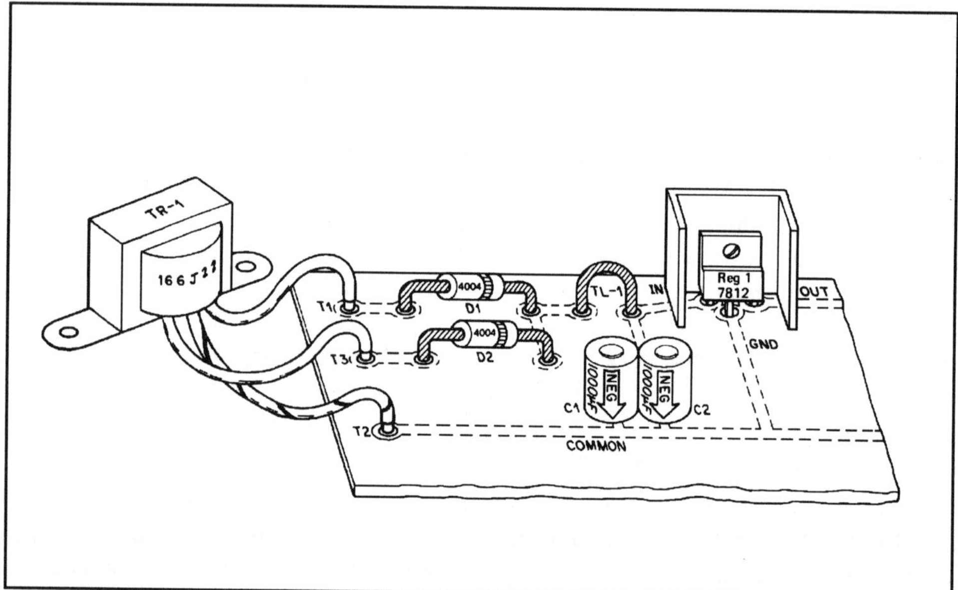

Figure 5-15 PCB pictorial of schematic reference.

Step 2 (Place the Components)

Noting how the various components are interconnected, we now have a better understanding of where the components should be placed. Figure 5-15 shows a pictorial placement of these five components, and Fig. 5-16 a top assembly view. It is this last view that is expanded until all components are placed within the confines of the board. Some juggling may be required to make them fit. It is the efficient placement of components at this stage that makes PCB draftspeople worth half their weight in gold.

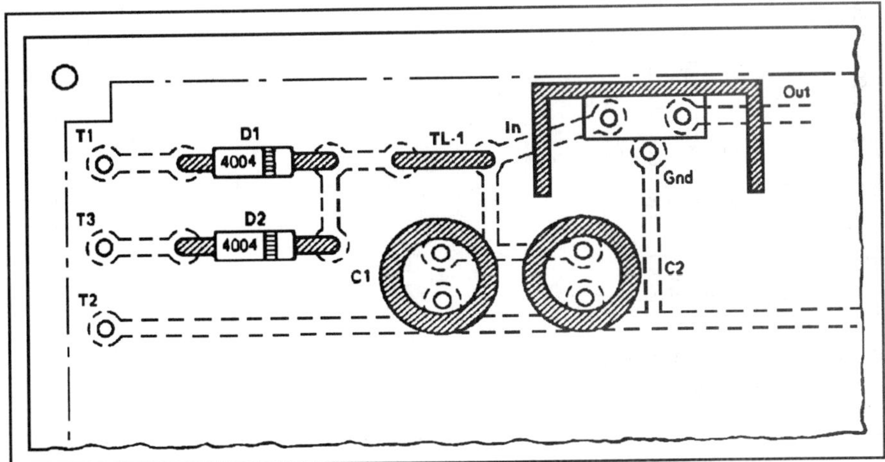

Figure 5-16 Assembly view of schematic reference.

Step 3 (Interconnect the Components)

Having the components correctly oriented, placed, and spaced, we can now interconnect them, using the schematic as a reference. As we join the lines on the assembly view, we should highlight them with red pen on a copy of the schematic to indicate that we have covered that area. Overcoming blocked paths to components will be your biggest problem in laying out single sided boards. We had to use one jumper (JPR1) in the layout of our dual voltage supply. It is in this area where PCB draftspeople are worth the other half of their weight in gold.

For the purpose of saving time on our project, an efficient component layout has already been developed for you. See Fig. 5-1. On the other hand, if you feel a challenge coming on, feel free to play!

EXERCISE 5-3. TRANSFERRING THE COMPONENT TEMPLATE

Discussion

To create the PCB Assembly Drawing and PCB Artwork Drawing, we'll first set up a symbol library of the components we will place on the board.

The components needed for this drawing are found on the Template disk that comes with this text, under the name ELECOMP.DWG (Fig. 5-17). To transfer these parts to the work disk, we will again use the Shell "Copy" command and the same procedure we used when transferring the electronic symbols to our work disk in Chapter 4.

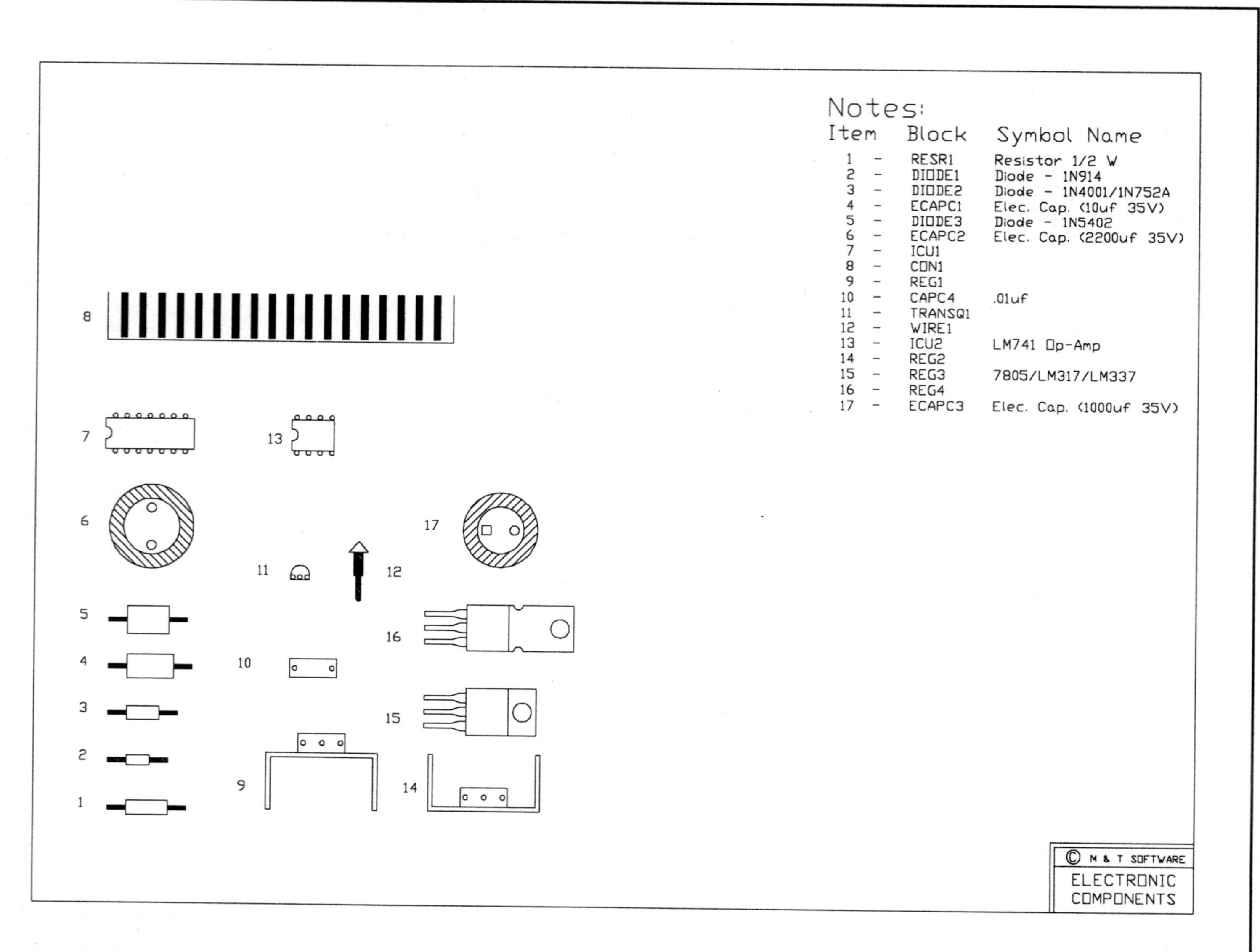

Figure 5-17 Component symbol template.

The components, when inserted, will be assigned to the layer "COMP" on our drawing.

To copy the data from the template disk on a single-drive system with a hard disk, refer to method 1 or 2 (Copying Files--A Floppy and Hard Disk) in Chapter 4.

On a two-drive system, place the Template disk in drive B and your work disk in drive A; and type the following:

```
Command: SHELL (RETURN)
```

You are now in DOS.

```
Dos Command: COPY B:ELECOMP.DWG A:ELECOMP.DWG (RETURN)
```

This will cause the drawing ELECOMP.DWG on the B drive to be copied to drive A under the same name. Upon completion of the copy process, you are placed back into AutoCAD. See Chapter 8 for more details on the Shell command.

Electronic Components and Wblocks

A list of components to be extracted from the ELECOMP drawing, suggested Block and Wblock names, and insertion points is shown in Fig. 5-17 and 5-18.

Housekeeping

To provide easy selection of the components, set:

Grid = 2
Snap = .5
Zoom so only two or three components are in view.
Use the Pan command to move about your drawing.

TRAP: Make sure that you are using layer 0 to develop your blocks!

Select these items from the component drawing (ELECOMP) as Blocks and then Save them to the work disk as Wblocks. Note that Wblocks are prefixed with a "W". Refer to Chapter 4, Exercises 4-2 and 4-3 if you require more detail on this procedure.

TIP: The selection of insertion points for the component blocks should be made with *ease of placement* in mind. *It is highly recommended that you use the insertion points suggested in Figure 5-18 as we will refer to them during the placement of the components in Exercise 5-4.*

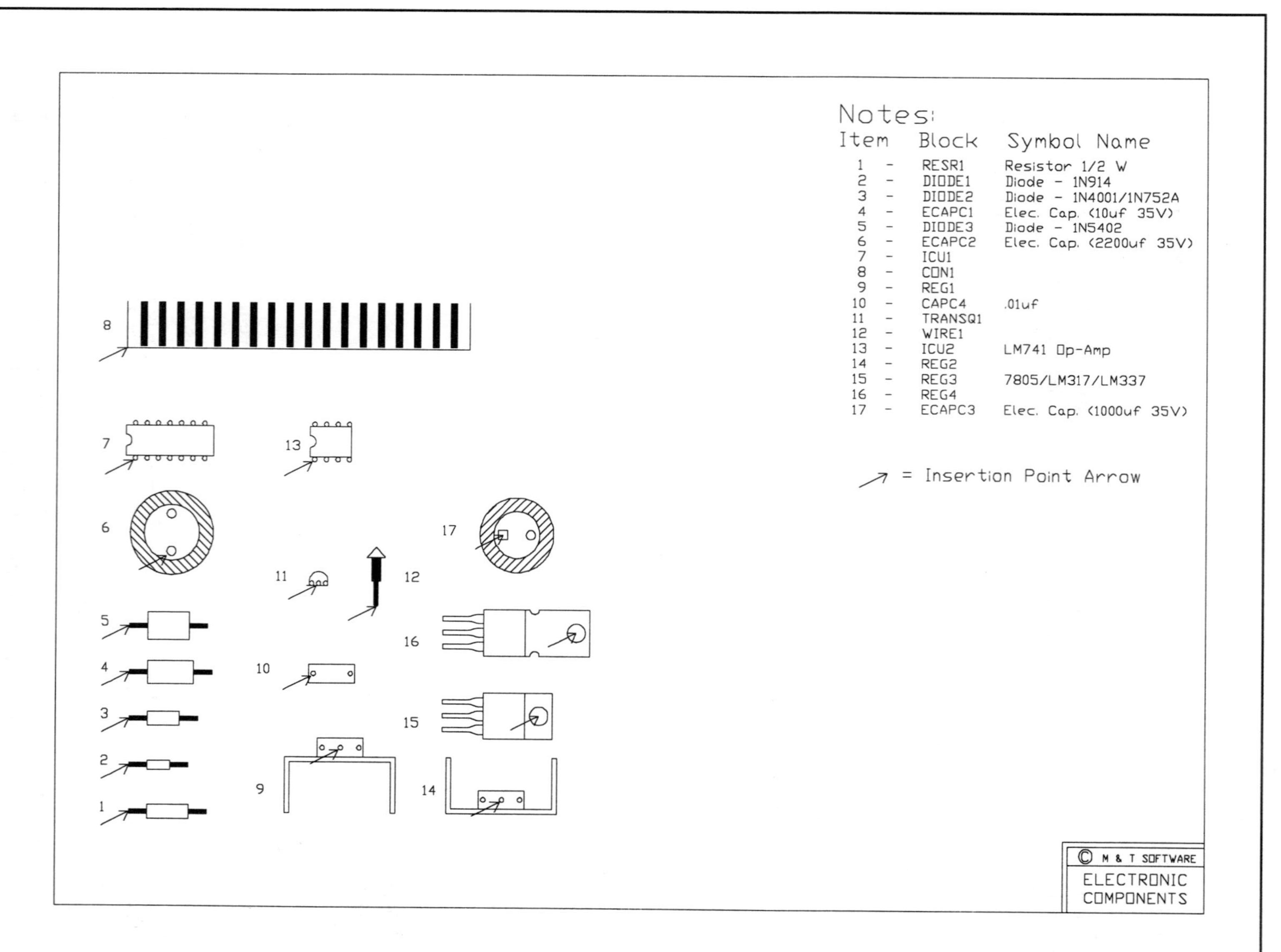

Figure 5-18 Component symbol template with insertion points.

TABLE 5-1

Electronic Components (ELECOMP)

Item	Symbol Name	BLOCK	WBLOCK Name
1	Resistor 1/2 W	RESR1	WRESR1
2	Diode (1N914/1N4004)	DIODE1	WDIODE1
3	Diode (1N752A/1n4001)	DIODE2	WDIODE2
4	Elec. capacitor (10 uF, 35V)	ECAPC1	WECAPC1
5	Diode (1N5402)	DIODE3	WDIODE3
6	Elec. capacitor (2200 uF, 35V)	ECAPC2	WECAPC2
10	Capacitor .01uF	CAPC4	WCAPC4
12	WIRE	WIRE1	WWIRE1
13	IC - U2 (741 op-amp)	ICU2	WICU2
15	Reg. 7805/LM317/LM337	REG3	WREG3
17	Ecap capacitor (1000 uF, 35V)	ECAPC3	WECAPC3

EXERCISE 5-4. POSITIONING COMPONENTS ON THE LAYOUT SKETCH

Discussion

Now that you have become familiar with the components and typical devices mounted on a PCB, let's start to place the component Wblocks on the drawing. Bring the bare board (PCB1A) back to the screen at maximum zoom so as to display the complete board. Use the block names of Table 5-1 as your reference when selecting the components.

We will set the current layer to "Comp" and establish a layout Grid on the board at 2-unit intervals (2 mm) and the Snap to 1 unit (1 mm). Remember, we are drawing at a one-to-one scale.

TIP: The components are placed so that they are parallel to the grid lines. All insertion points, terminal pads, donuts, and traces are placed on Snap intersection points or lines. Use Fig. 5-1 as your placement reference.

Note: The components will not require any scale factor, as they are actual size on the template (ELECOMP.DWG).

To get started, bring back the last drawing saved (A:PCB-1). Next we'll insert the Wblock (WECAPC3) of the four 1000-uF electrolytic capacitors (C1, C2, C3, and C4). I have given you the insertion coordinates to help in positioning these capacitors because finding the exact location on the first try is very difficult. Table 5-2 will also assist you in the placement of many of the major components.

Let's hope you have kept the same starting location recommended for the board (15,90). If so, your component placement will be the same as mine. If you have deviated from this, the following text will help you locate the insertion point for C1, C2, C3 and C4 with respect to the bottom left corner of the board:

C1 Insertion Point	= 35, 77 from lower left corner of board.					
C2 " "	= 35, 60	"	"	"	"	
C3 " "	= 20, 35	"	"	"	"	and rotate 180°
C4 " "	= 20, 18	"	"	"	"	and rotate 180°

The following insertion points are based on Table 5-2.

```
Command: LAYER (RETURN)
?/Set/New/On/OFF/Color/Ltype/Freeze/Thaw: SET
New Current Layer <Default>: COMP (RETURN)
?/Set/New/On/OFF/Color/Ltype/Freeze/Thaw: (RETURN)
```

To Insert C1:

```
Command: INSERT (RETURN)
Block name (or?): A:WECAPC3 (RETURN)
Insertion point: 50,167 (RETURN)
X Scale factor <1>/Corner/XYZ: 1 (RETURN)
Y Scale factor (default=X): 1 (RETURN)
Rotation Angle <0>:0 (RETURN)
```

The Wblock WECAPC1 now appears on our drawing.

To Insert C2:

```
Command: INSERT (RETURN)
Block name (or?): A:WECAPC3 (RETURN)
Insertion point: 50,150 (RETURN)
X Scale factor <1>/Corner/XYZ: 1 (RETURN)
Y Scale factor (default=X): 1 (RETURN)
Rotation Angle <0>:0 (RETURN)
```

To Insert C3:

```
Command: INSERT (RETURN)
Block name (or?): A:WECAPC3 (RETURN)
Insertion point: 35,125 (RETURN)
X Scale factor <1>/Corner/XYZ: 1 (RETURN)
Y Scale factor (default=X): 1 (RETURN)
Rotation Angle <0>:180 (RETURN)
```

To Insert C4:

```
Command: INSERT (RETURN)
Block name (or?): A:WECAPC3 (RETURN)
Insertion point: 35,108 (RETURN)
X Scale factor <1>/Corner/XYZ: 1 (RETURN)
Y Scale factor (default=X): 1 (RETURN)
Rotation Angle <0>:180 (RETURN)
```

The four electrolytic capacitors (C1 to C4) are now part of our Assembly Drawing.

The following table (Table 5-2) will help you place the major components if you have followed the coordinates used in this book.

TABLE 5-2 Major Component Locations Using Drawing 0,0 As Reference.

Item	WBLOCK	X	Y Location
C1	WECAP3	50	167
C2	WECAP3	50	150
C3	WECAP3	35	125 and rotate 180°
C4	WECAP3	35	108 and rotate 180°
Reg-1	WREG3	107	160
Reg-2	WREG3	107	142
Reg-3	WREG3	107	124
U1	WICU2	62	130

You're On Your Own Again

Using Fig. 5-1 (page 155) as a reference, continue with this exercise and place the remaining components on the Comp layer of your PCB. Then save your drawing under the name PCB1B.

```
Command: SAVE (RETURN)
File name <PCB1A>: PCB1B (RETURN)
```

EXERCISE 5-5. WIDE LINES AND THEIR USE

Discussion

In the development of the title block, we saw how the Line command drew a single thickness line. This next section discusses two special AutoCAD line types used in the layout of the tracks or traces on the printed circuit board. To handle the different electrical currents found on the PCB, the tracks must vary in width and eventually become the copper conductors that interconnect the various components on the board.

There are two methods of creating wide lines. One is to use the Trace command, which draws only straight lines, and the other is to use a Polyline or Pline command. The Polyline can produce straight as well as curved lines. Both line types may be drawn either solid (filled in) or left open, like a wire frame or skeleton. Let's explore these two commands and see how they behave.

TRACE Command

When you specify Trace, the width of the Trace line is requested, then the various From/To points. *Note:* Trace and Polylines lines are made up of "segments." You are actually drawing an "outline" of a wire-frame box which is equally spaced around the midpoint of the path you describe for it by the mouse. See Fig. 5-19A.

AutoCAD builds Trace segments as you enter each new From/To point. A segment is not displayed until the next segment is entered. This way the interconnecting angle between the two can be precisely calculated by AutoCAD to form a perfect match. Pressing the Return key will terminate the Trace command.

Let's try it. In a clear area of the drawing do the following:

```
Command: TRACE (RETURN)
Trace width <current>: 2 (RETURN)
From point: PICK A POINT
To point: PICK A POINT
To point: PICK A POINT
To point: PICK A POINT
To point: (RETURN)
```

See how Trace works? Try it again if you like. Okay, now Erase all the Trace Lines you added to your drawing.

Now that you have had a opportunity to play with Trace, *forget it!*

In my opinion, Trace is a throwback to the dinosaurs. Its limitations far outweigh its advantages. For example, the Trace line must be all the same width; therefore, you cannot make arrow heads from it. It cannot be used with the Dashed line type which gives us problems when creating the printed circuit board tracks. Likewise, you cannot Curve, Close, Continue, or Undo a Trace Line. A more elegant choice is the Polyline (PLINE). Use the Pline when creating the printed circuit board.

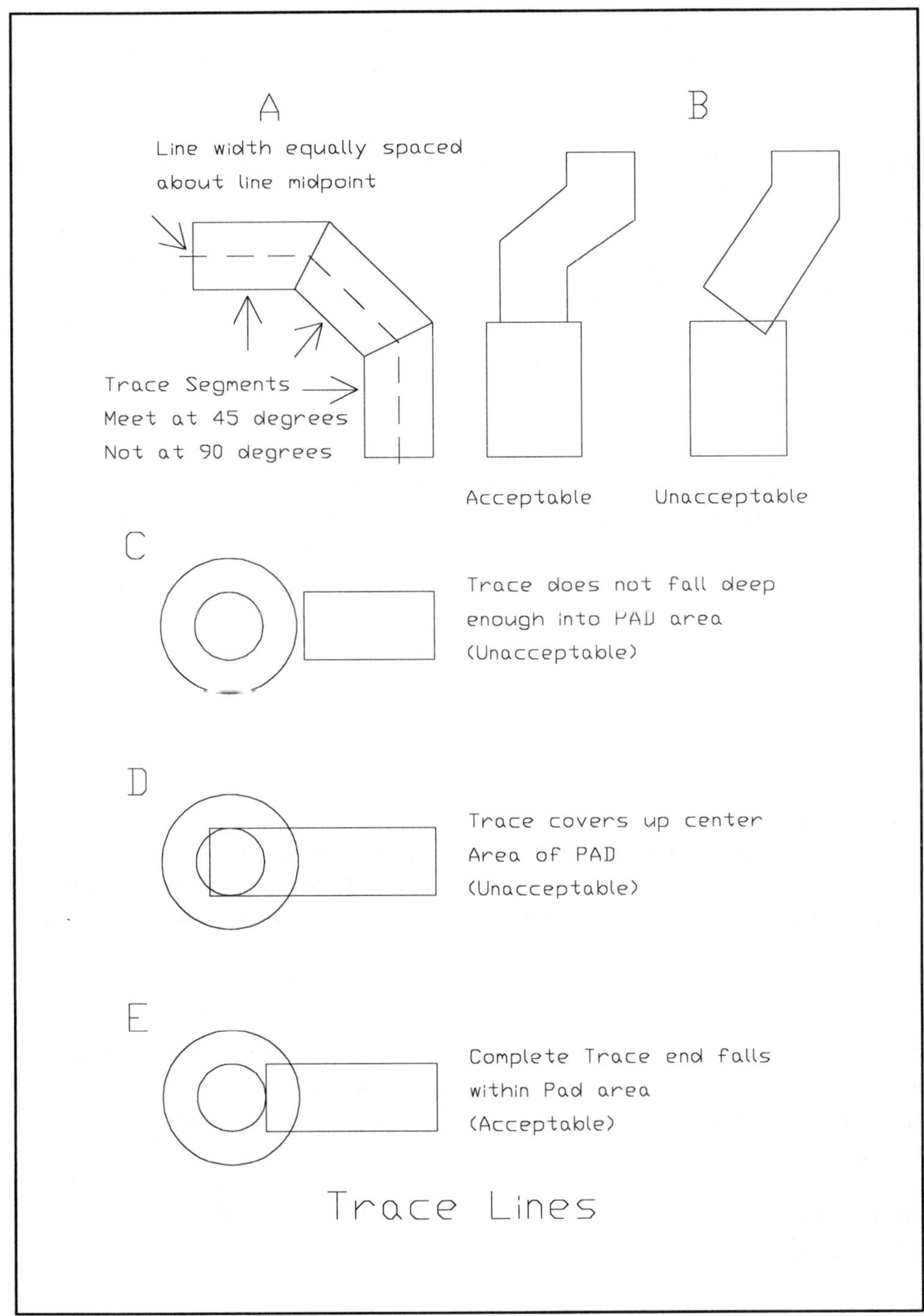

Figure 5-19A, B, C, D, E Trace lines.

TIP: To erase a trace or polyline, you must point to the edge of the line, because; like a wire frame, the center is effectively hollow. In the "object select mode," AutoCAD recognizes only outlines. Better still, use the Erase Last or Window command if you can.

TRAP AutoCAD starts the first trace segment and finishes the last trace segment with a flat end. These flat ends may become a problem if you do not meet the starting or termination points at the correct angle. See 5-19B. You must also be careful to ensure that the trace does not fill the center of the pad area, because this will make drilling the pad difficult at the manufacturing stage (see Figs. 5-19C, D, and E).

TRICK Printed circuit board tracks, when going around 90-degree corners, should be curved using a small radius of about five times the track width, or placed at a 45-degree angle. This prevents an etching problem that occurs during the manufacturing stage of the board. When traces make right angle bends, the corner becomes the last area to etch. When there are many traces grouped together, right angle bends will short out.

PLINE

Discussion

The Polyline is different from the Trace in that the width of the line may vary from its start point to its end point, such as in the development of an arrow head. A Pline can be a straight line, an arc or a circle. It's also used by the Donut command to make the pads used on printed circuit boards. See Fig. 5-20f on page 198.

Making Wide Lines with PLINE

In a clear area of the screen, let's practice drawing some polylines using the Pline command. Remember to erase these before you leave this exercise. Use the screen Draw menu, instead of the keyboard, and choose Pline. This will allow the Pline menu to be displayed.

The command first asks for the From point. You can select any screen location, or enter a set of coordinates from the keyboard.

```
Command: SELECT "PLINE" FROM DRAW MENU
From point: SELECT ANY POINT ON THE SCREEN
```

AutoCAD informs us of the default Pline line-width and prompts us for an input. We'll select the Width option and set the Start and End width to 2. Then we'll draw some practice lines. Type:

```
Current line-width <0.0000>
Arc/Close/Halfwidth/Length/Undo/Width/<Endpoint of line>: W (RETURN)
Starting width <0.0000>: 2 (RETURN)
Ending width <2.0000>: (RETURN)
```

Moving the input device to various locations on the screen and pressing the select button generates a series of Polylines segments 2 mm wide. Note that as you build up segments, they appear to be broken at the connecting points. This will correct itself when you finish drawing with Pline which causes a regeneration of the line to occur.

The definition for each of the prompt requests is:

(A)rc Selects the Arc mode.
(C)lose - Closes the Pline back to the beginning point.
(H)alfwidth - AutoCAD allows you to "show" it half the width you want the Pline to be, by using the input device.
(L)ength - The length of a segment may be specified.
(U)ndo - Undo the last polyline entered.
(W)idth - Allows the starting and ending widths of the line to be specified. The width is similar to trace line width, in that it is evenly spaced around the midpoint of the path you describe the Pline to follow.

Now to Create an Arrow Head:

```
Command: PLINE (RETURN)
From point: SELECT A POINT ON THE SCREEN
Current line-width <0.0000>
Arc/Close/Halfwidth/Length/Undo/Width/<Endpoint of line>: W (RETURN)
Starting width <5.0000>: 5 (RETURN)
Ending width <5.0000>: 0 (RETURN)
Arc/Close/.../Width/<Endpoint of line>: SELECT A POINT
Arc/Close/.../Width/<Endpoint of line>: (RETURN)
```

You will now see an arrow head, the length of the line selected, with a starting width of 5 mm going to a point (zero width) at the other end.

Having made a series of Plines on the drawing, don't forget to erase them before continuing with the next exercise.

Making an Arc with Pline

The greatest arc angle that can be created with the Arc option when using the Pline command is 359.99 degrees, as AutoCAD doesn't recognize 360 degrees in the Arc mode. This is a circle.

Let's explore the Pline command using the Arc option. The Pline width will be 2 mm wide, the Arc will be 90 degrees, with a radius of 4 mm. Again, select Pline from the Draw menu.

TIP Arcs are always drawn counterclockwise in AutoCAD unless you change the default settings.

This first step sets up the center point for the arc and then selects the Arc option:

```
Command: SELECT PLINE (RETURN)
From point: SELECT A CENTER POINT FOR THE ARC
Current line width 0.0000
Arc/Close/Halfwidth/Length/Undo/.<endpoint of line>:A (RETURN)
```

The Arc Angle menu, a long one, now appears on the screen. From this menu, we'll set the width parameters to 2 mm, the angle of rotation to 90 degrees, and the center radius of the Arc to 4 mm.

```
Angle/Center/CLose/Direction/Halfwidth/Line/
Radius/Second pt/Undo/Width/ <endpoint of arc>: W (RETURN)
Starting width:<0.0000>: 2 (RETURN)
Ending width:<2>: (RETURN)
```

The Arc Angle menu reappears. Now to set the Arc Angle to 90 degrees and the Radius to 4 mm.

```
Angle/Center/CLose/Direction/Halfwidth/Line/
Radius/Second pt/Undo/Width/<endpoint of arc>: A (RETURN)
Included angle: 90 (RETURN)
Center/Radius/<Endpoint>: R (RETURN)
Radius: 4 (RETURN)
Direction of chord <180>: (RETURN)
```

A Pline Arc of 90 degrees is now drawn counterclockwise around the original "From Point" selected at the beginning of this exercise.

Pressing the Return key gets us back to the Command prompt.

Okay, let's erase all the polylines we have just played with to clean up our drawing.

EXERCISE 5-5a. THE IC PAD

Discussion

In this exercise we will explore the Donut command and develop all of the pads used for the printed circuit board. This will include the large and small donuts (LPAD and SPAD) and the special pad for the 741 IC (ICPAD).

First, we will make a single IC pad. This will use a combination of the polyline and the polyarc commands to create the basic building block (ICPAD) Fig. 5-20a. This block is then made into an array to create the eight-pin IC pad (Fig. 5-20e) needed for the 741 op-amp. The same building block can also be used to build an IC pad for any size integrated circuit (IC).

Before we get started, one word of caution:

TIP: ***Develop all of your blocks on Layer 0.***

```
Command: LAYER (RETURN)
?/Set/New/On.OFF/Color/Ltype/Freeze/Thaw: SET (RETURN)
New Current Layer <Default>: 0 (RETURN)
?/Set/New/On.OFF/Color/Ltype/Freeze/Thaw: (RETURN)
```

We'll start by laying out a grid of 1 unit (mm) intervals, and then zoom in close to view a grid area of about 10 x 8 units. The snap is set to 0.5 unit.

Housekeeping Requirements

Set:
- LAYER to 0
- GRID to 1
- SNAP to .5
- ZOOM grid to display a 10 x 8 matrix.

Drawing the Basic Box

We are now ready to create the IC pad shown in Fig. 5-20a.

Note: Using the dimensions indicated, we will produce a pad twice its normal size. This will give us the opportunity to play with the Scale command when we "Insert" the IC pad.

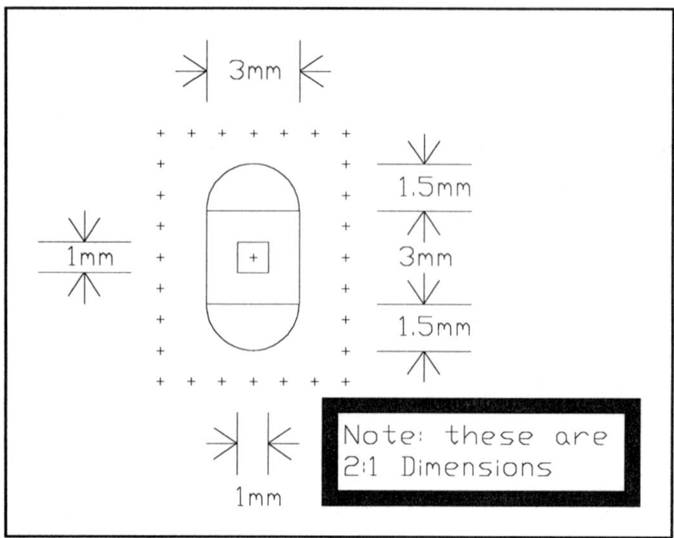

Figure 5-20a IC pad dimensions (shown at 2:1 scale).

TRAP You are about to develop an entity, the ICPAD, that could be used on many drawings. Make sure this is developed on Layer "0" so it can be inserted easily into other drawings in the future.

Using Fig. 5-20b as your drawing reference, the first part of the pad is developed by making a square box 3 mm by 3 mm on the outside perimeter. The width of the polyline used is 1 mm. Remember that the polyline is drawn about the center of the line used; therefore, the line lengths are only 2 mm long.

Getting started:

```
Command: PLINE (RETURN)
From point: SELECT POINT 1 (FIG 5-20b)
Current line width is <default>
```

Set up the line Width.

```
Arc/Close/Halfwidth/Length/Undo/Width/<Endpoint of line>: W (RETURN)
Starting width <default>: 1 (RETURN)
Ending width <1.000>: (RETURN)
```

Move the cursor to the right two units to create the first polyline. See point 2 of Fig. 5-20b.

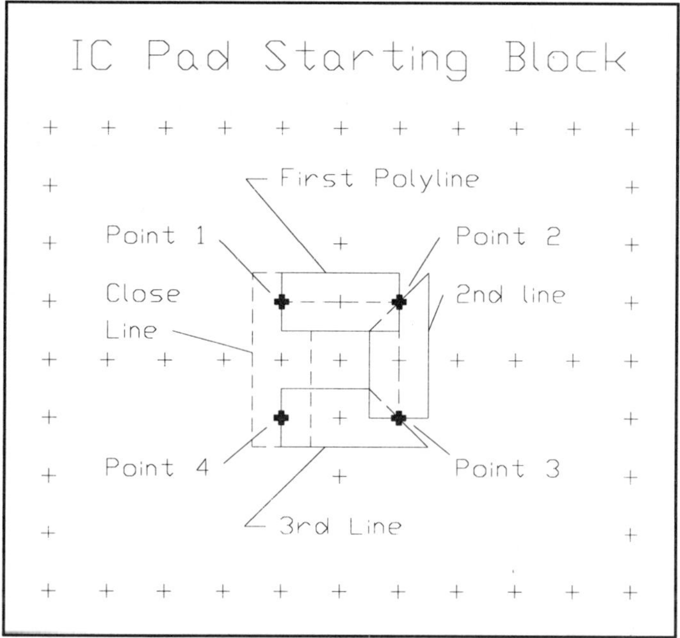

Figure 5-20b IC pad starting block.

```
Arc/Close/Halfwidth/Length/Undo/Width/<Endpoint of line>:
  SELECT POINT 2 @ 2.0000 <0
```

A Pline, 1 mm wide, is drawn between the points 1 and 2.

Now draw the second Pline.

```
Arc/Close/Halfwidth/Length/Undo/Width/<Endpoint of line>:
  SELECT POINT 3 @ 2.0000 <270
```

A second Pline is drawn between points 2 and 3.

```
Arc/Close/Halfwidth/Length/Undo/Width/<Endpoint of line>:
  SELECT POINT 4 @ 2.0000 <180
```

A third Pline is drawn. *Note:* This time, select Close (C) in order to complete the box.

```
Arc/Close/Halfwidth/Length/Undo/Width/<Endpoint of line>: C (RETURN)
```

The square is regenerated, and a box 3 mm x 3 mm is drawn with a 1 mm square hole in the center.

Adding the Pline Arc to the Top of the IC Pad

We'll now add a Pline Arc across the top of the box using Fig. 5-20c as a reference. The width of the polyarc line is first set to 1.5 units; then two locations are selected along the top edge of the box as the Starting and Center points for the arc. The Starting point A (Fig. 5-20c) is 0.75 unit in from the top right corner. The second point B is midway across the top of the box and becomes the arc Center. In order to locate these small spacings, you'll need to change the Grid and Snap spacing to .25.

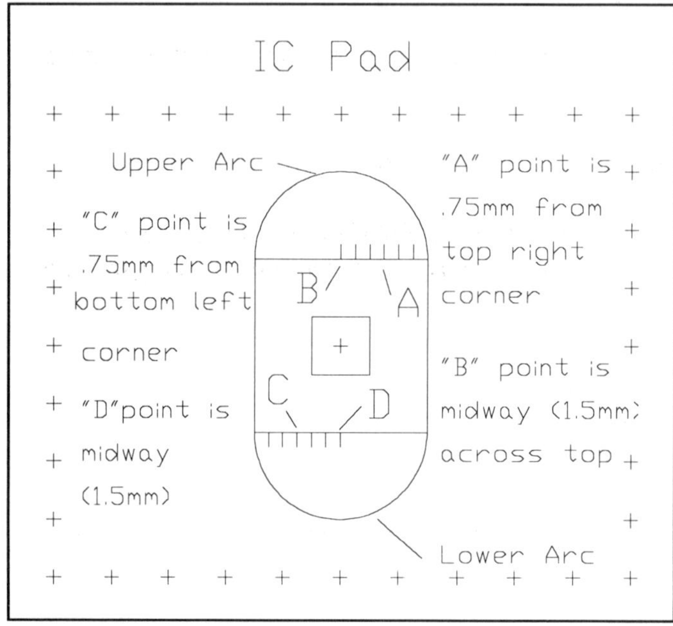

Figure 5-20c IC pad.

Housekeeping

Set:

GRID to 0.25
SNAP to 0.25

Drawing the Top Arc

```
Command: PLINE (RETURN)
From point: SELECT POINT A, FIG. 5-20c
```

Set Pline Width to 1.5 units.

```
Current width <default>
Arc/Close/Halfwidth/Length/Undo/Width/<Endpoint of line>: W (RETURN)
Starting width <1.0000>: 1.5 (RETURN)
```

```
Ending width <1.5000>: (RETURN)
```

Set up Arc mode.

```
Arc/Close/Halfwidth/Length/Undo/Width/<Endpoint of line>: A (RETURN)
```

Position Center of Arc.

```
Angle/CEnter/CLose...... <Endpoint of arc>: CE (RETURN)
Center Point: SELECT POINT B, (FIG. 5-20c)
Angle/Length <Endpoint>: A (RETURN)
Included Angle: 180 (RETURN)
```

The upper part of the arc is now drawn across the top of the box. Pressing Return gets us back to the command line.

Drawing the Lower Portion of the IC Pad Arc

Like the upper arc, we first select a starting point (Point C) to create this lower arc; then the Center point of the arc is chosen (Point D). The width of the polyarc line is left at 1.5.

```
Command: PLINE (RETURN)
From point: SELECT POINT C, FIG. 5-20c
Current line width is <1.5000>
```

Set up for Arc.

```
Arc/Close/Halfwidth/Length/Undo/Width/<Endpoint of line>: A (RETURN)
```

Choose Center.

```
Angle/CEnter/CLose ...<end point of arc>: CE (RETURN)
Center Point: SELECT POINT D, (FIG.5-20c)
 Angle/Length/<Endpoint>: A (RETURN)
Included angle: 180 (RETURN)
```

The lower arc is now drawn across the bottom of the box to complete the IC pad. Press Return to get back to the command menu.

Save It

You have now created an IC (integrated circuit) pad building block. Save the IC pad as a Block named ICPAD and a Wblock named WICPAD to your work disk. You might also consider adding it to your symbol library. It is suggested that you use the center point of the pad as your insertion point (see Fig. 5-20d). This block will be very useful if you are required to create the fourteen-pin IC pad in the future.

Zoom to All and erase any unwanted objects from your drawing.

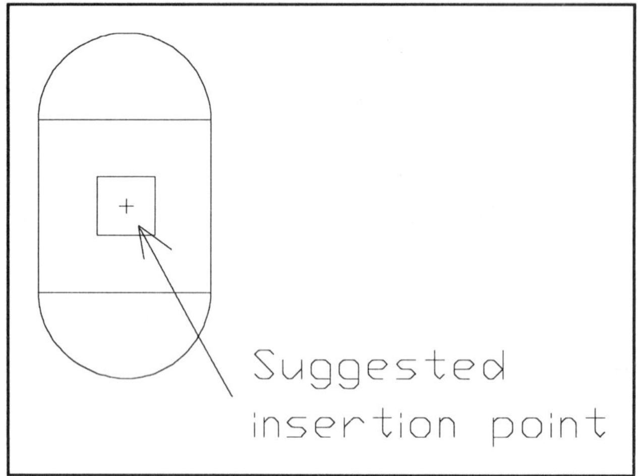

Figure 5-20d The single IC pad (ICPAD).

EXERCISE 5-5b. CREATING THE 8 PIN IC PAD

Discussion

In this exercise, we will make use of the smarts of AutoCAD and show you how it handles various grid ratios and scales, as well as the Array command. The building block ICPAD, just created, is used to produce the eight-pin IC pad layout required for the 741 op-amp.

In reviewing the component data sheets (Appendix D) for the center distance between the legs of the IC, or Fig. 5-20e, we see the center-to-center distance to be 0.1 inch apart and the distance between the rows to be 0.3 inch apart. "What is this 0.1 and 0.3 inch business? I thought we were working in millimeters!"

Note: Not all of the world has changed to the Metric System as yet; the standard dimensions for IC leg spacing is in tenths of an inch. Therefore; it will be good to know how to use both the Metric and Inch systems.

TRICK Let's use AutoCAD to automatically change the scaling from inches to metric for us. The conversion between inches and millimeters is 1 inch equals 25.4 millimeters. Therefore, 2.54 millimeters equals 0.1 inch. Remember this when it comes time to input our data.

To create the IC pad drawing, we'll use a grid based on 2.54 mm instead of 1 mm; that is, it is on 0.1 inch centers. Remember, the ICPAD building block is at double scale, so we'll place the pads on two grid intervals to maintain the scale of two to one (double scale). *We should also be using layer 0.*

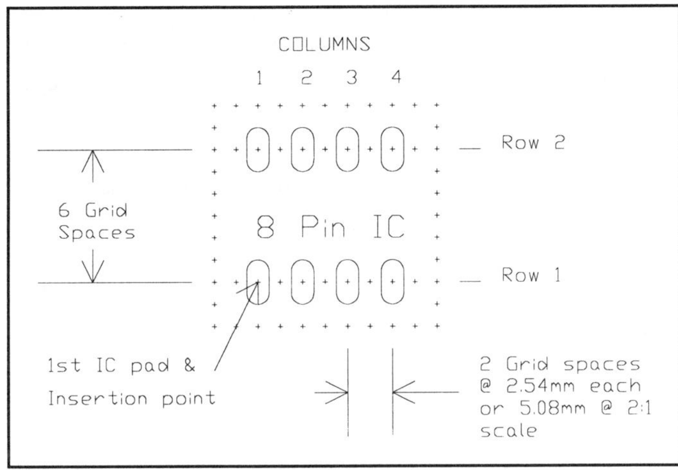

Figure 5-20e Eight-pin IC pad.

Housekeeping

Set:

Layer = 0
Grid = 2.54
Snap = 2.54
Zoom = display an area of 20 grid units by 16 grid units.

Creating the Pad

On your drawing, zoom to a location that will show a grid area of *about* 20 grid units by 16 grid units. Next insert the IC pad building block ICPAD at a grid location near the lower left corner. Note Fig. 5-20e.

```
Command: INSERT (RETURN)
Block name (or?): ICPAD (RETURN)
Insertion point: SELECT A LOCATION NEAR THE LOWER LEFT CORNER,
                 ON A GRID POINT
X Scale factor <1>: 1 (RETURN)
Y Scale factor <1>: (RETURN)
Rotation angle <0>: (RETURN)
```

The Array Command

The first ICPAD building block is now in place. Through the use of the Array command, we will now use this building block to create the 8-pin IC pads needed for our drawing.

The Array command allows us to create multiple copies of an entity. The copies may be placed in an arc or circle, called a "Polar Array," or in horizontal rows and vertical columns, called a "Rectangular Array."

To create our 8-pin IC pads, we will use the Rectangular Array method, and our ICPAD as the "reference" entity. For the 8-pin IC, our Array will have two rows, each containing four columns of IC pads. Each row is (3 x 2.54 = 7.62 mm @ 1:1 scale) 15.24 mm apart at 2:1 scale or six grid spaces, and each column is (2.54 mm @ 1:1 scale) 5.08 mm apart at 2:1 scale, or two grid spaces. Remember that the ICPAD building block was developed at 2:1 scale.

```
Command: ARRAY (RETURN)
Select objects: W (RETURN) PLACE A WINDOW AROUND ICPAD
1 selected, 1 found.
Select objects: (RETURN)
Rectangular or Polar array (R/P): R (RETURN)
Number of rows (---) <1>: 2 (RETURN)
Number of columns (III) <1>: 4 (RETURN)
Unit cells or distance between rows (---): 15.24 (RETURN)
Distance between columns (III): 5.08 (RETURN)
```

With the completion of the last command, AutoCAD creates an array for a 8-pin IC pad. Save this array as 8ICPAD on both your work disk and the Symbol Library disk as a Wblock. Note the suggested insertion points on Fig. 5-20e.

```
Command: BLOCK (RETURN)
Block name (or?) 8ICPAD (RETURN)
Insertion base point: SELECT CENTER OF LOWER LEFT PIN (#1)
Select objects or Window or Last: W (RETURN)
First point: SELECT POINT
Second point: SELECT POINT
(RETURN)
```

Don't bother to Oops the 8ICPAD that just disappeared as it is not necessary to use it now. We have it captured as a Block.

```
Command: WBLOCK (RETURN)
Wblock file name: A:W8ICPAD (RETURN)
Block name: 8ICPAD (RETURN)
```

For more information on Arrays, see Chapter 8.

The Copy Command of AutoCAD

An alternative method of reproducing the 8-pin IC pad would be to use the AutoCAD Copy command. This should not be confused with the DOS Copy command as described in Chapter 2.

The AutoCAD Copy command allows you to make a single copy of an existing entity by giving it a "Displacement," or "Multiple" copies by simply selecting new locations at which to place them. When you have made all the copies you need, just press the Return key. As an example, we'll play with the IC pad building block ICPAD and make copies of it.

Using the same housekeeping requirements as Exercise 5-5b, we will insert the ICPAD building block and then Copy it.

Housekeeping

Layer = 0
Grid = 2.54
Snap = 2.54
Zoom = display an area of 20 grid units by 16 grid units.

```
Command: INSERT (RETURN)
Block name (or?): ICPAD (RETURN)
Insertion point: SELECT A LOCATION NEAR THE LOWER LEFT CORNER,
                 ON A GRID POINT
X Scale factor <1>: 1 (RETURN)
Y Scale factor <1>:   (RETURN)
Rotation angle <0>:   (RETURN)
```

This places our ICPAD building block back on the drawing. Now to Copy it, using "Multiple" copy.

```
Command: COPY (RETURN)
Select objects: W (RETURN) PLACE A WINDOW AROUND THE ICPAD
<Base point or displacement>/Multiple:M (RETURN)
Base point: SELECT THE PAD POINT CENTER
Second point of displacement: SELECT A GRID POINT
Second point of displacement: SELECT A 2ND GRID
Second point of displacement: SELECT A 3RD GRID (RETURN)
```

Note that AutoCAD repeats the "Second point of displacement" message, allowing you to place copies of the entity until you press Return.

EXERCISE 5-5c. DONUTS (Since AutoCAD 2.5)

Discussion

One of the best instructions AutoCAD has added for those of you who lay out printed circuit boards is the Donut command. AutoCAD introduced this command with release 2.5 and it became a dream come true for many of us. It allows you to make Polyline type IC pads with only one instruction: *Donut*.

The simplest way to make a pad for your printed circuit board is through the use of the Donut command found under the Draw menu. The Donut command generates either circles with a hole in the center, that is, a donut (Fig. 5-20f), or circles that are completely filled in, that is, a solid donut (Fig. 5-20g). In the creation of these circles, AutoCAD requires to know the inside and outside diameter dimensions of the circle. If the inside diameter is made zero, you have a solid circle.

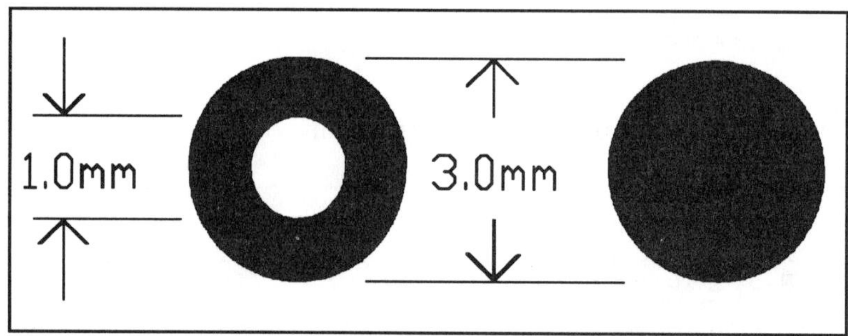

Figure 5-20f "Donut" **Figure 5-20g** "Solid donut"

There are two sizes of Donut pads used on our printed circuit board, a small size pad (SPAD), 3 mm in diameter with a 1 mm hole and a large size pad (LPAD), 4 mm in diameter with a 1.5 mm diameter hole. See Fig.5-20h and Fig.5-20i

Let's experiment with the Donut for the small pad.

Housekeeping

Set:

GRID to 1
SNAP to 0.5
ZOOM to an appropriate scale to see the pad.

```
Command: DONUT (RETURN)
Inside diameter <default>: 1 (RETURN)
Outside diameter <default>: 3 (RETURN)
```

At this point, a dotted donut appears in the center of the cross-hairs of the screen cursor. AutoCAD requests a position at which to place the donut.

```
Center of doughnut: SELECT A POINT ON SCREEN
```

Note that the dotted donut transforms into a solid donut having an inside diameter of 1 mm and an outside diameter of 3 mm. A request to position another donut for a second selection is now made. Pressing the Return key will terminate the command.

```
Center of doughnut: SELECT YOUR SECOND POINT
Center of doughnut: (RETURN)
```

Now let's create a solid circle (Fig. 5-20g). Enter a value of zero (0) as the inside diameter.

```
Command: DONUT (RETURN)
Inside diameter <default>: 0 (RETURN)
Outside diameter <default>: 3 (RETURN)
Center of doughnut: SELECT A POINT ON SCREEN
Center of doughnut: (RETURN)
```

We now have a solid Donut or circle. Erase any unwanted entities and continue with Exercise 5-5c.

Fill On/Off

Having drawn Plines and Donuts that always look like solid objects, you probably have not noticed that Plines sometimes appear as outlines of boxes and the donuts as two circles that are not filled in. This is because the default for the Fill command is always ON, causing the polylines to be filled in. To see these entities in their other form, the Fill must be switched to OFF. Fill operates like a switch that can be turned on and off through the Fill command. To switch the Fill off, type FILL at the command line and reply to the prompt with OFF.

```
Command: FILL (RETURN)
ON/OFF: OFF (RETURN)
```

Note: You may have to regenerate the drawing to see the results of this operation.

```
Command: REGEN (RETURN)
```

One should be careful with using Fill in the on mode, for it gobbles up precious computer time when regenerating the drawing. To save time, it is better to leave Fill in the off mode when you are developing a drawing and only turn it on when you are ready to print or plot it.

Before we leave this exercise, erase all the unwanted polylines and donuts from your drawing if there are still some remaining.

During this exercise, you have added two more Blocks and Wblocks to your work disk. These are the ICPAD and the 8ICPAD pad drawings. *Did you create them on layer "0" ? I hope so!*

Getting Ready for the Pads (Donuts) and Traces

Discussion. You are now familiar with the two ways to draw wide lines and design pads. Let's use this knowledge to: (1) place pads (Donuts) at each of the component leads where they are located on the board and (2) interconnect these pads to trace out the circuit described by the schematic. This will be done on the layers called "DONUT" and "TRACES." During this exercise, we will leave Fill off in order to save drawing time. Use Polyline to draw the traces.

EXERCISE 5-5d. LAYING DOWN THE PADS (DONUTS)

Discussion

To get started, we will change the Grig and Snap spacing, and the Layer to DONUTS; then create a small donut (OD = 3, ID = 1), as shown in Fig. 5-20h, and place it at the termination points (end points) of each component.

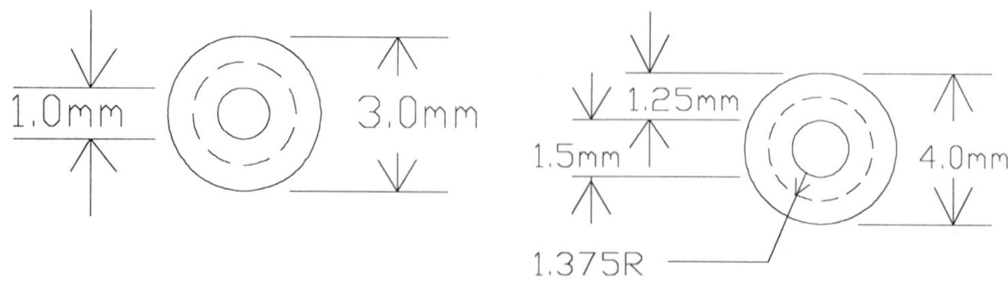

Figure 5-20h Small Donut

Figure 5-20i Large Donut

Housekeeping

SET:
GRID = 2
SNAP = .5
LAYER = DONUTS

Now, develop the small Donut (SPAD) and place it at the end of all components on the board. Use Fig. 5-1 or Fig. 5-21 as a reference.

```
Command: DONUT (RETURN)
Inside diameter <default>: 1 (RETURN)
Outside diameter <default>: 3 (RETURN)
Center of doughnut: PLACE A DONUT AT THE ENDS OF ALL COMPONENTS
```

Next, we'll position a LPAD (Large Donut, Fig. 5-20i) at all locations where wires from the outside world attach to the board, and then insert the 8ICPAD over U1, the 741 op-amp socket. Use Fig. 5-1 or Fig. 5-21 as a reference.

```
Command: DONUT (RETURN)
Inside diameter <default>: 1.5 (RETURN)
Outside diameter <default>: 4 (RETURN)
Center of doughnut: PLACE THE LARGE PADS AS PER TABLE 5-3
```

To help you with the placement of the large pads, refer to Table 5-3.

TABLE 5-3 Large Donut Locations

Item	Comment	X	Y Location	
T1	Large Donut	22	103	(From lower left corner of drawing)
T3	" "	27	97	(0,0 location)
T2	" "	32	97	
T7	" "	37	97	
T8	" "	80	97	
T6	" "	89	97	
T5	" "	95	97	
T9	" "	108	103	
T10	" "	108	109	
T11	" "	108	115	

You're On Your Own Again.

Note: Because we created the ICPAD and its big brother, 8ICPAD on layer "0" of this drawing, and then saved them as blocks, they do not have to be transferred back to this drawing as Wblocks, since they already exist. To use these blocks, just insert them where needed. Let's do that now:

If all is well, an insertion point of 62, 130 of the 8ICPAD should correspond to the center of pin 1 of the 741 op-amp. If not, adjust your location to fit.

```
Command: INSERT (RETURN)
Block name (or?): 8ICPAD (RETURN)
Insertion point: 62,130
X scale factor <1>/Corner/XYZ: .5 (RETURN)
Y scale factor <default=X): .5 (RETURN)
Rotation Angle <0>: 0 (RETURN)
```

By now, you should have placed pads at the end of each component, over the op-amp pins, where test points and jumpers should go, and along the edge of the board where wires from the outside world attach.

Now, check that the space between components and pads is at least 1 mm apart. Using the Move command will allow you to move any entity without changing layers. Also; note Appendix-E, Exercsie 7, Modifying lines with "Stretch" and "Grips." When all pads and components are in place, save the drawing as PCB1C.

EXERCISE 5-6. ADDING THE TRACES

Discussion

We will now add the traces that interconnect the various components, using Figs. 5-21 and 5-1 (pages 206 and 155) as references, to the drawing PCB1C. We'll use a Pline trace width of 1 mm for the non-power conductors and a trace width of 1.5 mm for the Common conductor and the three regulator conductors. This falls within the minimum trace width of 1 mm and 1.5 mm our PCB-1 specification called for under the PCB layout design criteria. Remember not to cover up the center hole of the pad when laying down the traces (see Figs. 5-19C, D, and E, page 185).

TIP: You may have noted that the Assembly view of the traces are shown as a dashed line. This is *not* a Fill option, but a procedure used in the plotting process of the drawing and will be discussed in Chapter 7. For now, place the traces as normal, with Fill off.

Housekeeping

SET:
- GRID = 2
- SNAP = .5
- LAYER = TRACES
- ZOOM to an area that represents about one-quarter of the board.

It is preferable to select Pline from the Draw menu rather than using the keyboard, as the Draw menu displays all the Pline options.

```
Command: PLINE (SELECT MENU)
From point: SELECT THE INNER EDGE OF A PAD
```

Set Width to 1 mm, and using Fig. 5-1 complete the Pline.

```
Arc/Close/Halfwidth/Length/Undo/Width/<end point of line>: W (RETURN)
Starting width <0.0000>: 1 (RETURN)
Ending width <1.0000>: (RETURN)
Arc/Close/Halfwidth/Length/Undo/Width/<end point of line>: SELECT THE
PAD TO BE CONNECTED TO.
```

A Pline 1 mm wide is drawn between the two selected points. Press Return to fix the Pline in place.

Pressing Return again returns the Pline command.

Add the remaining traces using Fig. 5-1 or Fig. 5-21 as a reference. You will have to zoom or pan to various areas of the drawing in order to complete the trace layout. Should you encounter areas where the placement of the trace becomes too tight, the next few commands may become good friends.

MOVE Command

Discussion. In the process of laying down a track, you may find a location where the track runs too close to another track or pad. Sometimes the object can be moved or shifted enough to allow the track to pass by without violating the design "conductor spacing" criteria of a minimum distance of 1 mm. The Move command is one method to shift objects around in that it allows the moving of objects as whole entities by putting a window around them.

Note: A few words before you use the Move command.

TIP1: If you are working on the Donut layer and want to Move an object, such as a resistor that was originally inserted on the component layer, it is not necessary to switch back to the Comp layer before making the move. Move moves objects on their respective layers with any current layer setting.

TIP2: Move does not cause the tracks and pads associated with a component to be stretched or rubber banded, when you move the component. You must lay new tracks to the Moved object. To rubber-band entities, see the next two commands, Stretch and Grips.

TRAP If it becomes necessary to reinsert a component, remember to do so on the appropriate layer. Forgetting, and inserting the component, such as a resistor, on the Donut layer will produce problems when the Donut layer is mirrored to create the Tape side of the drawing. The reinserted resistor would then appear as an object in the middle of the artwork. You could erase it, but it is better to keep it out of the tape artwork initially.

The following is the command sequence to Move an object created on the Component layer while you work on the Donut layer, using a window. Type:

```
Command: MOVE (RETURN)
Select object: W (RETURN)
First point: START WINDOW SELECTION
Second point: COMPLETE WINDOW SELECTION
```

The object becomes dotted and you are asked to select more objects. If you are satisfied, press Return.

```
Select objects: (RETURN)
```

You are now asked to select a Base point to start the move, and then to indicate the Displacement you want the object to be moved. Selecting a Base point is similar to selecting an insertion point. Choosing a location on the object that can be used as a reference point makes it easy to select its final resting place.

```
Base point or displacement: SELECT A POINT ON THE OBJECT
```

followed by,

```
Second point of displacement: SELECT A POINT WHERE YOU WANT THE BASE
POINT TO END UP, AND PRESS THE SELECT BUTTON.
```

The object is now Moved to the new location. If you are satisfied with the move, press Return.

Stretch and Grips for Editing Traces

Two new commands you should become familiar with for editing, are the Stretch and Grips features of AutoCAD.

In addition to editing an entity with the Move and Extend commands which we played with in Chapter 3, you can also "stretch" it using the Stretch command. Stretch provides a rubber-banding effect on the entities connection points, like a rubber band; whereas Extend just extends the end of the line, and Move moves the whole entity.

In version 12, AutoCAD introduced another unique feature for editing entities called Grips. These are small square areas (Grip boxes) on each entity that allow you to "grip the entity" and manipulate it, such as: Move, Stretch, Rotate, Scale, Mirror, Copy and Undo.

The Grip boxes are usually located at object snap points and become active by: a) clicking on the object with the cursor grip window (the small square at the cross-hairs of the cursor) or b) by wrapping a window around the entity using the cursor.

At this time, you should review Appendix-E, Exercise 7, *Modifying Lines With "STRETCH" and "GRIPS,"* to become more familiar with these two commands.

Note: The cursor grip, as you have probably already found out, is present all the time in AutoCAD version 12. You can't get rid of it.

Review and PCB1D

At this point, you have come a long way, using AutoCAD as a design tool. Your drawing should now look very much like Fig. 5-21. We have almost completed the Assembly Drawing except for four areas:

(1) To add the representative input wires from the outside world.

(2) To add the test links (TL1 & 2) and the jumper (JMPR1) now that the pads are laid down. This is done on the Component layer.

(3) To add the Text, and

(4) To add the right-angle Pline targets and corner tabs that are used as reference marks when the board is cut to its final size.

Next: Add the wires, test links and jumper to the component layer "COMP", and then, add the text to the layer TEXT. Add only the text that is within the perimeter of the PCB board. The remaining text will be inserted after the artwork is complete. Use Fig. 5-1 (page 155) as a reference.

Finally: Add the Pline corner tabs to layer TRACE, if they are not there already.

Before going any farther, it might be wise to save the drawing under another version name, such as PCB1D. You should also make a complete backup of your disk as well. By now, some of you may have already experienced some problems with trying to recover lost files or drawings. Backing up your complete disk now will certainly do no harm.

```
Command: SAVE (RETURN)
File name <default>: A:PCB1D (RETURN)
```

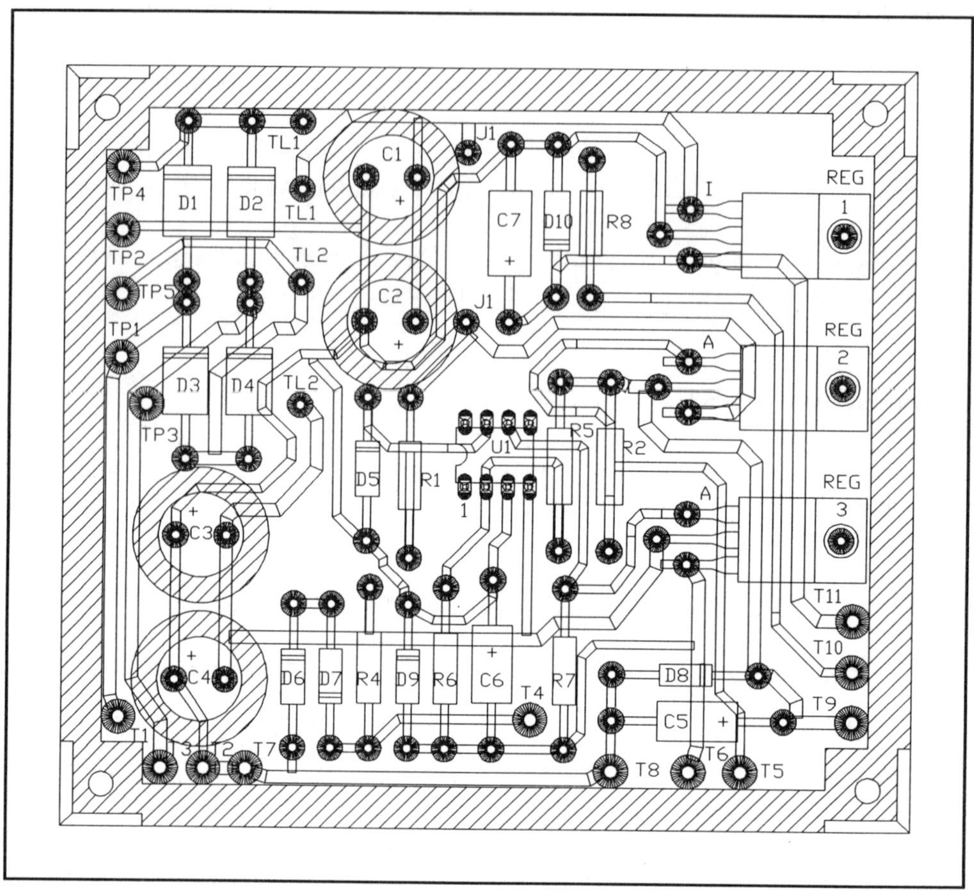

Figure 5-21 PCB-1 assembly view.

EXERCISE 5-7. CREATING THE ARTWORK

Discussion

You are about to see one of the magical features of AutoCAD, the making of th Artwork view of the drawing with one single command, Mirror.

Reference Arrows

To prepare our drawing to be mirrored, we will add a reference arrow at the bottor right corner of the Assembly Drawing. This will become a common point on both th Assembly and Artwork portions of the drawing. Without the reference arrow, tryin to orient the two drawings becomes very difficult during manufacturing. The referenc arrow is drawn on the Traces layer and may be placed at either the top or bottor right corners of the Assembly View. See Fig. 5-22.

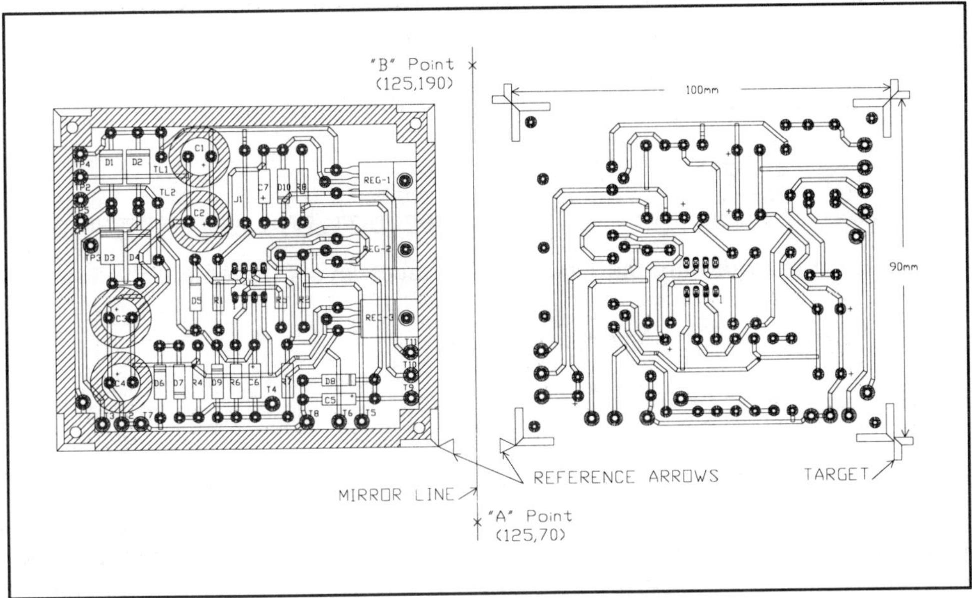

Figure 5-22 Assembly and Artwork with reference arrows, targets and mirror line.

The reference arrow is drawn on the Assembly View side of the page, using a variable-width polyline with a starting width of 0 mm and an ending width of 4 mm.

```
Command: PLINE (RETURN)
From point: SELECT A CORNER
Current line width <default>Arc/Close/Halfwidth/
  Length/Undo/Width/<endpoint of line> W (RETURN)
Starting width <default>: 0 (RETURN)
Ending width: 4 (RETURN)
```

Move the input device four units to the right of the selected corner and press its Select button.

This creates a 4 mm arrow head adjacent to the Assembly drawing.

Creating the Artwork View (Tape Side)

If you placed the Assembly drawing correctly on your page, there will be plenty of room to mirror the artwork on the right side of it.

Mirror requires you to select a plane or line about which the image of the Donuts and Traces layer will be mirrored. The mirror line will be on the exact center of the page, 125 mm in from the left edge. The two points we will use for the line are 125,70 (Point A) and 125,190 (Point B).

This is done with all layers turned off except the Donut and Trace layers. You will be asked if you want to "delete old objects." To this we will reply "No."

Turn off the unwanted layers. Use the modify layer dialogue box or:

```
Command: LAYER (RETURN)
?/Set/New/On/OFF/....../Thaw: OFF (RETURN)
Layer name(s) to turn off: 0,HATCH,COMP,TEXT,DIM,TAPE (RETURN)
```

This should leave us with only the Donuts and Traces layers turned on.

Now to use the *magic* and produce the Artwork.

```
Command: MIRROR (RETURN)
Select objects or Window or Last: W (RETURN)
```

Place a window around the complete Assembly drawing. Include the arrow heads.

We are now asked to establish the mirror plane or line. The mirror line must be either vertical or horizontal. Turn Ortho *on* (F8), and use Fig. 5-22 as a reference. We will select point A at 125,70 just right of the arrow head and slightly below the Assembly drawing. Our second point (B) is located at 125,190 (up 120@<90) which is directly above point A.

```
First point of mirror line: SELECT POINT A (125,70)
Second point: SELECT POINT B (125,190) Move Up 120<90 degrees
Delete old objects? <N>: N (RETURN)
```

The mirror image of the Donut and Trace layers is now drawn to the right of the mirror line. This simple step saves us many hours of drawing time that hand-taping the board would require.

The remaining things for us to complete before we are finished with this drawing are

1. The remaining text for the Assembly view.

2. A set of targets, dimensions lines and text to use as a reference when we photographically reduce the artwork.

EXERCISE 5-8. ADD REMAINING TEXT

Using Fig. 5-1 as your reference, turn on all layers and insert any remaining text on the Text layer. Use a text height of 2 mm for normal text and 3mm for titles. Have you saved your drawing lately?

EXERCISE 5-9. DATUM LINES AND TARGETS

Discussion

Board Scale Factors: Printed circuit boards, in addition to being schematically correct, must be very accurately drawn, because components will eventually be mounted on them. An error of only .1 millimeter could make the difference between a board's acceptance or rejection by a company's quality assurance department. In the hand layout method of designing a PCB, the error factor is reduced by laying out the drawing at two or three times its normal size. When the board is finally reduced for manufacturing, the error is reduced at the same time by the scale factor. This problem does not exist when working with AutoCAD because of the accuracy of the system. Because of this accuracy, we have designed the board at a one to one scale.

Datum Lines/Dimension Lines

To ensure that the physical size of the board is accurate, two "Datum" lines are placed on the artwork between three Target points. One Datum line is placed in the horizontal direction and the other in the vertical. These references allow shifts in the paper size to be noted should the paper become unstable dimensionally with time. The use of Mylar instead of paper is very popular in industry, because it resists shrinking and stretching due to temperature and humidity problems. It is, on the other hand, very expensive.

Datum lines and Targets provide a reference distance for the photographer to use when the drawing is converted to photographic artwork. The Datum line may double as a dimension line if it represents a length on the PCB. We will place our Datum lines across the top and right side of the tape view and also use them as dimension lines.

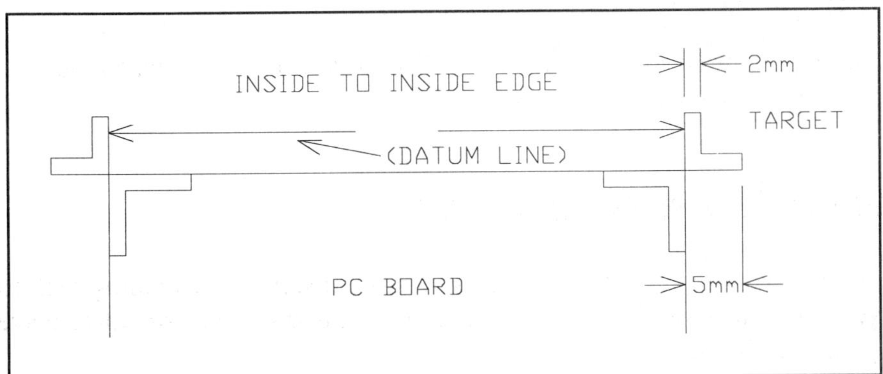

Figure 5-23 Targets and datum lines.

Targets

To establish the outer limits of the board, we will mark the corners with a "Target" (Fig. 5-22). The Datum/Dimension lines are placed between the target points. The dimension placed between these targets must represent the actual board dimensions when reduced to its final size. The targets may eventually become cutting lines when the board is cut to its final size.

Usually only three targets are placed on a drawing to minimize the error. See Fig. 5-22. A common target is used for both the horizontal and the vertical dimensions.

From Fig. 5-23, note that the inside-to-inside edge of the targets represents the outer edge of the board.

Draw your targets as a right angle so that they will meet the corner markers of the board. Use the Pline command to draw the targets 2 mm wide by 5 mm long. Be careful in drawing the targets so that they do not cover any of the board area. By now, your skills with lines and polylines should be well developed.

Now add your Datum/Dimension lines between the Targets. Make sure that the horizontal datum line is exactly 100 mm long, and that the vertical datum line is exactly 90 mm. Add the text "100 mm" to the horizontal datum and "90 mm" to the vertical datum lines, as shown in Fig. 5-22. Remember, these dimensions will be used as a reference when the artwork is photographically reduced during the manufacturing stage.

Fill in the Title Bock with your name, the date, and a drawing revision number of "0."

From Fig. 5-1, you will note that the Component side of the drawing shows the trace lines as a dashed line, while the Tape side shows the trace lines as a solid line. This is accomplished in the plotting process and is covered in Chapter 7, Exercise 7-2. To make a hard copy of your drawing, refer to Chapter 7.

Now that you have completed the board, it is a good time to:

(1) Save the drawing as PCB1E .

(2) Erase any of those old files that may be cluttering your disk.

(3) Make a backup copy of your complete disk and store it at some other location than that of your master.

QUIZ: PRINTED CIRCUIT BOARDS

Name ____

Circle the most correct answer or fill in the blanks.

1. A Trace line may:

   a. Be drawn solid or filled in.
   b. Be drawn in an arc.
   c. Have different starting and ending widths.
   d. All of the above.

2. A Polyline (Pline) may:

   a. Be drawn solid or filled in.
   b. Be drawn in an arc.
   c. Have different starting and ending widths.
   d. All of the above.

3. A Trace or Pline, when meeting a donut, should:

   a. Just touch the outer edge of the donut.
   b. Cover the complete pad.
   c. Fall completely within the pad area but not cover the center hole.
   d. All of the above.

4. The Fill command:

   a. May be turned on and off at will.
   b. Should be left off until you plot.
   c. Allows solid polylines and traces.
   d. All of the above.

5. What are the three types or groups of PCBs?

   a.____

   b.____

   c.____

Name:________________________________

6. Hatching:

   a. Is the process of filling in an area with dotted and/or dashed lines.
   b. Has at least 53 patterns available in AutoCAD.
   c. Allows you to construct your own pattern.
   d. All of the above.

7. Before designing in a component, you should:

   a. Consult the manufacturer's data sheet.
   b. Test its resistance.
   c. Get the cheapest unit.
   d. All of the above.

8. Some components are polarized and should:

   a. Never be used.
   b. Have their orientation verified.
   c. Be mounted vertically.
   d. All of the above.

9. Electrolytic capacitors:

   a. Will explode if mounted backwards.
   b. Are nonpolarized.
   c. Come with only axial leads.
   d. All of the above.

10. The input leads on the LM317 and LM337 regulator:

    a. Are the same.
    b. Are different.
    c. Have axial leads.
    d. Should be heat-sinked.

11. To manually lay out a PCB, you should:

    a. Observe the schematic.
    b. Check component orientation.
    c. Interconnect components as in the schematic.
    d. All of the above.

chapter 6

Main Assembly Drawing and Front Panel

DISCUSSION

In this chapter, we will look at how to dimension a drawing by using one of AutoCAD's most powerful features; its ability to dimension things automatically. Up to this point, we dimensioned our printed circuit board by hand, made the arrowheads by hand, and applied the text by hand. When you have completed this chapter, you will understand dimensioning standards and know how you can use AutoCAD to automatically dimension a drawing. Note Appendix E, Exercise 2 - *A Practice Session On Dimensioning*.

It is suggested that a new disk be used for the drawings developed in this chapter because these exercises and drawings will quickly fill up a large chunk of your disk.

This chapter will lead you through the following steps:

Exercise 6-0. Understanding Dimensioning.
Exercise 6-1. Dimensioning - A Practice Session.
Exercise 6-2. Setting Up The Main Assembly Drawing.
Exercise 6-3. Front Panel and Graphics View.
Exercise 6-4. Inserting Components On The Front Panel.
Exercise 6-5. Drawing The Top Chassis View.
Exercise 6-6. Drilling/Hole Chart.

DIMENSIONING

Dimensioning uses a system of conventional lines, symbols, numbers, and notations developed by the American National Standards Institute (ANSI) and the International Standards Organization (ISO). When these conventions are followed, the drawing may be used universally and understood by everyone.

Dimensioning information must be precise and clearly understood, so that the device can be completely fabricated from the data given. It becomes your responsibility, as a designer, to ensure that the drawing represents what the product really looks like.

EXERCISE 6-0. UNDERSTANDING DIMENSIONING

Discussion

In this exercise you will be introduced to the AutoCAD terms and commands of :

DIMENSION LINE	TEXT
EXTENSION LINE	DIM
DIMENSIONAL VALUE	DIM1
LEADER LINE	EXIT
CENTER LINE	UCS
CENTER MARK	UCSICON
TOLERANCES	WCS

The User Coordinate System (UCS)

Until version 10 of AutoCAD, the method of defining the location and point of view of an object on a drawing was the Cartesian Coordinate System, known to AutoCAD users as the World Coordinate System (WCS). This system uses the reference axes of X, Y and Z. In version 10, AutoCAD introduced another reference system called the User Coordinate System (UCS). This system allows you to work more easily with three-dimensional objects and to define the *point of view of the object* in any of the three axes. It has also introduced some confusion to the first-time user of AutoCAD, as most people understand two-dimensional (2D) objects better than 3D objects.

The UCS system is ideal for the graphics person who must think of objects in three-dimensional space because it allows you to change the origin point (0,0 reference) as well as shift and start the axes to suit the drawing application. To understand the position and orientation of objects in the User Coordinate System you should look at the UCS icon found in the lower left corner of the screen, Fig. 6-1. Its representation will help you visualize the orientation of your object with respect to the World Coordinate System (WCS).

As each object on your drawing can be looked at from a different point of view, several UCS icons may appear on a drawing. The X and Y plane of the UCS icon always indicates the positive direction of the X and Y axes. See Fig. 6-1.

The control of the UCS icon is through the Ucsicon Command which allows you to turn the icon ON or OFF, make changes to all the active viewports, place the icon in the lower-left corner of the viewport, or forces the icon to be displayed at the origin (0,0,0).

```
Command: UCSICON (RETURN)
ON/OFF/All/Noorgin/Origin <current ON/OFF state>:
```

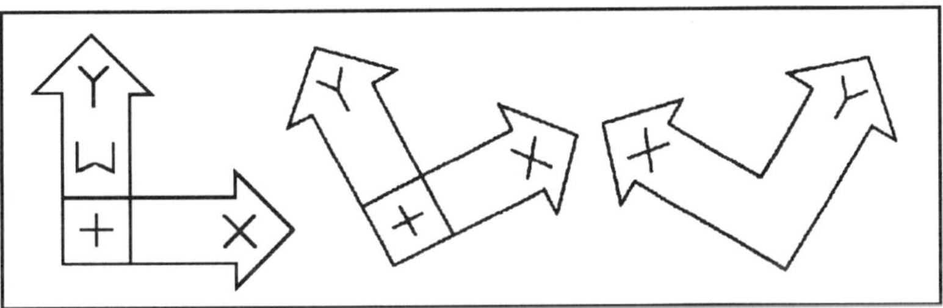

Figure 6-1 The UCS ICON

If a "+" appears in the UCS icon, the view is at the origin established by you.

If a "W" is in the Y arm of the icon, the UCS coordinates are the same as the World Coordinate System (WCS).

A box in the lower left corner of the icon indicates you are seeing the view as a top view.

No box in the lower left corner means you are seeing the view from the bottom.

Dimension Information

From Fig. 6-2, we see that there are three basic parts to dimension information:

1. The Dimension Line, which is a thin solid line terminated with an arrowhead.
2. The Extension Line (Witness Line), used to extend a visible line to the object.
3. The Dimensional Value or Text, representing the actual length of the line or text associated with the line.

Other symbols and notations include:

The Leader Line, which is used to point to a curved feature of a drawing or an area where the dimensional value will not fit.

The Center Line and Center Mark, used as reference points to highlight the center of a circle or arc.

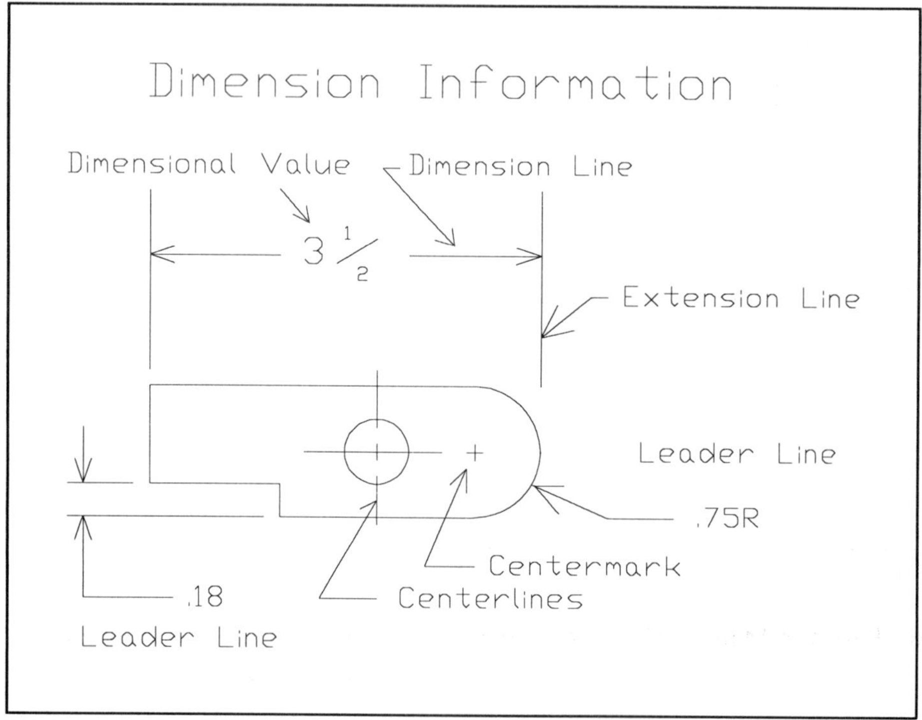

Figure 6-2 Dimension Information.

Dimension lines should not cross. Since this is not always practicable, you should strive to use as few crossings as possible.

Tolerances

As we can see from Figs. 6-3A and B, there are two basic methods to show the Tolerance of a dimension:

1. To use the plus (+) and minus (-) symbols.
2. To use "limit dimensioning," which shows the upper and lower limits of the dimension. As shown in Fig 6-3B, the higher value is usually above the line.

Both of these figures (6-A and 6-B) achieve the same results.

All dimensions are subject to a Tolerance. When a dimension line has no Tolerance, the implied Tolerance is specified in the title block.

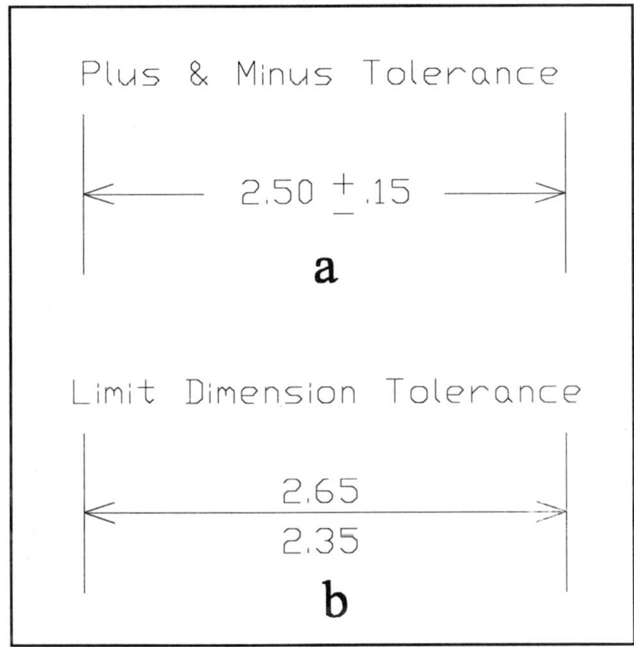

Figure 6-3a Plus and Minus Tolerance. **Figure 6-3b** Limit Dimension Tolerance.

Some typical Tolerances for general surfaces are:

Tolerance for Dimensions in Decimal Inches:

One-place decimals	+/- 0.04	(inch tolerance)
Two-place decimals	+/- 0.01	(inch tolerance)
Three-place decimals	+/- 0.005	(inch tolerance)

Tolerance for Dimensions in Metric (mm):
All dimensions unless otherwise specified are: +/- 0.25 mm

Angle Tolerance of Machine Parts (both surfaces):
+/- 0.0 degrees, 30 min.

DIMENSIONING AND AUTOCAD

Discussion

In previous version of AutoCAD up to and including 12, we use the DIM Command Set to dimension a drawing. This command set placed us in a whole new world within AutoCAD. In the world of dimensioning, we cannot use the normal AutoCAD command set nor can we use the standard editing features. This sometimes presents a problem. In the DIM command set, the prompt is changed to display "DIM:" instead of "COMMAND:." To get back to the normal operation of AutoCAD, we must use the Exit command or the CTRL-C option.

If you want to quickly dimension a single line, AutoCAD in version 2.5 introduced the DIM1 command. This command allows you to draw a single dimension line without becoming too involved with the many DIM command set options. When you have finished with this single instruction, you are automatically brought back to the AutoCAD Command line.

Because of the limitations of editing while in the dimensioning world, version 13 of AutoCAD, has replaced the DIM command set with the DIM prefix command set. To do Linear dimensioning, for example, you type DIMLINEAR instead of LINEAR. In the DIM prefix command set, you can now do full editing on your drawing as you are working in the normal AutoCAD mode. Note Appendix E, Exercise 3: *A Practice Session on Dimensioning Using AutoCAD Version 13* for a detailed review of DIMLINEAR commands.

TIP: It should be noted that although version 13 now has a new DIMLINEAR command set, it still supports the older DIM command set. Therefore the following discussion still holds true.

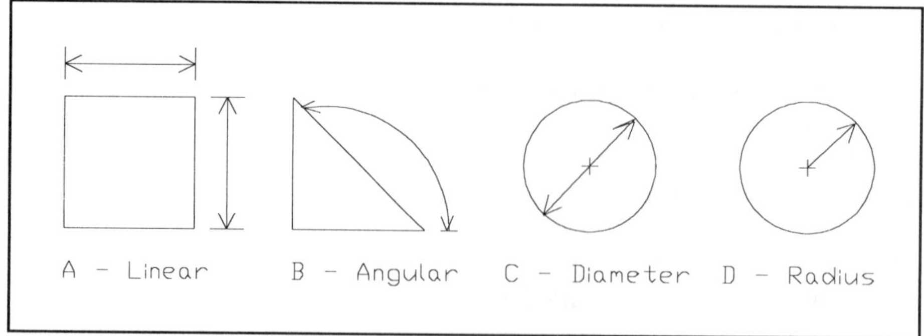

Figure 6-4A, B, C, D Dimensional methods.

There are five types of dimensioning methods available when using AutoCAD: Linear, Angular, Diameter, Radius and Ordinate, as well as a Dimensioning Utility command. Figure 6-4 shows four basic methods. Ordinate uses a 0,0 reference point on the drawing. Some of the methods can be further subdivided.

Linear

The Linear Dimensioning method has six subheadings: Horizontal, Vertical, Aligned, Rotated, Baseline, and Continue. Let's examine these one at a time. See Fig. 6-5.

1. Horizontal Dimension produces a horizontal dimension line between two vertical extension lines. Text is in the horizontal plane.
2. Vertical Dimension produces a vertical dimension line between two horizontal extension lines. Text may be horizontal or vertical.
3. Aligned Dimension produces a dimension line parallel to the specified extension line origin points.
4. Rotated Dimension produces a dimension line similar to the Aligned dimension, except that the angle of rotation of the line is specified.
5. Baseline Dimension produces a series of parallel dimension lines, all starting from a common extension line.
6. Continue Dimension produces a series of linear dimension lines in a contiguous line between extension lines.

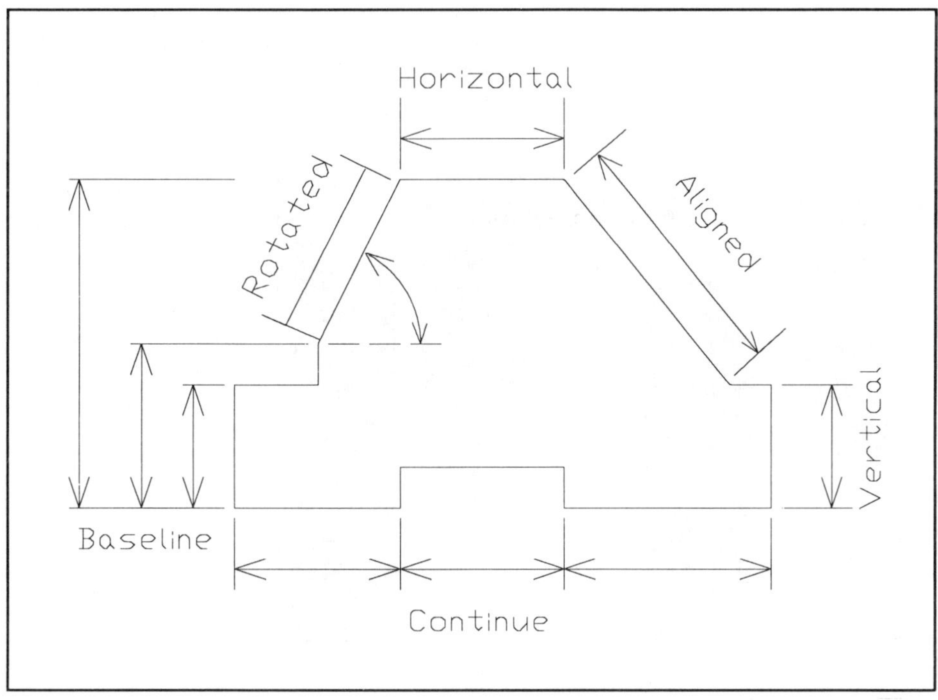

Figure 6-5 Linear dimensioning methods.

Angular

Angular Dimensioning produces an arc instead of a straight line between the two nonparallel lines specified. See Fig. 6-4B.

Diameter

Diameter Dimensioning dimensions the diameter of a circle or arc. See Fig. 6-4C.

Radius

Radius dimensions the radius of a circle or arc with the option of having a center mark or center lines. See Fig. 6-2 or Fig. 6-4D.

Ordinate

Ordinate Dimensioning is based on a 0,0 reference point and displays the X or Y coordinate feature along with a leader line. It is suggested that Ortho be turned ON.

Dimensioning Variables

AutoCAD 12 contains a series of 42 Dimensioning Variables, and version 13 now has 58. These allow you to change the look of the dimension information and become the preset values or defaults used by the DIM commands. Table 6-1 shows version 12 variables with their default settings, and version 13 is shown in Appendix G.

The two default settings that you should immediately become familiar with are "DIMTXT," which sets the dimensional text height, and "DIMASZ," which sets the arrowhead size. See Fig. 6-5a and Exercise 2 in Appendix E.

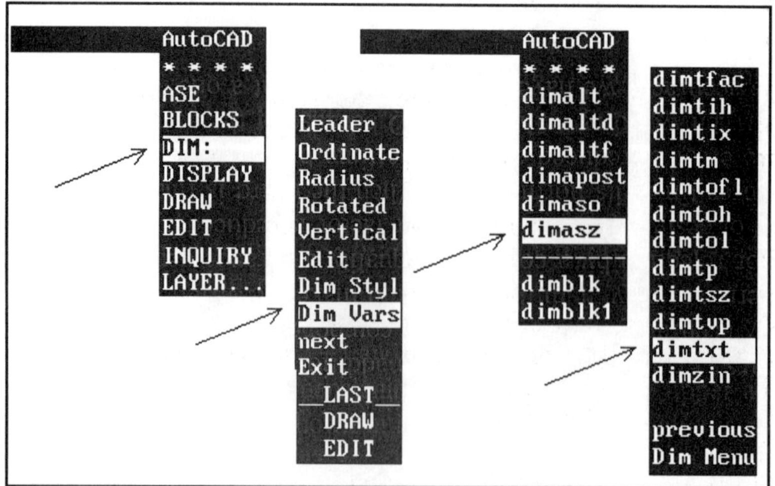

Figure 6-5a Selecting the DIM command and the arrow head and text size.

TABLE 6-1 Dimension Variables and Default Settings

Name of Variable	*Default Value*	*Functional Description*
DIMALT	Off	Alternate dimensions adjacent to standard dim
DIMALTD	2	Decimal places used by DIMALT
DIMALTF	25.4	Multiplier factor for DIMALT
DIMAPOST	<"">	Alternate units text suffix
DIMASO	On	Associative dimensioning
DIMASZ	0.1800	Arrowhead size
DIMBLK	None	Customized arrowhead block name
DIMBLK1	None	Separate arrow block 1
DIMBLK2	None	Separate arrow block 2
DIMCEN	0.0900	Center mark size
DIMCLRD	Byblock	Dimension line color
DIMCLRE	Byblock	Extension line color
DIMCLRT	Byblock	Dimension text color
DIMDLE	0.0	Dimension line extension
DIMDLI	0.38	Dimension line continuation increment
DIMEXE	0.1800	Distance extension line is beyond dimension line
DIMEXO	0.0625	Extension line origin offset
DIMGAP	0.0900	Gap from dimension line to text
DIMLFAC	1.0000	Linear unit scale factor
DIMLIM*	Off	Add limit dimension tolerance
DIMPOST	None	Default suffix for dimension text
DIMRND	0.0000	Dimension rounding tolerance
DIMSAH	Off	Separate arrow blocks
DIMSCALE	1.0000	Overall scale factor
DIMSE1	Off	Suppress the first extension line
DIMSE2	Off	Suppress the second extension line
DIMSHO	On	Update dimensions while dragging
DIMSOXD	Off	Suppress outside extension dimension
DIMSTYLE	UNNAMED	Current dimension style (read-only)
DIMTAD	OFF	Place text above the dimension line
DIMTFAC	1.0000	Tolerance text height scaling factor
DIMTIH	On	Text inside extension is horizontal
DIMTIX	OFF	Place text inside extension lines
DIMTM	0.0000	Minus sign tolerance
DIMTOFL	0.0000	Text outside, force line inside
DIMTOH	1.0000	Horizontal text outside extension lines
DIMTOL*	Off	Adds dimension tolerance text
DIMTP	0.0000	Plus sign tolerance
DIMTSZ	0.0000	Tick size
DIMTVP	0.0000	Text vertical position
DIMTXT	0.1800	Text height
DIMZIN	Off	Suppress zero inches after feet

* DIMTOL and DIMLIM cannot both be on at the same time.

EXERCISE 6-1. DIMENSIONING - A PRACTICE SESSION

Discussion

To better understand the operation of dimensioning, complete Exercises 2 and 3 in Appendix E before going any further. This will hone your skills when it comes time to dimension the various views in the Main Assembly drawing.

EXERCISE 6-2. SETTING UP THE MAIN ASSEMBLY DRAWING

Discussion

Figure 6-6 is the Main Assembly Drawing of our project. It consists of three views and a table. They are:

1. Front Panel View.
2. Front Plate Graphics.
3. Top Chassis View.
4. Drilling Table.

I have decided not to show the side view of the chassis in order to make room for the Front Plate Graphics from which you will make the front plate overlay during the manufacturing of this project.

Housekeeping

The three views are drawn at one-to-one scale and evenly spaced on an Architectural C size sheet (550,400) to provide a balanced look. Note Fig. 6-6.

An enlargement of the Front Panel View, Fig. 6-7, shows the location of the voltage adjust knob, power ON/OFF switch, and the light (LED). It also contains the information necessary to dimension the Front Plate.

The Top Chassis View shows the location of the printed circuit board, the transformer T1, the switch, knob, binding posts, and the associated wire cable used to connect the board to the front panel.

We will use the Front Plate Graphics as the "Artwork" drawing to develop a silk screen or light-sensitive overlay sheet during the manufacturing stage of our project. This will give our multi-output variable Power Supply a professional look when it is finally assembled.

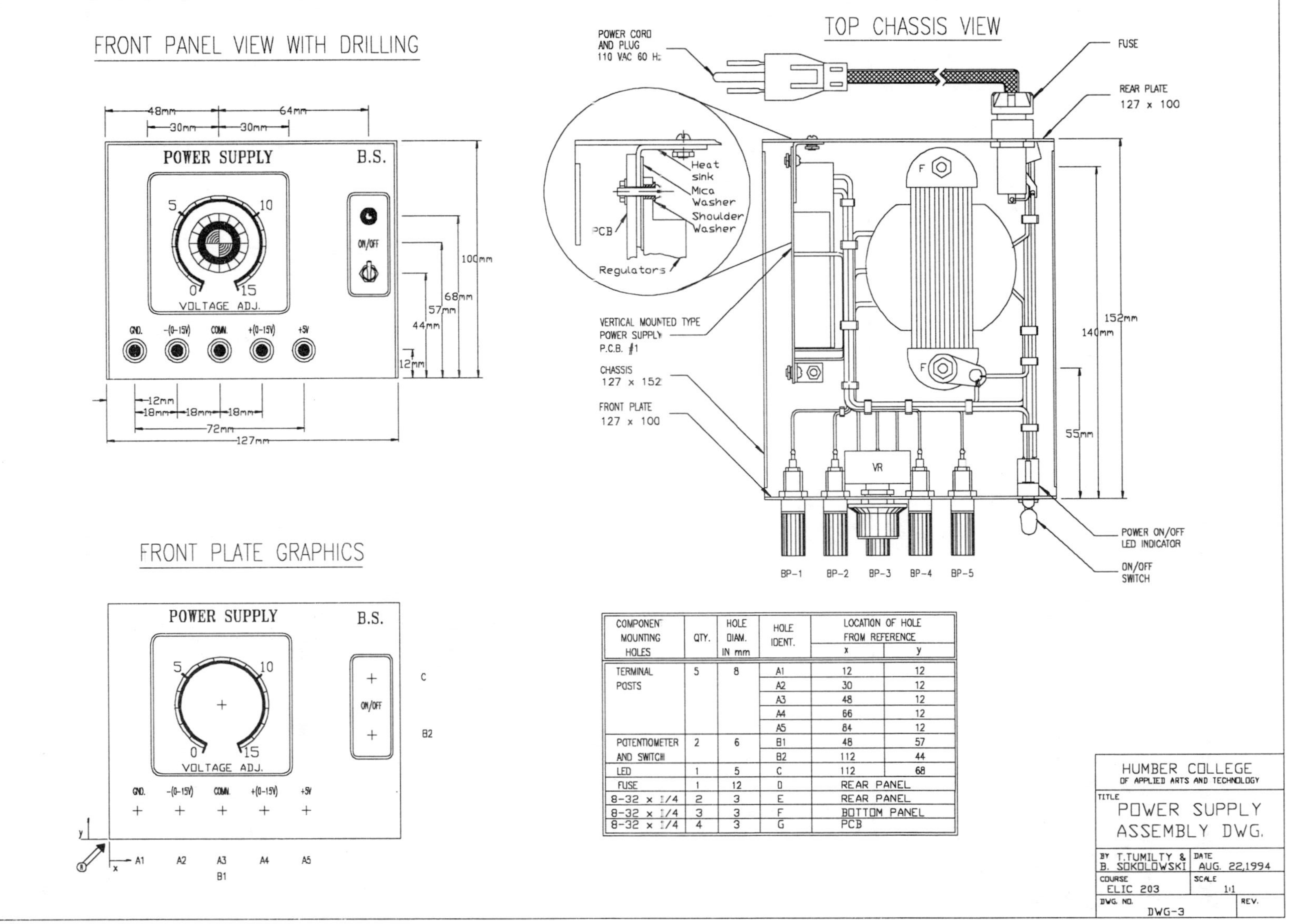

COMPONENT MOUNTING HOLES	QTY.	HOLE DIAM. IN mm	HOLE IDENT.	LOCATION OF HOLE FROM REFERENCE x	y
TERMINAL POSTS	5	8	A1	12	12
			A2	30	12
			A3	48	12
			A4	66	12
			A5	84	12
POTENTIOMETER AND SWITCH	2	6	B1	48	57
			B2	112	44
LED	1	5	C	112	68
FUSE	1	12	D	REAR PANEL	
8-32 x 1/4	2	3	E	REAR PANEL	
8-32 x 1/4	3	3	F	BOTTOM PANEL	
8-32 x 1/4	4	3	G	PCB	

Figure 6-6 Power supply main assembly drawing.

Now to set up the ASSEM drawing

Drawing Name: ASSEM
Title Block Name: P.S. MAIN ASSEMBLY
Limits = C size (C-SHTA) Architectural paper: 550,400
Grid = 10 units
Snap = 5 units
Coordinates = ON
Zoom = ALL
UCSICON = OFF

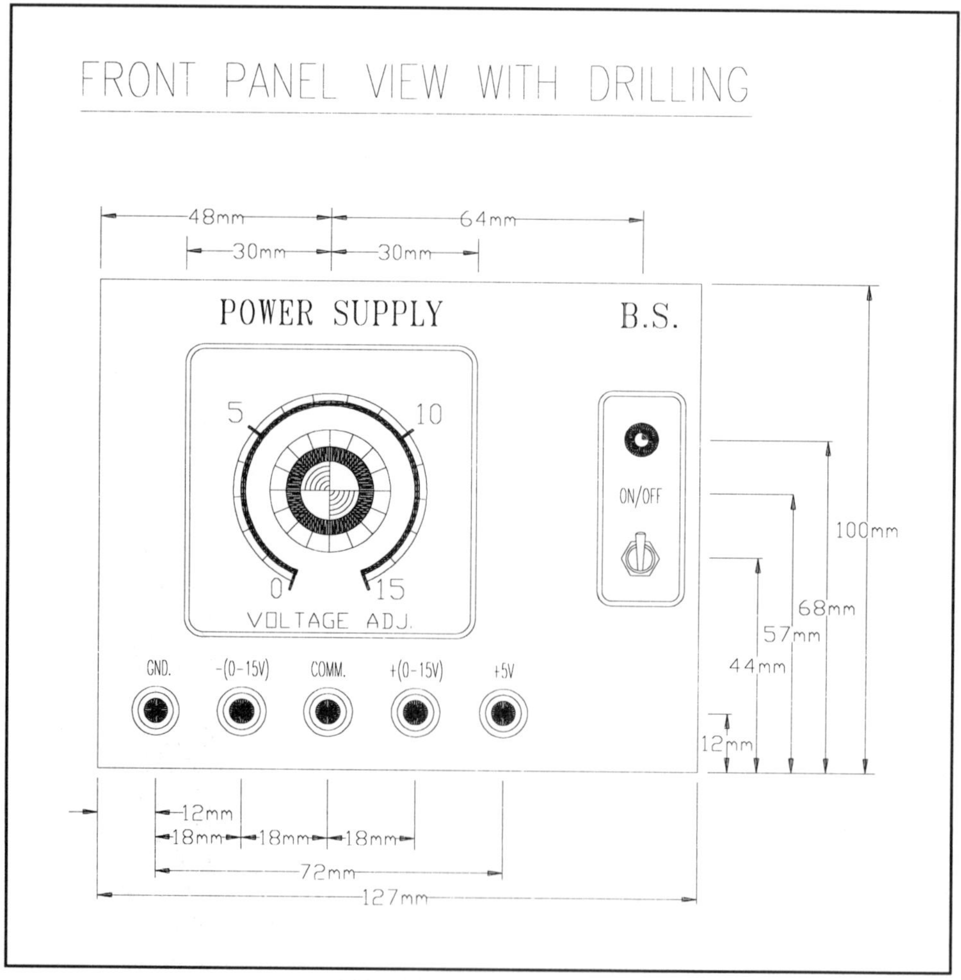

Figure 6-7 Front panel view.

Establish the following Layers:

Layer	Comment	Color	Linetype	Pattern Used
0	Main Dwg.	White	Continuous	Continuous
Hatch1	Hatched Area	White	Continuous	ANSI31
Hatch2	Hatched Area	White	Continuous	ANSI37
Comp1	Front Plate Graphics	White	Continuous	Continuous
Comp2	Basic Comp.	White	Continuous	Continuous
Text1	Front Plate Text	White	Continuous	Text
Text2	General Text	White	Continuous	Text
Dim	Dimensions	White	Continuous	Continuous

Where Should We Start to Draw?

Discussion

Using Figs. 6-6 and 6-7 as a references, you will see that the Front Panel and Front Plate Graphics are approximately 125 mm (127 mm) by 100 mm. This becomes approximately 160 mm by 145 mm when the dimension and text information is added around it. The Top Chassis View is approximately 125 mm (127 mm) by 150 mm (152 mm) and when you include the area for the knob, fuse, power cord, dimensioning and text, it becomes approximately 150 mm by 245 mm.

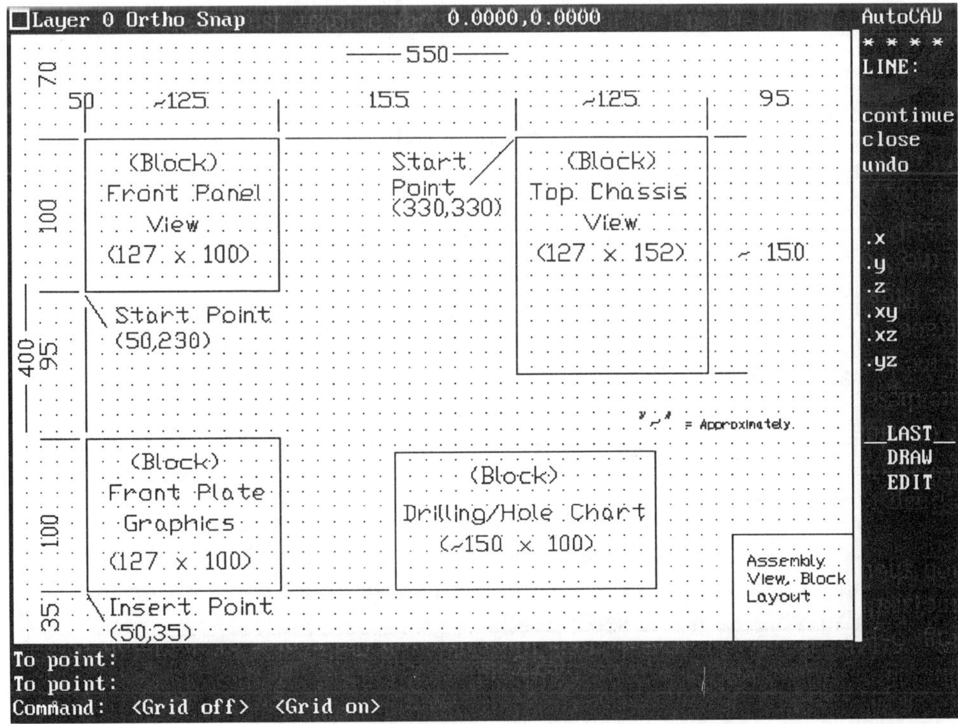

Figure 6-8 Assembly view layout.

If you allow for a border area around each view, you can start the drawing by placing the lower left corner of the Front Panel View at a location 50 mm from the left edge and 230 mm from the bottom, as shown in Fig. 6-8. The *upper left corner* of the Front Panel View thus becomes 50 mm and 330 mm as it is 100 mm high.

Next, we'll place the top left corner of the Chassis View on a line parallel to the top of the Front Panel View, and to the right of it. Choosing the location of 330 mm across, and up 330 mm will provide plenty of working space around the Top Chassis View to include the external components and text.

We have chosen the upper left corner of the Top Chassis View instead of the lower left corner as our starting point because: (1) the chassis is slightly longer than 150 mm (152 mm), thus our starting point would not be on a standard grid/snap point; and (2) it provides symmetry with the top of the Front Panel View.

Our Drilling/Hole Chart, Fig. 6-13 (page 233), is positioned below the Top Chassis View and is approximately 150 mm by 100 mm. Near the end of this exercise, the table will be moved so the lower left corner is parallel to the lower Front Plate Graphics.

EXERCISE 6-3. FRONT PANEL AND GRAPHICS VIEW

Discussion

In this exercise, we will save time by developing both the Front Panel View and Front Plate Graphics at the same time as they are very similar.

Basic Setup - *Drawing the Front Panel View outline on your own.*

On layer 0, zoom to the upper left side of the drawing and draw the *"outline"* of the Front Panel View starting at 50,230. It is 127 mm long by 100 mm high. Use Figs. 6-6, 6-7 and 6-8 as references.

Establishing the Center Marks on the Front Panel View

Now, set the layer to COMP1 and, using Fig. 6-9 as a reference, draw the eight center marks at each of the hole locations on the Front Panel View. The center marks are 4 mm by 4 mm cross-hair-lines. You can draw these by either blocking them, copying them, or making each one separately.

Dimensioning the Front Panel View

Next, using Figs. 6-5a and 6-9 as references, switch to the DIM layer and dimension the Front Panel View using the knowledge gained when you worked through the dimensioning exercises 2 and 3 in Appendix E. Change the Dimension Variables of the arrowhead DIMASZ to 3 mm and the text height DIMTXT to 3 mm. When dimensioning this view, choose several combinations of the Linear Dimensioning options to gain experience working with them. Your drawing should look similar to Fig. 6-9 when complete.

Developing the Front Panel Graphics

Next, we'll develop the graphic for the Voltage Adjust variable resistor (R3), the Power On/Off switch, and add the Front Panel text. Note Fig. 6-10a.

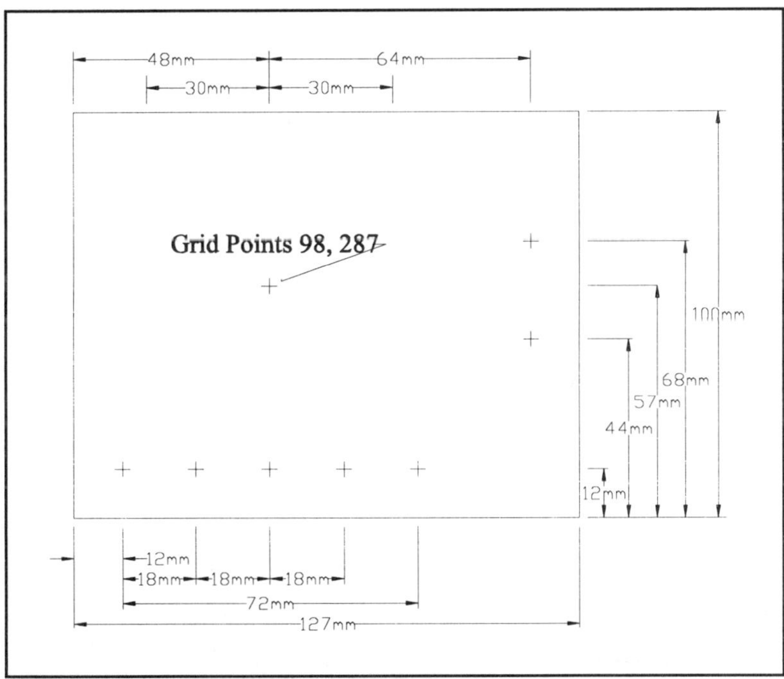

Figure 6-9 Front panel with center marks.

Voltage Adjust Graphics: Discussion

To develop the Voltage Adjust graphics, use Fig. 6-10a as a reference. If you have accurately positioned the center mark for this control, you should find it at the grid junction points of 98, 287. Zoom into this area of your drawing so you can easily work on the graphic. Set the Snap to 1 unit and the Grid to 7 units.

The outside box. Using Fig. 6-10a as a reference, start by drawing a 60 mm by 60 mm box around the Voltage Adjust center mark. You may want to use a radius at the corners or keep them square. The outside of the box may be a thin polyline or a double line. The choice is yours.

Voltage Adjust Control Graphic: How to Develop It!

The Voltage Adjust Control Graphic (Fig. 6-10a). The shaft of the Voltage Adjust potentiometer (R3) rotates through 280 degrees. This leaves a 80 degree segment at the bottom of the rotation that is not used. This is the arc area between 230 degrees and 310 degrees in Fig.6-10b. In the rotational area of R3, we'll develop the 0 to 15 voltage scale and split it into 15 equal spaces.

In developing this graphic you will discover several unique steps to create it. These same steps will later be use to develop the graphic for the knob which covers the shaft of R3.

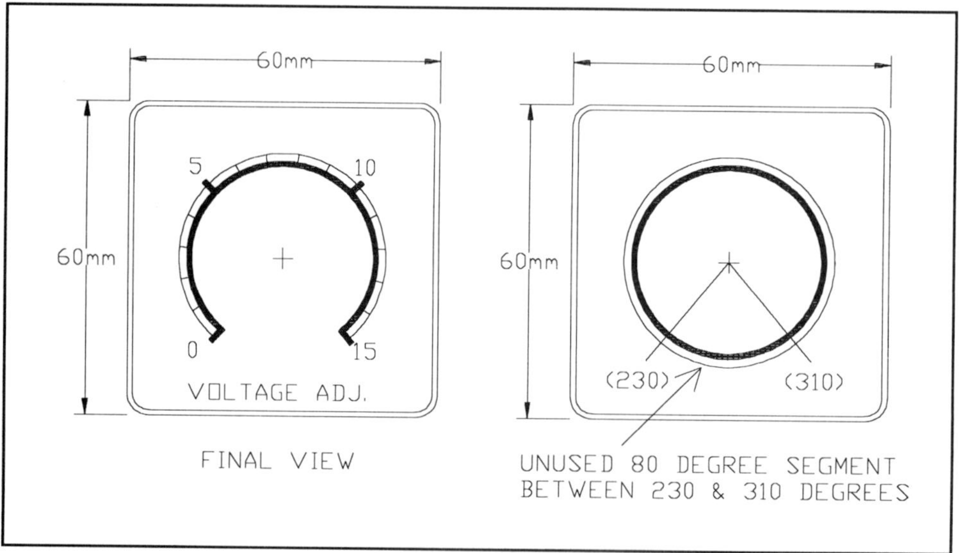

Figure 6-10a **Figure 6-10b**

The Voltage Adjust Graphic is created from a circle and a Pline arc. See Fig. 6-10b.

First, we'll draw a circle with a 20 mm radius at location 98, 287. This establishes the outer ring of our graphic. The circle will be trimmed to size, after we have added the wide polyline arc.

The Outer Circle

```
Command: SELECT DRAW
Command: SELECT CIRCLE
Command: SELECT CEN,RAD
Command: Circle 3P/2P/Centerpoint: SELECT LOCATION 98,287 AND DRAG
                                   A 20 mm RADIUS CIRCLE.
```

Now we'll draw a Pline arc using the S,C,E (Start, Center, End) option, starting at 116, 287. The Pline starts 2 mm to the left of the circle and its width is 1 mm.

Polyline

```
Command: PLINE (Return)
From point: SELECT THE POINT (116, 287) WHICH IS 2 mm TO THE LEFT OF
            THE CIRCLE AT THE "0" DEGREE POINT.
```

AutoCAD displays the current line width; set it to 1 mm.

```
Current line-width is 1
Arc/Close/Halfwidth/Length/Undo/Width/<Endpoint of line>:W (Return)
Starting width <0.1000>: 1 (Return)
Ending width <1.0000>: 1 (Return)
```

```
Now select an ARC with an Angle of 359.9 degrees.
```

```
Arc/Close/Halfwidth/Length/Undo/Width?<Endpoint of line>: A (Return)
Angle/CEnter/CLose/......../Undo/Width/<Endpoint of Arc>: A (Return)
Angle/<Endpoint>:
Included Angle: 359.9 (Return)
Center/Radius/<Endpoint>: C (RETURN)
Center Point: Click on the center point (98,287)
(Return)
```

Your Voltage Adjust graphic should look similar to Fig. 6-10b.

Removing the unwanted segment.

Now we'll remove the unwanted segments at the bottom of the circles with the Trim command. See Figs. 6-10b and 6-10c. To establish the cutting edge for the Trim command, we'll draw a line from the graphic center mark (98,287), through the Pline and terminate it just outside the circle. This is done at the angles of 310 degrees and 230 degrees to create our two trim lines.

```
Command: Line (Return)
From point: SELECT THE CENTER POINT OF THE GRAPHIC (98,287)
To point: DRAG A LINE AT 310 DEGREES, PAST THE CIRCLE.
```

Draw a second line, again from the center point past the circle, at 230 degrees.

Now to Trim the segment.

```
Command: TRIM (Return)
Select cutting edges(s): SELECT THE TWO LINES AT THE ANGLE 310 AND
                         THE ANGLE 230. (Return)
Select object to trim: SELECT THE PLINE AND OUTER CIRCLE LINE IN THE
                       UNUSED SEGMENT AREA. (Return)
```

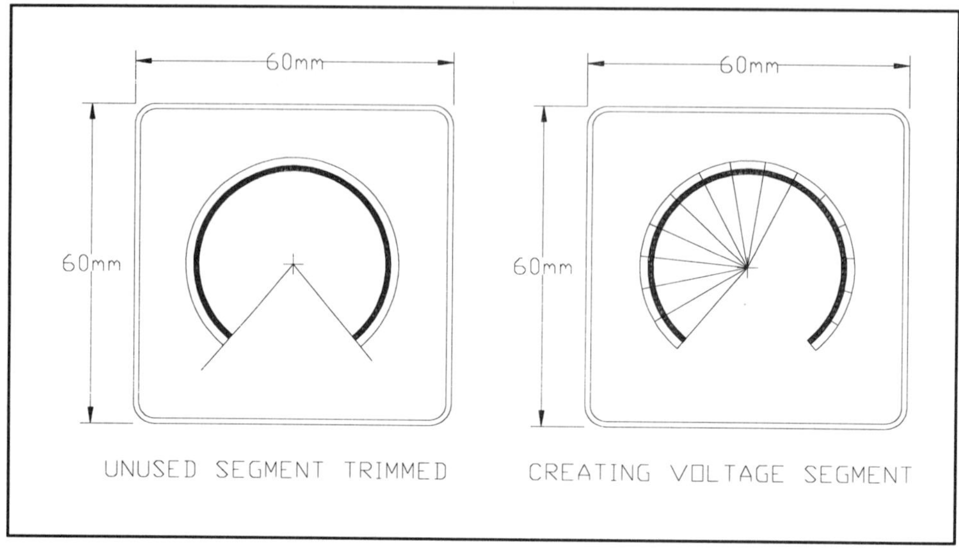

Figure 6-10c **Figure 6-10d**

Like magic, the Pline and Circle segment are removed. Your drawing should be similar to Fig. 6-10c.

Creating the Voltage Adjust Segments: Discussion

Now to create the 15 small voltage adjust segments using Fig.6-10d. The potentiometer rotates through 280 degrees, while the voltage is adjusted between 0 and 15 volts. Therefore, each volt out is equivalent to a rotation of 280/15 which is 18.6 degrees. We will use Table 6-2 which shows each angle used in creating the voltage adjust segments.

Table 6-2 Voltage Adjust Segment Angles

Segment	Previous Angle	Increment	Drawing Angle (in degrees)
0	0	310	310
1	310	18.6	328
2	328	18.6	347
3	347	18.6	366 (6 degrees)
4	6	18.6	25
5	25	18.6	43
6	44	18.6	62

7	63	18.6	80
8	81	18.6	100
9	100	18.6	118
10	118	18.6	136
11	137	18.6	155
12	156	18.6	173
13	174	18.6	192
14	192	18.6	211
15	211	18.6	230

Now to draw the Voltage Adjust Segments.

TIP: Use the F6 function key (Coordinates toggle) to cycle the coordinates display through the ON/OFF, X-Y Coordinates reference and Angular reference.

TIP: To create each short segment, we must reference them to the center mark of the graphic (98,287). Each segment is developed by drawing two lines. The first establishes the correct angle, and thus the spacing between segments, while the second line creates the small segment line itself. The first line is drawn from the center mark to the outer circle. The second line traces over top of the first line, from the outer circle, back to the area covered by the Polyline. The original line is then erased leaving the short segment line in place. See Fig. 6-10d.

Using Table 6-2, draw the 16 segments lines using the specified Drawing Angle. Draw the line from the graphic center to the outer circle and back to the polyline. Note Fig 6-10d.

Power On/Off Switch and LED GRAPHIC

Using the same line style as that of the box around the Voltage Adjust control (R3), draw a similar graphic around the LED and On/Off switch. It is important to keep the two graphics looking similar to provide uniformity in the design. Note Fig. 6-11.

Lettering the Front Panel

We are now ready to apply the text to the Front Panel. Change the layer to Text1 and zoom into the Front Panel view. Use Fig. 6-11 as your reference.

The small text height should be 3 mm, and the large text 5 mm. It is recommended to use the Center option of the Dtext command and the center line of the hole as a reference point when inserting text. Your graphic should look similar to Fig. 6-11.

When you have completed typing all the text, we are now ready to use the Copy command on the Front Panel View to create the Front Plate Graphics.

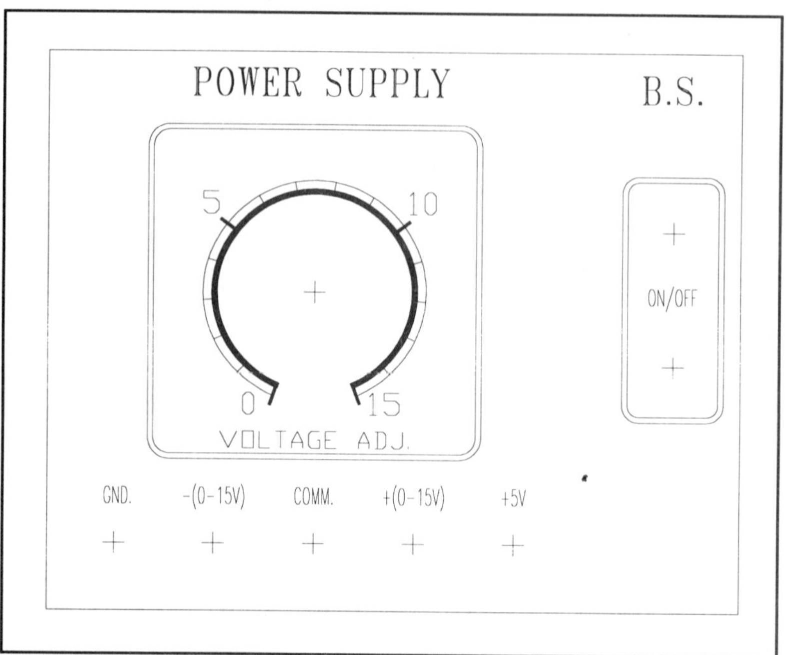

Figure 6-11 Front Panel with graphics.

Creating the Front Plate Graphics

The present view of the Front Panel is essentially the Front Plate Graphics. Copy this view using the lower left corner as the insertion point and reinsert it at location 50, 35 on your drawing. This can be done with ease if the Grid and Snap are set to 5 and a window is used to capture the Front Panel.

```
Command: COPY (Return)
Select objects: W (Return) (Capture Front Panel with a window)
<Base point or displacement> /Multiple: SELECT THE LOWER LEFT CORNER
                                OF THE FRONT PANEL VIEW (Point 50,230).
Second point of displacement: SELECT LOCATION 50,35 (Return)
```

Front Plate Graphics Text

Using a text height of 3 mm, the Text2 Layer, and Fig. 6-12 as a reference, insert the text around the Front Plate Graphics View. This text is associated with the Front Plate Hole Chart, Fig.6-13. The dimensions shown in the "X" and "Y" locations on this chart are referenced to the lower left corner of the Front Plate Graphics. The circled "R" shows this point along with its X and Y coordinate lines.

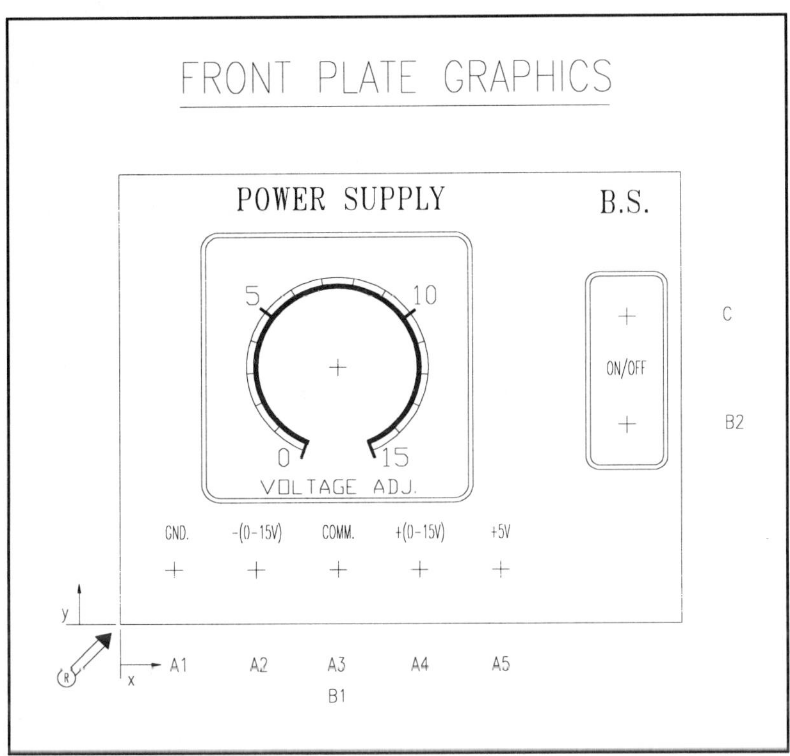

Figure 6-12 Front Plate Graphics View

COMPONENT MOUNTING HOLES	QTY.	HOLE DIAM. IN mm	HOLE IDENT.	LOCATION OF HOLE FROM REFERENCE x	y
TERMINAL POSTS	5	8	A1	12	12
			A2	30	12
			A3	48	12
			A4	66	12
			A5	84	12
POTENTIOMETER AND SWITCH	2	6	B1	48	57
			B2	112	44
LED	1	5	C	112	68
FUSE	1	12	D	REAR PANEL	
8-32 x 1/4	2	3	E	REAR PANEL	
8-32 x 1/4	3	3	F	BOTTOM PANEL	
8-32 x 1/4	4	3	G	PCB	

Figure 6-13 Drilling/Hole Chart

With reference to Fig.6-12 and on the Comp2 Layer, draw the X and Y coordinate lines, and the associated reference circle and arrow in the lower left corner of the Front Plate Graphics. When you have finished, your drawing will look similar to the Front Plate Graphics view of Fig. 6-12. This completes Exercise 6-3.

EXERCISE 6-4. INSERTING COMPONENTS ON THE FRONT PANEL

Discussion

We are now ready to add the components such as the knob, power switch, LED, and the binding posts to the Front Panel View. Note Fig. 6-7. Much use is made of graphics and blocks in this exercise to save time. Furthermore, you have the freedom to create the graphics of the Front Plate View with your own personal style as we do not want your project to look exactly like your neighbor's.

Designing the Knob

Let's create the knob together. In this exercise we will use the tricks we learned in developing the Front Panel graphics to establish angular segmented lines. When the drawing is complete, we'll save the knob as a block under the name KNOB.

Using Fig. 6-14a as a reference, note the outside diameter of the knob is 25 mm. On the inside area is a solid 3-mm-wide polyline ring, and finally a series of 90-degree arcs at the center. The outer ring is divided into 16 segments, 22.5 degrees apart.

First we'll draw the outer circle. Set the Grid to 5 and the Snap to 2.5. Select a clear area of the drawing and zoom to a screen size of about 40 units by 40 units.

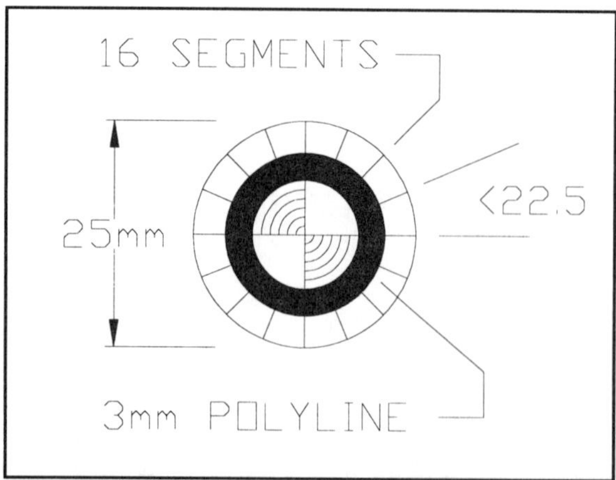

Figure 6-14a Knob layout.

The Outer Circle

```
Command: SELECT DRAW
Command: SELECT CIRCLE
Command: SELECT CEN,RAD
Command: Circle 3P/2P/<Centerpoint>: SELECT THE CENTER POINT AND
                                     CREATE A 25 mm CIRCLE.
```

Polyline

```
Command: PLINE (RETURN)
From point: SELECT A POINT 7.5 mm TO THE RIGHT OF CENTER
```

AutoCAD displays the current line width; set it to 3 mm.

```
Current line width is 0.0000
Arc/Close/.....Width/<Endpoint of arc>: W (RETURN)
Start width <0.0000>: 3 (RETURN)
Ending width <3.0000>: (RETURN)
```

Now select an Arc with an Angle of 359.9 degrees.

```
Arc/Close/....Width/<Endpoint of line>: A (RETURN)
Angle/CEnter/..Width/<Endpoint of arc>: A (RETURN)
Included Angle: 359.9 (RETURN)
Center/Radius/<Endpoint>: C (RETURN)
Center point: SELECT CENTER OF THE KNOB
```

The 3 mm wide polyline is drawn. Your drawing should look similar to Fig.6-14b.

Press Return to terminate Polyline mode.

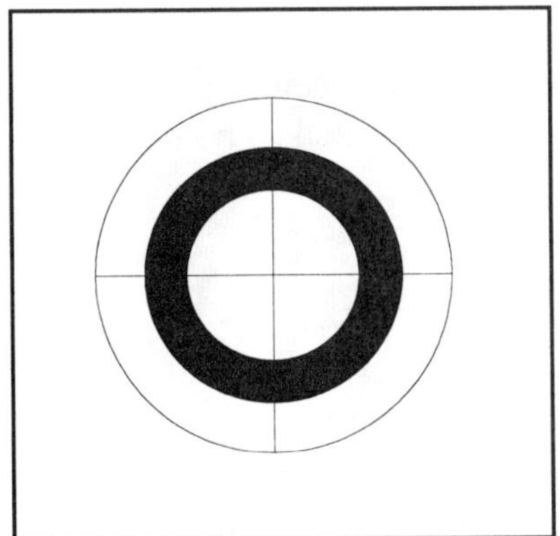

Figure 6-14b Knob polyline.

The 16 Segments and a Bit of Trickery. Now we'll add the 16 segments to the outer ring. Set the coordinates ON (F6 Function Key).

Using Fig. 6-14b, draw a set of cross-hair lines through the center of the knob. Start your line at the outside circle and draw from top to bottom and from side to side. Our knob should look similar to Fig. 6-14b at this point.

Set the Snap to OFF.

The bit of trickery comes into play when we draw the short 22.5-degree segments. To create these segments we will draw two lines. The first line starts from the center point of the knob to the outside circle. The second line is a continuation of the first, and retraces it back to a point midway into the polyline area. This is our segment line. See Figs. 6-14c and 6-14d. This procedure is similar to drawing the segment lines in the Voltage Adjust Graphic.

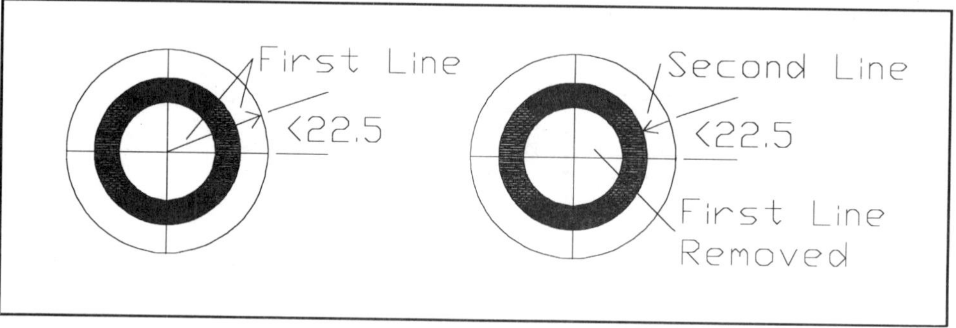

Figure 6-14c First Line. **Figure 6-14d** Second Line

The first line establishes the angle we want with respect to the center point. It uses the coordinates measurement technique, and is eventually erased. The second line becomes the segment line, "the keeper."

Now to draw the first segment, a 45-degree line:

```
Command: SELECT LINE
Line from point: SELECT THE CENTER POINT
To point: SELECT OUTER CIRCLE AT  45 DEGREES
```

The coordinates display will show a distance and angle. It may be similar to 12.4967 <45. Make sure you just touch the outside circle and don't overshoot it. Take Snap OFF if you need to.

Draw the next "To point" back to the polyline, retracing the first line. The angle should be 225 degrees (45 + 180).

```
To point: SELECT A POINT IN POLYLINE AREA
```

The coordinates display may show: 4.0025 <225.

Press Return to terminate the line command.

We now have two lines drawn, one at 45 degrees from the center point to the edge of the outer circle, and the second at 225 degrees from the outer circle over the first line and back to the wide polyline.

Now select Erase and remove the first line. Select a point close to the center point area because we want to remove only the first line, not both.

```
Command: ERASE (RETURN)
Select objects or window or last: SELECT THE FIRST LINE DRAWN, IN
                                  CENTER POINT AREA.
```

When we press Return, both lines disappear. What happened? Now toggle the GRID OFF and ON again. Note that the segment line, our "keeper," reappears.

You may want to remember this technique for future applications.

You are on your own again. Now, use this method to make the remaining 15 segments, each at approximate multiples of 22.5 degrees apart.

The Center 90-Degree Arcs. Inside the center of the knob is a series of 90 degree arcs. Note Fig. 6-14a. To draw these, first set the Snap to 5, then select the Arc command and use the Start, Center, and End (S, C, E) option.

TIP: Remember Arcs are drawn counterclockwise.

```
Command: SELECT ARC
Command: SELECT S,C,E
Command: START 1 UNIT ABOVE CENTER, THEN THE CENTER POINT,
         AND THEN THE END POINT, 1 UNIT LEFT OF CENTER.
```

This process is repeated until you complete the center portion of the knob. When you have finished, your drawing should look similar to Fig. 6-14a.

Block the Knob

As you may want to use the knob in several other drawings, add it to the symbol table on your disk. Block it under the name KNOB and use the center mark as the insertion point. See the Wblock command in Chapter 4 if you need help.

Completing the Front Panel View

Use your creative skills to develop the remaining components on the Front Panel View. If you hatch any areas, do so on the appropriate layer.

THE PARTS CATALOG

All of the components shown on the Front Panel and Top Chassis view may be found in the data sheets in Appendices C and D. The data sheets contain electrical as well as mechanical information about the parts. Use these sheets, as you would in industry, to find the dimensions needed to draw the remaining views. The parts catalog is organized with an index for quick reference to the components. Appendix C is the major parts list for this project.

Using the parts catalog, verify the dimensions of the following parts in Appendix C before you use them.

ON/OFF Switch = 8 mm parallel sides (Item SW1)

ON/OFF Lamp = 7 mm diameter (LED-1)

5 Binding Posts = 10 mm diameter (Item BP 1-5)

1 Potentiometer = 1/4 inch dia. shaft, 28 mm x 30 mm (R3)

EXERCISE 6-5. DRAWING THE TOP CHASSIS VIEW

Discussion

Helpful Hints

This exercise is straightforward, but requires some thought when you visualize the knob, switch, and binding posts from a top perspective.

When drawing the components that are mounted on the front panel, first draw those items that appear closest to the eye, such as the knob, LED and power switch. Then draw the next layer of items, such as the binding posts.

When possible, Block or Copy an item to save time in redrawing it.

The basic chassis is comprised of a single sheet of metal, bent to form a front and rear panel. (See Fig. 6-6 Top Chassis View). The printed circuit board is mounted vertically on the top left side of the chassis. It is held in place by a heat sink, with a 90-degree bend. One flap of the heat sink is mounted under the three power regulator tabs while the other is mounted to the back chassis to allow further heat dissipation.

To begin the Top Chassis View, draw the basic outline of the bottom on layer 0 starting at location 330,330. The bottom is 127 mm wide by 152 mm deep. Note Fig. 6-8.

Use the parts catalog in the back of the text (Appendix D) to determine the size and shape of the various components. Then, on the Comp2 layer, draw the various components to scale. When this is complete, switch to the Hatch layer and hatch those areas indicated. Finally, insert the text on layer Text2.

EXERCISE 6-6. DRILLING/HOLE CHART

Discussion

The Drilling/Hole Chart contains the drilling information you require during the manufacturing stage of this project. To design the chart, call upon your knowledge to make lines, polylines, and dotted lines. Using Fig. 6-13 as a reference, insert the text on the Text2 layer. Choose a text size that is appropriate for the space available. It should not be less than 2 mm.

To complete the drawing, title each of the views, using a 7-mm text height.

Completing the Drawing

As you recall, we are using a C size architectural sheet. Fill in the title block. Do you recall the name of the title for this drawing from our setup information? It is "P.S. MAIN ASSEMBLY."

REVIEW

Looking back over the last six chapters, we see that you have gained a tremendous amount of practical experience and a vast amount of knowledge. You have learned many of the basic AutoCAD commands, how to set up a drawing, the technique to make a simple printed circuit board, the method to lay out an electrical schematic and wiring diagram, and how to develop an assembly drawing with several views. Most of all, *you* did the work of thinking through the layout. All I did was walk you through the process and guide you through the steps to accomplish it.

This is no simple task, and you should feel proud of your accomplishments to date! But before we pat ourselves on the back too much, we have one last task to accomplish. All this work is of no use if we can't get a hard copy of it. Continue on to Chapter 7 where you will learn several ways to make a hard copy of your work.

Notes:

QUIZ: ASSEMBLY DRAWINGS

Name ____

Circle the most correct answer or fill in the blanks.

1. Dimensional Information:

   a. Must be precise and clearly understood.
   b. Use Dimension Lines and Extension Lines.
   c. Always use tolerances.
   d. All of the above.

2. The UCS system uses:

   a. The Cartesian Coordinate System.
   b. Only the World Coordinate System.
   c. The X, Y coordinate system.
   d. A method that allows you to work more easily with three-dimensional objects.

3. The WCS system uses:

   a. The Cartesian Coordinate System.
   b. Only the World Coordinate System.
   c. The X, Y coordinate system.
   d. A method that allows you to work more easily with three-dimensional objects.

4. Tolerances use:

   a. The plus (+) and minus (-) symbols.
   b. The Limit Dimensional System.
   c. Upper and Lower Limit Dimensions.
   d. All of the above.

5. In version 12, the Prompt when the Dimension Command is active, is:

   Answer: ____

6. The two methods to get out of the Dimensional mode are:

   Answer: ____

7. The Extension Line is also known as:

   a. Witness Line
   b. Leader Line
   c. Center Line
   d. Dimension Line

8. The DIM1 Command:

   a. Is now obsolete.
   b. Provides a Horizontal Array.
   c. Only uses the + and - symbols.
   d. Allows you to draw a single Dimension Line.

9. How many types of Dimensioning methods are there in AutoCAD?

   a. 1 type of Dimensioning.
   b. 3 types of Dimensioning methods.
   c. 5 types of Dimensioning methods.
   d. 7 types of Dimensioning methods.

10. The term "DIMTXT" changes what Dimensional Variable?

    Answer:____

11. The term "DIMASZ" changes what Dimensional Variable?

    Answer:____

12. What AutoCAD Command is used to remove a segment from a circle?

    Answer:____

chapter 7

Using the Printer and Plotter

DISCUSSION

It is said that we are in the *"Information Age,"* and that computers are supposed to save us paper. *Don't believe it!* Computers are the leading cause for an increase in the use of paper, and this chapter will show you how. Computers allow you to become more productive and with an increase in productivity comes more documentation and eventually more paper. Just think, without paper, you would need a $2500 computer just to read this document, ...my humor for the day.

Information and documentation is useless if we cannot gain access to it. Let's be realistic, all of the work of developing the drawings in this book would be in vain if we cannot produce "several hard copies" of it; first a review copy to allow us to see how the product will look in its final form, and finally, as a set of working drawings to be used on the production shop floor.

We are at an interim stage when it comes to computerization of the manufacturing process. The time is coming when the output from the development computer in the engineering office will directly drive numerical or computer controlled machines on the shop floor. In fact, some modern companies already have advanced to this stage of manufacturing. In the meantime, the hard copy is still necessary.

In version 12 of AutoCAD, the Prplot (printer plot) command was dropped in favor of the Plot command. In version 13, AutoCAD introduced the Print command which now replaces the Plot command. Print and Plot are found in the Files pull-down menu and evoke the Plot Configuration dialogue box which allows you to select various output devices such as a printer or plotter. Note Figs. 7-3 an 7-4 on page 251.

This chapter deals with the two methods to transfer your drawings into a printed output; first from a printer through the a printer plot routine (Prplot), and next from a pen plotter using the Plot/Print command. You may choose to try all of these exercises or only those necessary to produce the results you desire.

These exercises are:

Exercise 7-1. Printer Plotting SCHEMA1 with Versions 10 and 11.
Exercise 7-2. Printer Plotting SCHEMA1 with Versions 12 and 13.
Exercise 7-3. Preparing the PCB-1 Drawing for Plotting.
Exercise 7-4. Doing a Pen Plot with Dashed Lines on PCB-1.
Exercise 7-5. Doing a Printer Plot with Dashed Lines on PCB-1.

TIP: Before plotting a drawing that contains Solids, Traces, or Polylines, remember to turn the Fill ON and Regenerate the drawing; then End or Save it. If you don't, those entities required to be filled will not be plotted as a solid. You should also remember to Zoom to All before you End a drawing session so the full drawing will appear when you recall it.

PRINTED OUTPUT

The quickest and easiest way to get a "reference or hard copy" of your drawing is to use a dot matrix printer, or a laser printer. When working with dot matrix printers, the emphasis is on "reference," as the resolution or fine detail of the drawing is limited by the number of dots per inch the printer can make. When using laser printers, the output is often comparable to that of plotters as many lasers will produce 600 dots per inch (600 dpi) resolution. This difference can be noted in Fig. 7-1a, a dot matrix printer plot and Fig. 7-1b, a laser printer plot. Note how the text in Fig. 7-1b of the Printed Circuit Assembly View is much clearer.

If you need the hard copy only as a reference, a printer plot is usually quite adequate for the job, but is limited to the size of paper the printer can handle. Depending upon the capabilities of your printer, you can even get a printer plot in several colors.

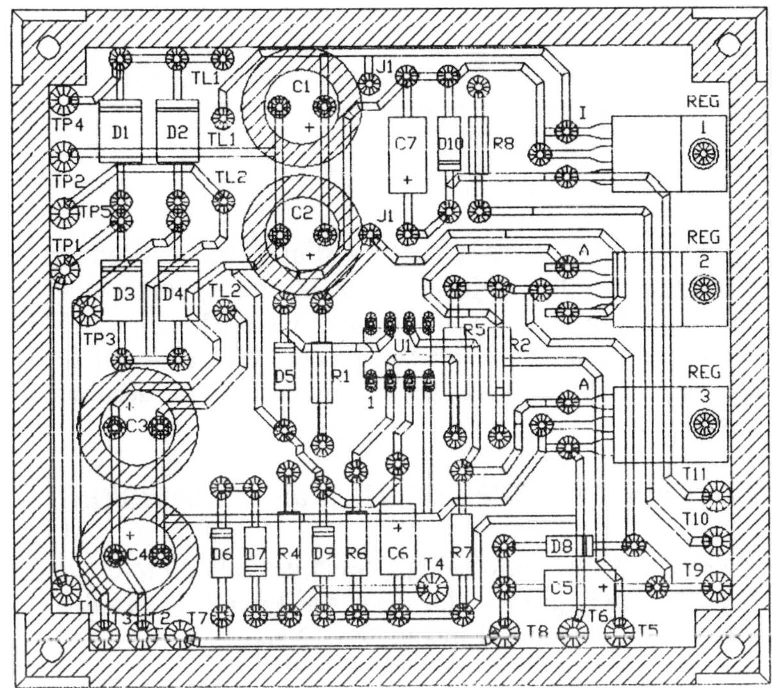

Figure 7-1a
Dotmatrix Printer Quality Output.

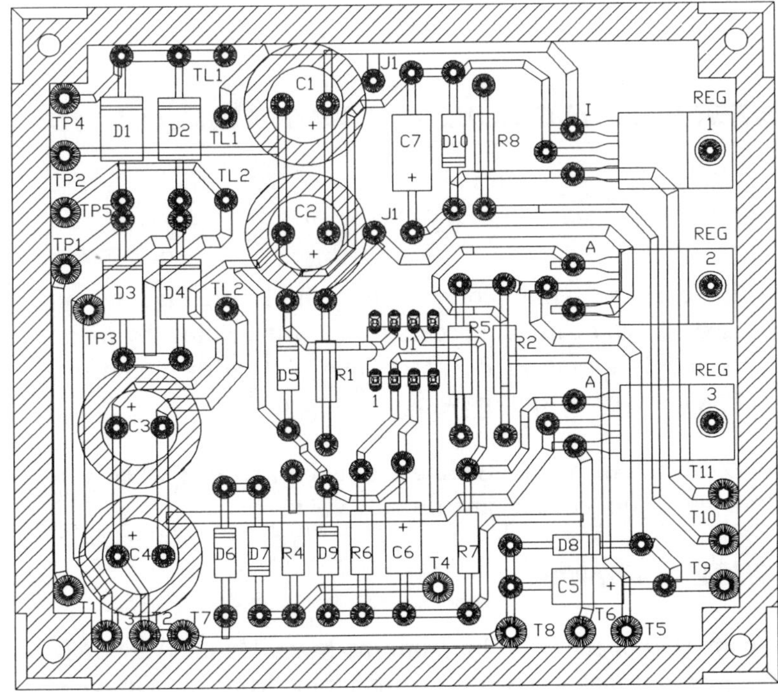

Figure 7-1b
Laser Printer / Plotter Quality Output.

AutoCAD gives you the option to rotate the drawing 0, 90, 180 or 270 degrees when performing a plot. When the drawing is rotated 90 degrees using a printer plot, the height of the drawing is limited by the paper and platen width. The drawing width or length is virtually unlimited and depends upon the length of paper available. This is ideal for making long narrow drawings, such as banners.

PEN PLOTTERS

Pen plotters are very accurate, use various sizes of paper, and may use multicolored pens. Coloring is desirable if you want to highlight various layers on the drawing. To achieve the highest resolution and quality of your drawing, the final copy should be done using a pen plotter. With a pen plotter, circles come out looking like circles and lines look straight, not a series of short, jagged dashes. The quality and accuracy of the plot are reflected in the price of the plotter.

ACCESSING PLOT ROUTINES WITH DIFFERENT AutoCAD VERSIONS

There are several methods to access the two AutoCAD plot routines called Prplot and Plot. We will first look at a "simple" printer plot using the Prplot command in versions 10 and 11, and then the Plot command in versions 12 and 13.

If you are using version 10 or 11, continue with this exercise. If you are using version 12 or 13, read this next section to understand the definitions of the plot elements and then move to Exercise 7-2, Printer Plotting SCHEMA1 With Versions 12 & 13 (page 250).

EXERCISE 7-1. PRINTER PLOTTING SCHEMA1 WITH VERSIONS 10 & 11

Discussion

When you have finished this exercise, your plot will look similar to Fig. 4-3 on page109. From versions 10 and 11, you can gain access to the printer plot, by using either the Main Menu or by simply typing PRPLOT (printer plot) at the command line. This will plot the *last view saved* of any drawing on your disk.

TIP: It is important to Zoom to All before Ending a drawing, so that when you plot, all of the drawing is plotted, not just the last view.

Printer plotting our SCHEMA1 drawing.

When you type the Prplot command or select Prplot from the Main Menu, AutoCAD directs you to the printer plot routine. Let's do a printer plot from the Drawing

Editor Screen of our SCHEMA1 drawing. The starred items (*), 3 and 4 below, are used for the plot routines from the Main Menu.

Insert your disk with the SCHEMA1 drawing on it into drive A, and retrieve it.

```
MAIN MENU

     0. Exit AutoCAD
     1. Begin a New drawing
     2. Edit an Existing drawing
*    3. Plot a drawing
*    4. Printer Plot a drawing
        .
     8. Convert old drawing files

Enter Selection 2 (Return)

Enter NAME of drawing (default A:PCB1): A:SCHEMA1 (Return)
```

With your SCHEMA1 drawing on the screen (Fig. 7-2), at the Command Line type:

```
Command: PRPLOT (Return)
```

AutoCAD next requests which portion of the drawing is to be plotted:

```
What to plot: Display, Extents, Limits, View, or Window <D>:
```

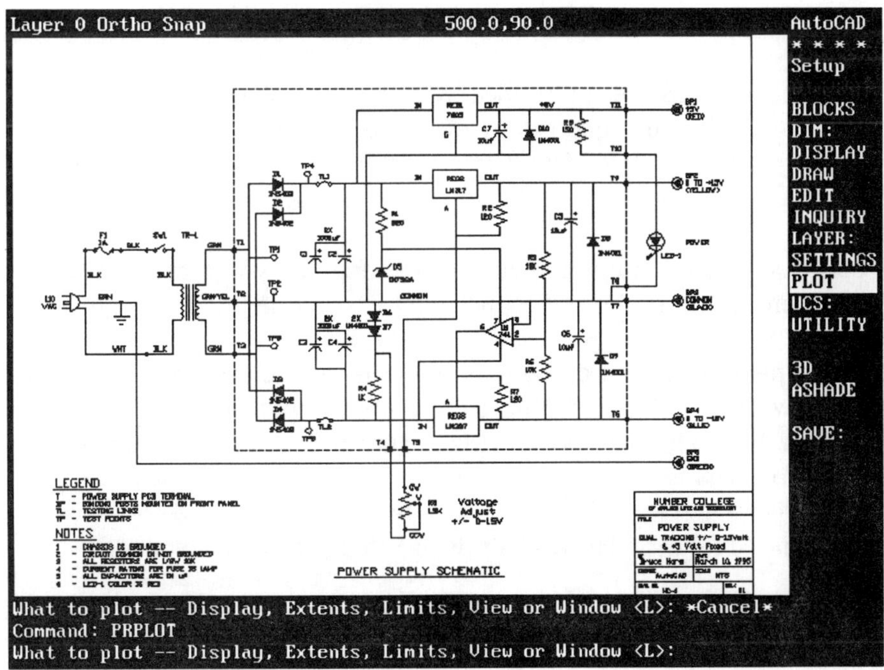

Figure 7-2 Command Line printer plot selection "PRPLOT"

The following options are available:

D (Display)

The Display option will plot the last display on the monitor whether it was obtained from the disk as a result of a Save, End, or EndSave, or obtained through the edit command in AutoCAD, such as Plot or Prplot.

E (Extents)

If your drawing has no border, the Extents option will plot to the extents of what you have drawn. If there are large areas within the Limits of your drawing that contain no entities, these areas will be eliminated and the plot is re-sized to fill the area available on the drawing sheet. If you have a border and title block, the complete drawing is plotted.

L (Limits)

The Limits option plots the entire drawing area, including that area which may contain no entities.

V (View)

A View is a stored zoom that has been identified by name. The View option will plot that stored view. AutoCAD will request the view name to plot.

W (Window)

The Window option allows you to plot an area of the drawing you specify, using a window as the reference. A First Point/Second Point is requested.

TIP: Note, when plotting from a drawing on the screen, use the input device to select your window. When plotting from a disk stored drawing, you will need to have the coordinate points. Make sure that you have these points available before you enter the plot routine.

We'll plot to the Limits of our SCHEMA1 drawing. Reply with L, Return.

```
What to plot: Display, Extents, Limits, View, or Window <D>:
L (RETURN)
```

You are now taken to the Plot Routine. AutoCAD next displays the existing Default conditions. They will be similar to the following, but the units may be in millimeters instead of inches. Let's walk through it:

```
Plot will NOT be written to a selected file.
Sizes are in inches.
Plot origin is at (0.00,0.00)
Plotting area is 14.00 wide by 9.80 high (MAX size)
Plot is NOT rotated 90 degrees
Pen width is 0.025
Area fill will not be adjusted for pen width
Hidden lines will NOT be removed
Plot will be scaled to fit available area

Do you want to change anything? <N>:
```

Decision Time!

A "Y" response will allow you to change the default values. "N" will continue with the plot.

Let's change inches to millimeters and set the printer plot to 200 mm x 250 mm (8 inches x 10 inches) so it will fit on a standard size page. Note that your system may already be in metric. If so, re-enter the data just to get more experience.

```
Do you want to change anything? <N> Y (RETURN)
```

A new printer plot setup table appears, similar to the following:

```
Write the plot to a file? <N>: N (Return)
Size units (Inches or Millimeters)<I>: M (RETURN)
```

The plot origin for the printer is at the top left-hand side of the page. This may be moved to the right and/or down the page, depending upon the response to this next AutoCAD request. Let's keep the plot origin where it is and rotate our drawing by 90 degrees, and Fit it on the page.

```
Plot origin in millimeters <0.00,0.00>: (RETURN)
Standard values for plotting size
Size   Width   Height
MAX    225.38  279.40
USER    80.00  180.00

Enter Size or Width, Height (in Millimeters)<USER>: 200,250 (RETURN)
Rotate 2D plots 90 degrees clockwise?<Y>: Y (RETURN)
Remove hidden lines? <N>: (RETURN) (There are none on this drawing)

Specify scale by entering:
Plotted mm = Drawing Units or Fit or? <F>: F (RETURN)
Effective plotting area: 181.82 wide by 250.00
Position paper in printer.

Press RETURN to continue: (RETURN)
```

At this point the printer plot begins and you will see the vector count being incremented. Within a minute or so, you will see a printer plot of your SCHEMA1 drawing. This should look similar to Figure 4-3 in Chapter 4.

Note that AutoCAD checked the size of the drawing we requested (200.00 by 250.00) and realized it would not fit on a standard 8-1/2 by 11 inch sheet of paper. It then set the effective plotting area to 181.82 mm wide by 250.00 mm long to keep the same aspect ratio.

Had we not rotated the drawing, the maximum drawing width would have been only 190 mm instead of 250 mm. Rotating allowed us to use the full paper length. When the plot is complete, you get the message "Press Return to continue," and the Main Menu reappears.

```
Plot complete, press RETURN to continue (RETURN).
```

Let's look at some other plot characteristics!

TIP: Near the beginning of this plot sequence, we were asked if the "Plot Origin in millimeters is at <0.00, 0,00>." Do not get this confused with the "Drawing Origin" coordinates we learned about when we first set up the drawing. The plot origin allows us to shift the plot on the page with respect to the page edges. If you see the plot is too close to an edge, increase the numbers in the plot origin to move it across and down the page.

Drawing Rotation and User Width/Height Selection

Note that the Drawing was rotated by 90 degrees. This allowed our schematic drawing to fit sideways on the page, rather than across it. When the drawing is rotated, the X and Y Plot axes also change. The X drawing axis is now the plot Y axis, and the Y drawing axis becomes the plot X axis. Because of this, the USER Width and Height becomes 200,250 instead of 250,200 (X,Y), which is how you normally think of the drawing limits.

EXERCISE 7-2. PRINTER PLOTTING SCHEMA1 WITH VERSIONS 12 & 13

Discussion

When you have finished this exercise, your plot will look similar to Fig. 4-3 on page 109. Using AutoCAD version 12 or 13 to do a printer plot is quite different than the steps used in previous versions. First, the software uses a "raster graphic" plot routine when printing from the screen. This speeds up the printer plot time tremendously. Second, it uses a new dialogue box for the selection of the plot, see Fig. 7-3; and third, there is no Prplot command; it has been eliminated. Earlier versions of AutoCAD used a "vector graphic" software routine when printer plotting.

This caused the program to calculate every vector, making it slow, because of the many math calculations.

To create a printer plot, first we'll open the SCHEMA1 drawing. In AutoCAD 12 select Plot (Print in version 13) from the Files pull-down menu.

```
Select OPEN from the FILES pull-down menu
Choose the A: Drive, and click on SCHEMA1, and OK
```

Before you start a printer plot, make sure that the Fill is turned ON

```
Select PLOT/PRINT from the FILES pull-down menu
```

This brings up the Plot Configuration dialogue box, Fig. 7-3.

AutoCAD 12 and 13 use the same dialogue box to access both the printer and pen plot functions. The dialogue box appears when you select Plot from the Files pull-down menu or by typing PLOT at the keyboard. Through the "Device and Default Selection..." button, you can select all the output devices presently configured for the system. See Fig 7-4.

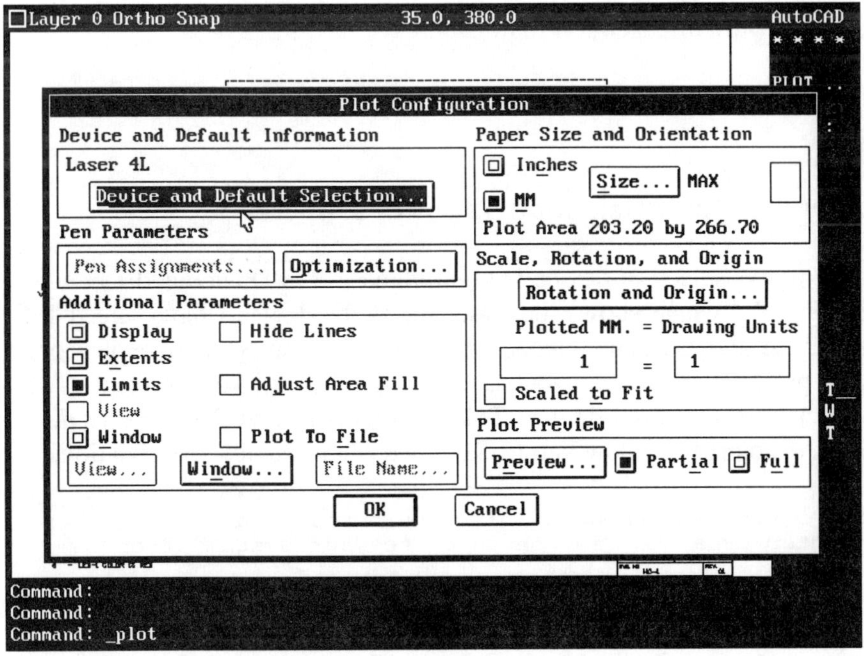

Figure 7-3 Versions 12 and 13 Plot Configuration dialogue box (DOS).

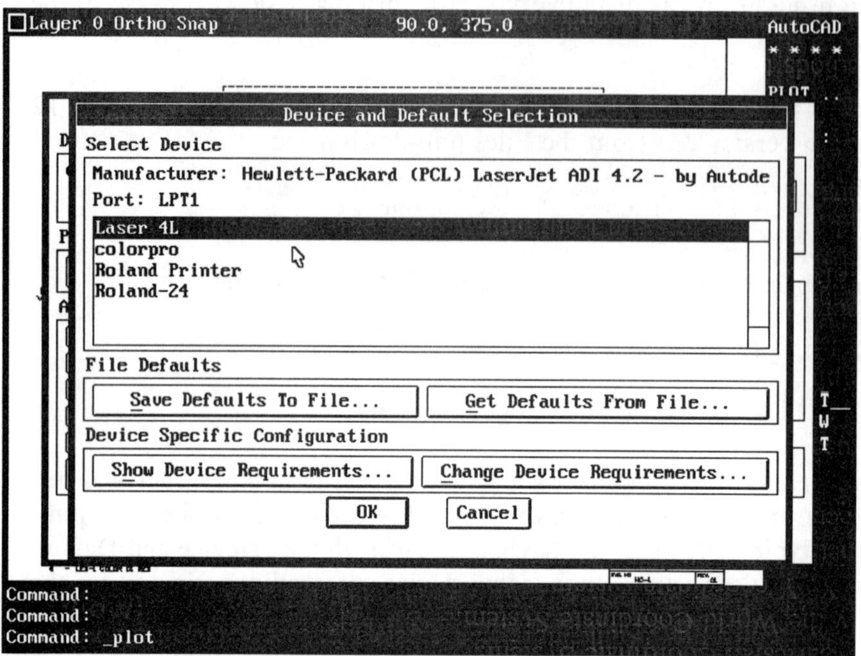

Figure 7-4 Device and Default Selection dialogue box.

The Plot Configuration dialogue box has six areas which control the plotting of the drawing.

1) The Device and Default Selection Area (Button).

The current printer/plotter selection device is highlighted in the Device and Default Selection area. See Fig. 7-4 which shows the Laser 4L as the device I have selected. Clicking on the Device and Default Selection button will allow you to review or change any of the printers or plotters you have configured into your system. To change a device, click on the desired device. This highlights it, and then select OK. Other options in this dialogue box allow you to Save and Show device specifications.

```
Click on the Device and Default Selection button and
Choose a printer type that is connected to your system, then
Select OK
```

2) The Pen Parameters / Optimization Area (Button).

With reference to Fig. 7-3, the Pen Parameters button is not active when a printer is selected because you can't change the type of print head used by the printer. Pen Parameters only becomes active when a pen style plotter is selected.

With a pen plotter selected, the Pen Parameters button allows you to choose a variety of pens, pen widths, pen colors, linetypes and plotter speeds which are used by the device selected. See Fig. 7-5.

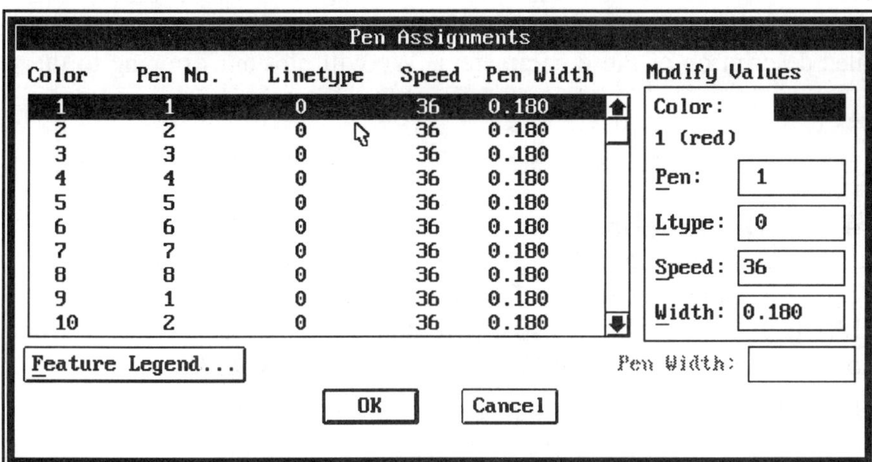

Figure 7-5 Pen Assignments dialogue box.

Picking the Feature Legend button in the Pen Assignment dialogue box will show the Linetypes that are available through the plotter. See Fig 7-6. Note, these are different from the Linetypes used in your drawing.

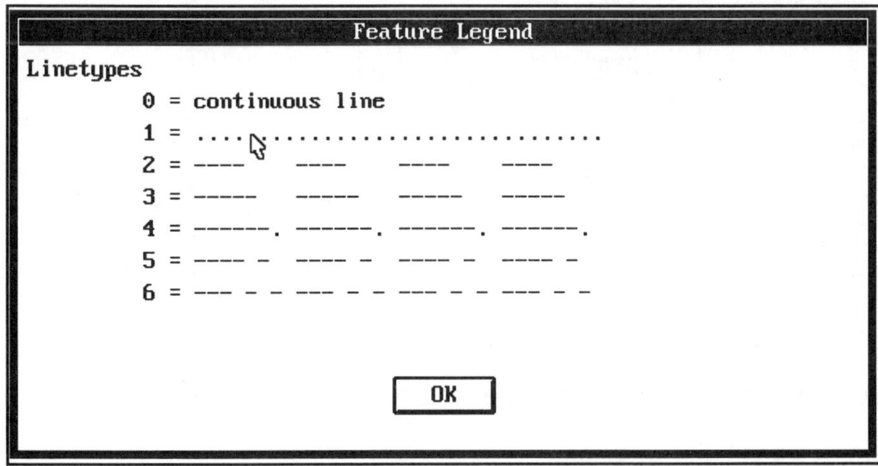

Figure 7-6. The Feature Legend subdialogue box displays the linetypes.

The Optimization... button (Fig. 7-3) allows you to change and control the movement of the pen. With these features you can optimize pen motion, thus increasing productivity.

3) Additional Parameters.

Additional Parameters of the Plot Configuration dialogue box give you control over the plotting of such things as: Display, Extents, Limits, View, and a Window, through the use of click-on/click-off radio buttons. Hide Lines, Fill Area, and Plotting to a File is also controlled from this subdialogue box. Refer to Exercise 7-1 for a more detailed description of these parameters. We will plot our drawing to the outside Limits.

```
Click on the LIMITS Radio Button, it should go dark when selected.
```

4) Paper Size and Orientation.

The Paper Size and Orientation area of Fig. 7-3 (page 251), will allow you to change from Inches to Metric and the submenu Paper Size dialogue box (See Fig. 7-7) will allow you to choose the sheet sizes (A, B, C, D) or establish a USER sheet size. We want the drawing plotted in millimeters.

```
Click on the MM button in the Paper Size and Orientation area (Fig.
7-3).
```

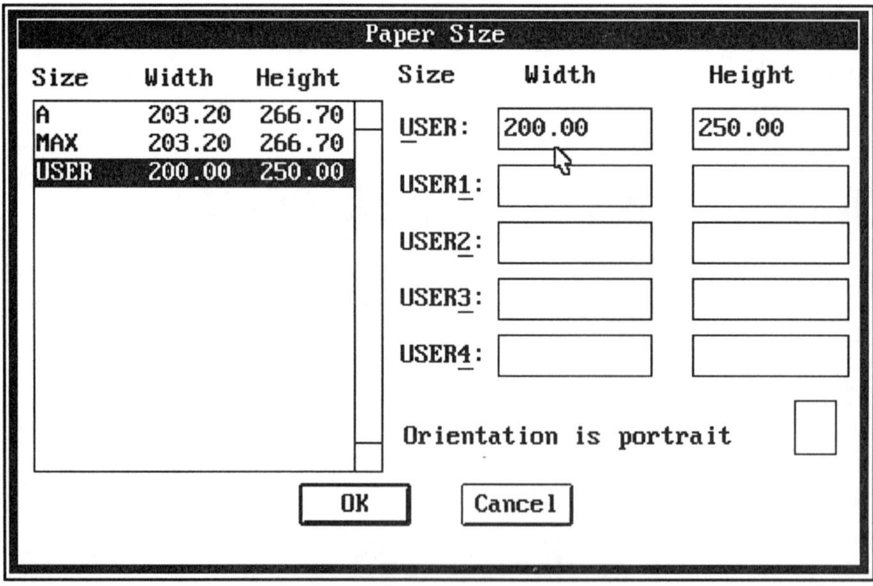

Figure 7-7. Paper Size subdialogue box.

5) Scale, Rotation, and Origin

With reference to Fig 7-3, the Scale, Rotation and Origin buttons allow you to change the plot rotation to 0, 90, 180, and 270 degrees as well as the origin or starting point of the drawing. See Fig. 7-8. Scale allows you to "Fit" the drawing on the page or change the "Plotter Scale" and "Drawing Scale" ratios through the two Drawing Unit edit boxes. For our drawing, we will rotate it 90 degrees, set the drawing size to 200 by 250, and select Scale to Fit.

```
Click the Rotation and Origin button and Select 90 degrees and OK,
then Click on the Scale to Fit button.
```

6) Plot Preview.

This is a new feature to version 12 which will allow you to quickly check what the plot will look like before printing or plotting. Preview Plot will save re-plot time and material should you find there is a problem with the plot setup.

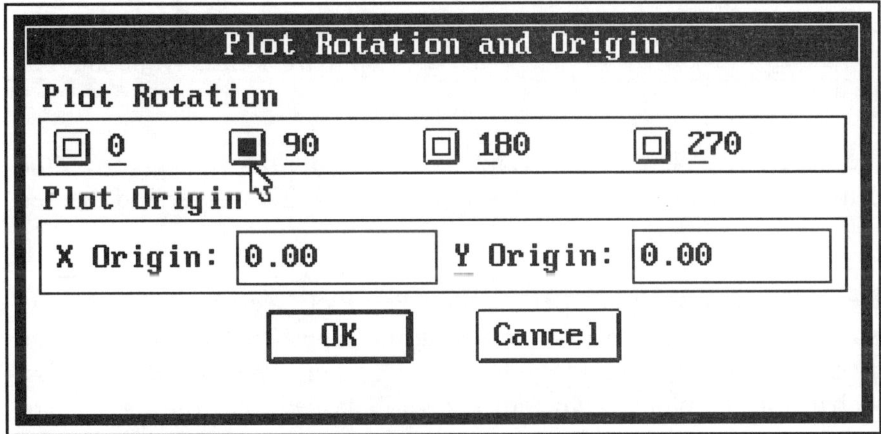

Figure 7-8. The Plot Rotation and Origin subdialogue box.

Doing the plot!

TIP: In Fig. 7-3, Make sure the "Plot To File" button is NOT selected as we want to send the plot to the printer, not to a file.

To plot the SCHEMA1 drawing,

```
Click on OK at the bottom of the Plot Configuration dialogue box.
```

At the Command line, press Return to start the plot

```
Command: Plot effective plot area: 190 wide by 250 high.
Position paper in plotter.
Press RETURN to continue or S to Stop for hardware setup: (RETURN)
```

You will see that a regeneration is done and the (3150) raster data scan lines are calculated and eventually sent to the printer. After several seconds, the printer generates the printed output which should look similar to Fig. 4-3 in Chapter 4.

EXERCISE 7-3. PREPARING THE PCB-1 DRAWING FOR PLOTTING WITH DASHED TRACE LINES IN THE ASSEMBLY AREA

Discussion

In this exercise, we will prepare our printed circuit board drawing so we can plot it. See Fig. 7-15 on page 260. To have the trace lines of the printed circuit board Assembly View appear as if they are on the underside of the board, we will plot them as *unfilled dashed lines* instead of solid. To do this, they must be changed to a new layer called TAPE. Presently, all of the Assembly View traces are on the layer called TRACES along with the Plines that make up the Artwork Side (right side) of our PCB-1 drawing.

Separating the Assembly View Plines from the rest of the drawing gives us the flexibility needed to show these lines as dashed lines in both a printer and the pen plots.

To achieve the unfilled dashed line effect, with a Pen Plot (see Fig. 7-15 on page 260), the drawing is plotted twice, without removing the paper from the plotter. This assures both plots align themself exactly on top of each other. The dashed line effect is created by changing the plot linetype attributes to "dashed line" during the plot operation.

TIP: The printer plot uses a single pass through the printer as it is too difficult to assure the paper is correctly aligned with the print head during a second pass. To achieve the dashed line effect on a printer plot, the Tape layer is switched to a *dashed line* as part of the drawing attributes, not as part of the plot attributes. The dashed lines in this case will appear filled (solid) as you cannot segregate filled and unfilled lines on single pass printing. See Fig. 7-16 on page 266. This can only be achieved by equipment that allows two passes, such as pen plotters where the paper is not moved after plotting.

Switching The Traces To The Tape Layer

From the Layer Control dialogue box, we will change the Traces to the layer called Tape. A layer we have not used yet. Next, we'll set the current layer to Tape, and freeze all layers except TRACES and TAPE, then return to the drawing editor screen. If you have been following this text, the only lines that should be visible are the Polylines associated with the traces. See Fig. 7-9. This setup should NOT show donuts, text, dimensions, or anything else except those Plines associated with the Tape layer. If it does, don't worry, we'll fix that in a moment.

```
Select Layer Control from the Settings pull-down menu.
Set the Current Layer to TAPE
Freeze all layers except TAPE and TRACE, and Click OK.
```

Note: The layer color for the Traces layer is Magenta, and White for the Tape layer. If yours is different, change it now. Your drawing should look similar to Fig 7-9.

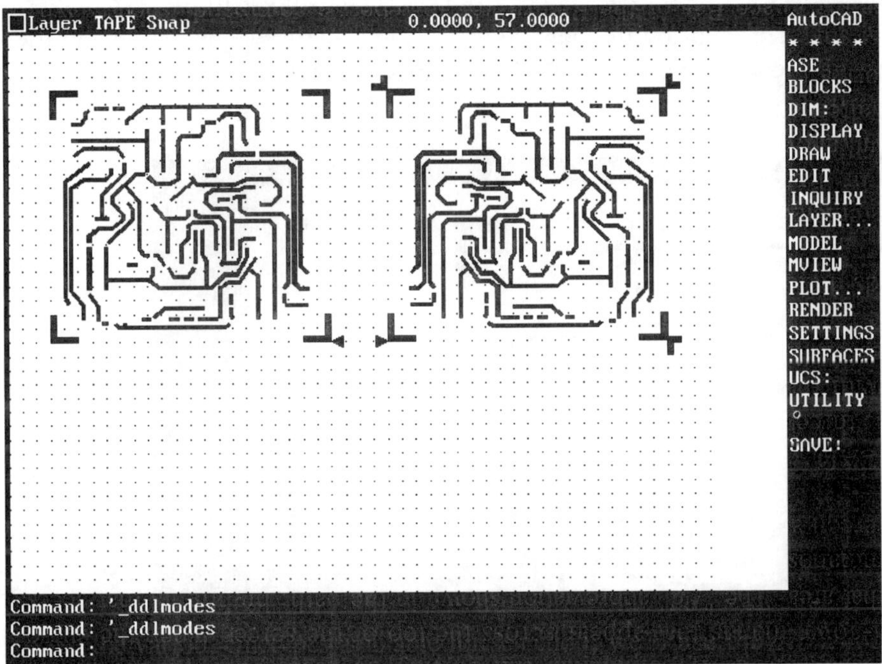

Figure 7-9. The TAPE and TRACES layer view.

Zoom into the Assembly View side (left side) of the drawing to make the selection of the window easier to capture during the next step. See Fig. 7-10. Then we'll change the traces to the layer Tape, using the Change command and a window.

```
Command: CHANGE (RETURN)
Select objects: W (RETURN)
```

Wrap a window only around the traces. *If there are other objects selected, de-select them by clicking on them with the cursor after your selection.* Press the Return key for the next selection.

```
First corner: Other corner: 67 found
Select objects: (RETURN)
```

Your drawing may contain a different number of found objects than shown above.

We will change the Properties and select the Layer to change using TAPE as the New Layer.

```
Properties/< Change point> P (RETURN)
Change properties (Color/Elev/LAyer/LType/Thickness)?: LA (RETURN)
New Layer <TRACES>: TAPE (RETURN)
Change properties (Color/Elev/LAyer/LType/Thickness)?: (RETURN)
```

AutoCAD now moves all the selected Plines from the Traces layer to the Tape layer. You should recognize them by the new color, as the Tape Layer is White and the Traces layer is Magenta, if you selected these colors originally.

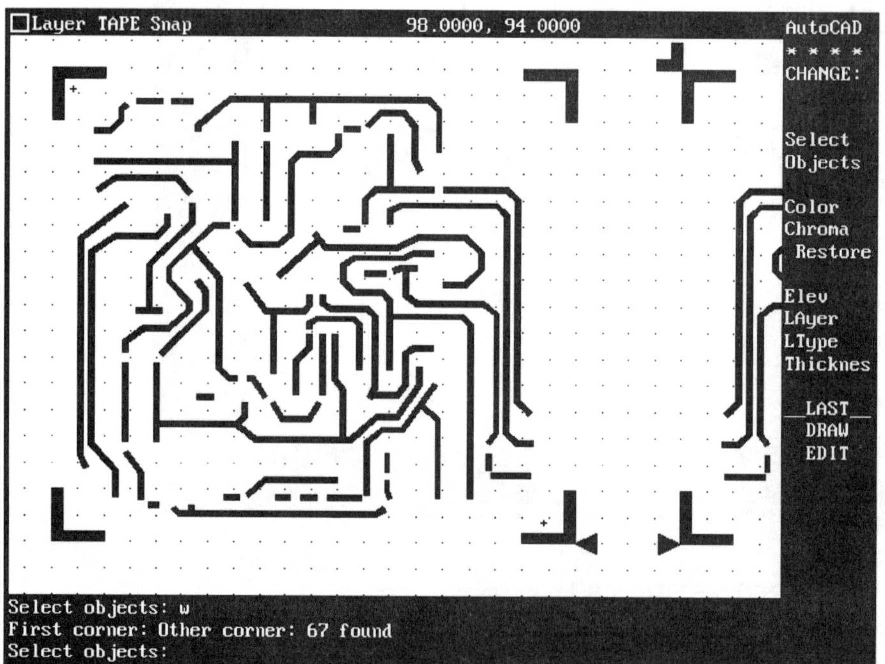

Figure 7-10. Window view for selection of TRACES.

From the Modify Layer/Layer Control dialogue box, thaw all layers we froze before we changed the traces.

```
Select Modify Layer/Layer Control from the Settings pull-down menu.
Thaw all Layers.
Set the Current Layer to 0, and Click OK to exit the box.
```

We are now ready to start a pen or printer plotting.

EXERCISE 7-4. DOING A PEN PLOT WITH DASHED LINES ON PCB-1

Discussion

The following are the steps to produce a Pen Plot of our PCB-1 drawing that will show the Assembly View trace lines as *unfilled dashed lines*. See Fig. 7-11 This is a two pass plotting process on the plotter, and uses version 12 or 13 of AutoCAD. In the second pass, the plotted line style is changed from Continuous to Dashed through the "Plot Features Legend."

Step 1. Turn on all layers except the Tape layer.
Step 2. Freeze the Tape layer.
Step 3. Set Fill to ON.
Step 4. From the Plot Configuration dialogue box (Fig. 7-3 page 251), plot the drawing using the following attributes. (Refer to Exercise 7-2 should you need help.)

Select the plotter connected to your system.
Plot to the Limits.
Use Millimeters.
Drawing Size = 250,200
Plot Origin = 0.00,0.00
Plot is NOT rotated (Set to 0).
Pen width = .25
Area Fill "IS" adjusted for pen width.
Hidden Lines are NOT removed.
Scale is 1=1 or 1:1

TIP: Because the PCB-1 drawing is used to produce the final printed circuit board, and as it is drawn at full scale, it is therefore plotted at 1:1 scale rather than Scaled to Fit. This allows us to use the drawing to directly manufacture the board.

When the plot is complete,

Step 5. Thaw the Tape Layer.
Step 6. Freeze all other layers.
Step 7. Set Fill to OFF.
Step 8. From the Plot Configuration dialogue box (Fig.7-3), select "Pen Assignment" under the Pen Parameters area. This displays the Pen Assignment dialogue box (Fig. 7-5, page 253). Next select the "Feature Legend" in the lower left corner of this dialogue box, and then select the Linetype as item "3 dashed lines" of the Features Legend dialogue box (Fig 7-6). Do this for all the pen numbers 1 through 16 in Fig. 7-5.

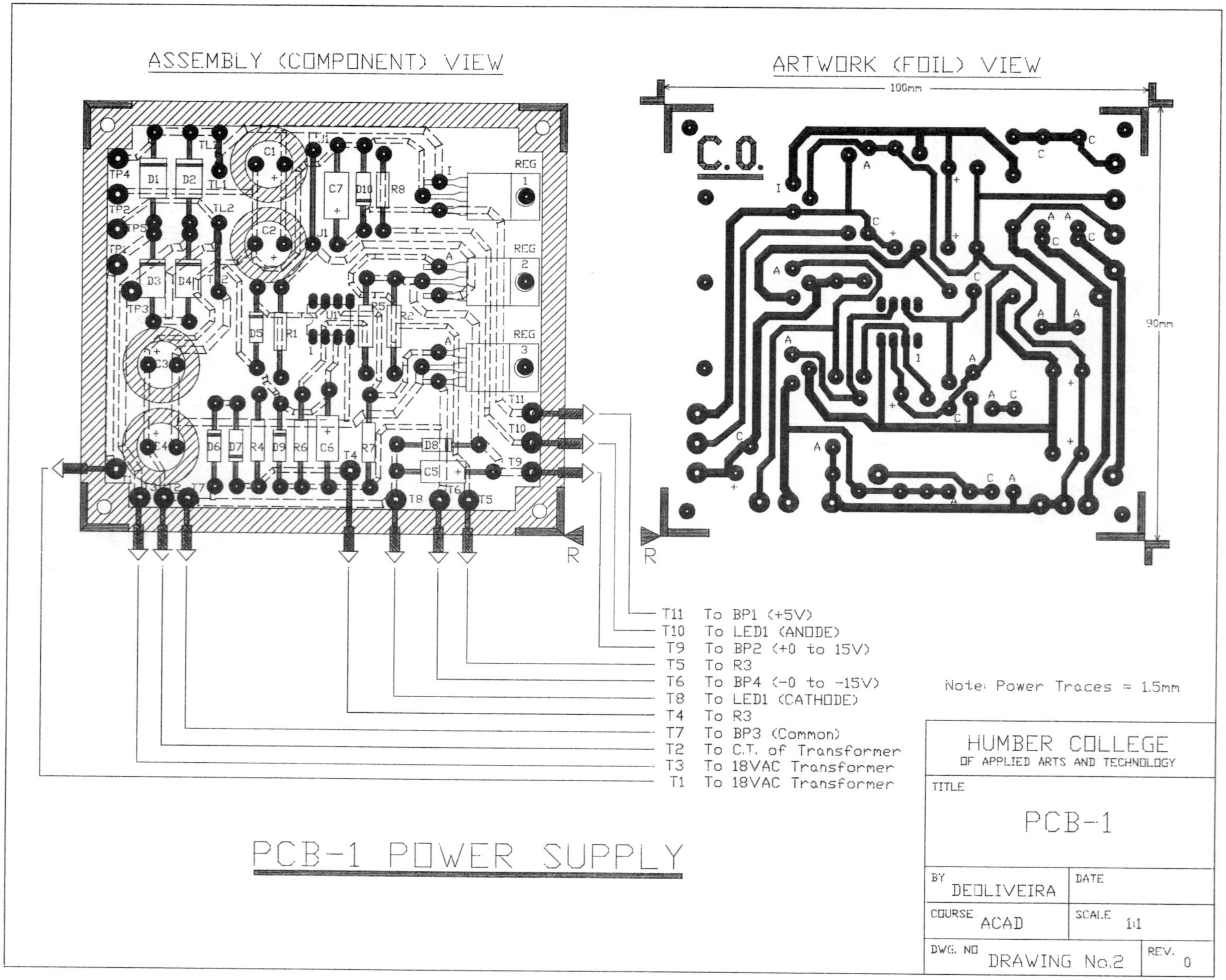

ASSEMBLY (COMPONENT) VIEW

ARTWORK (FOIL) VIEW

100mm

90mm

C.O.

T11 To BP1 (+5V)
T10 To LED1 (ANODE)
T9 To BP2 (+0 to 15V)
T5 To R3
T6 To BP4 (-0 to -15V)
T8 To LED1 (CATHODE)
T4 To R3
T7 To BP3 (Common)
T2 To C.T. of Transformer
T3 To 18VAC Transformer
T1 To 18VAC Transformer

Note: Power Traces = 1.5mm

PCB-1 POWER SUPPLY

HUMBER COLLEGE OF APPLIED ARTS AND TECHNOLOGY	
TITLE PCB-1	
BY DEOLIVEIRA	DATE
COURSE ACAD	SCALE 1:1
DWG. NO DRAWING No.2	REV. 0

Figure 7-11 Pen plot with unfilled dashed lines in the Assembly View.

Step 9. Replot the drawing using Step 4 setup.

Step 10. When the plot is finished, you will have the Artwork side (right side) of the drawing with solid filled-in traces and the Assembly View (left side) with dashed lines. Your drawing will look similar to Fig. 7-11.

EXERCISE 7-5. DOING A PRINTER PLOT WITH DASHED LINES ON PCB-1

Discussion

The following are the steps to produce a Printer Plot of our PCB-1, which will show the Assembly View Trace Lines (left side) as solid dashed-lines. This is a one-pass process, therefore the dashed lines will show up as filled dashed lines rather than non-filled. See Fig. 7-16 on page 266. Before we start, assure that the Assembly View Trace Lines have been moved to the new layer Tape. See Exercise 7-3.

In this exercise (7-5), we will change the PCB-1 drawing Linetype of the Tape layer to a dashed line called Dashed and then printer plot it.

Loading The Linetype Dashed Into AutoCAD.

Bring the PCB-1 drawing into AutoCAD then, at the command line, type:

```
Command: LINETYPE (RETURN)
```

AutoCAD responds with,

```
?/Create/Load/Set:
```

To see the various linetypes available for selection, we'll use the "?" as our response. At the prompt type:

```
?/Create/Load/Set: ? (RETURN)
```

The Select Linetype File dialogue box is displayed (Fig. 7-12), if the FILEDIA switch is set to 1. If it is set to 0, the "?/Create/Load/Set:" prompt is displayed. The dialogue box indicates that the ACAD.LIN file is where AutoCAD will look to find the Dashed linetype. This should be in the C:\ACAD12\SUPPORT Directory or the C:\ACAD13\SUPPORT Directory.

```
Click on OK or press the RETURN key.
```

AutoCAD now displays a list of the various linetypes available to Load. Note the Dashed style linetype.

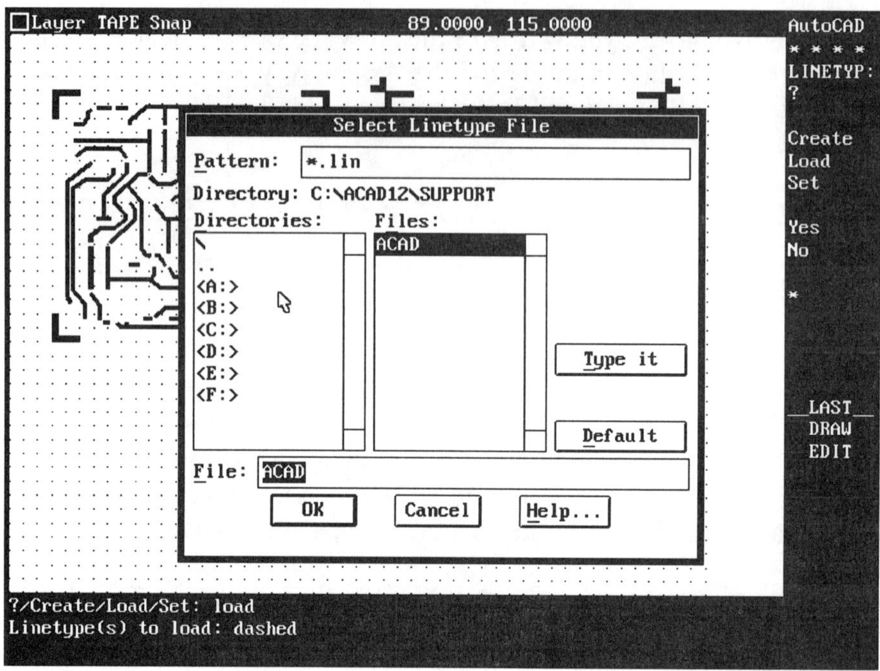

Figure 7-12. Select Linetype File dialogue box.

Press the Return key to advance, and then the F1 function key to get back to the Drawing Editor screen.

```
Press the RETURN KEY
Press F1 Function key.
```

Now to Load the Dashed linetype into our drawing attributes. Type:

```
Command: ?/Create/Load/Set: LOAD (RETURN)
Linetype(s) to Load: DASHED (RETURN)
```

The Selected Linetype File dialogue box (Fig. 7-12) again appears with ACAD as the reference file.

```
Click on OK or press the RETURN key.
```

AutoCAD responds with,

```
Linetype DASHED loaded.
```

You are questioned again if you want to load another linetype. We are finished, so press the Return key.

```
?/Create/Load/Set: (RETURN)
Command:
```

This has loaded the linetype Dashed into the linetype menu of the Layer Control Select Linetype dialogue box. After setting the Fill to ON, we will activate this dialogue box and set the linetype of layer Tape to Dashed.

```
Command: FILL (RETURN)
ON/OFF <ON>: ON (RETURN)
```

Now to activate the Layer Control dialogue box, Fig. 7-13:

From the **SETTINGS** pull-down menu, select **LAYER CONTROL** (Version 12)

or

From the **DATA** pull-down menu, select ***LAYERS*** and **LAYER CONTROL** (version 13)
Click on layer **TAPE** to activate it.
Click on the **Set Ltype...** button in the lower right corner.

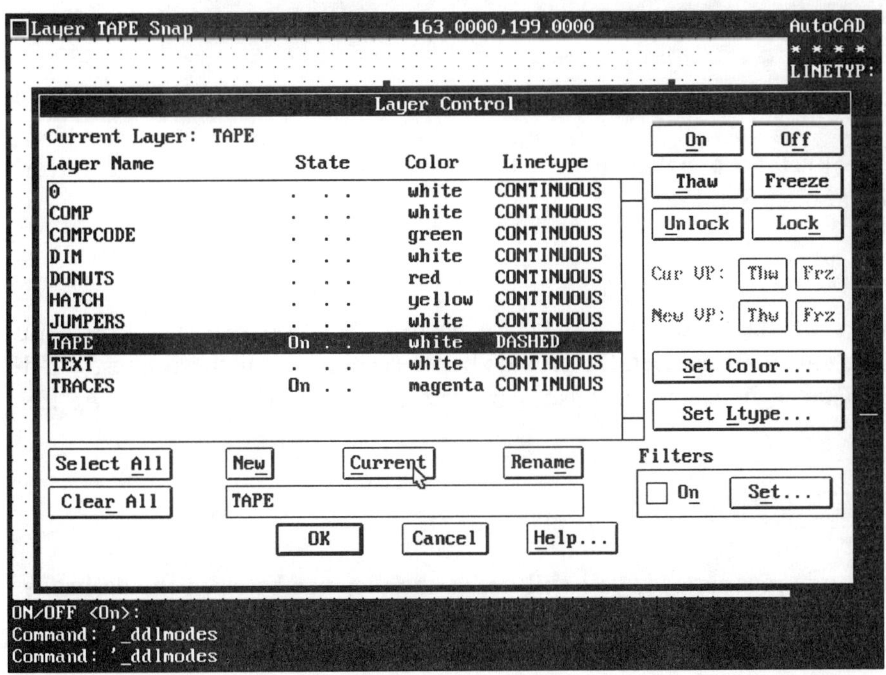

Figure 7-13. Layer Control dialogue box.

This displays the Select Linetype dialogue box, which now contains the linetype Dashed. See Figure 7-14.

Click on the **DASHED** linetype and **OK** to import it into the TAPE layer.

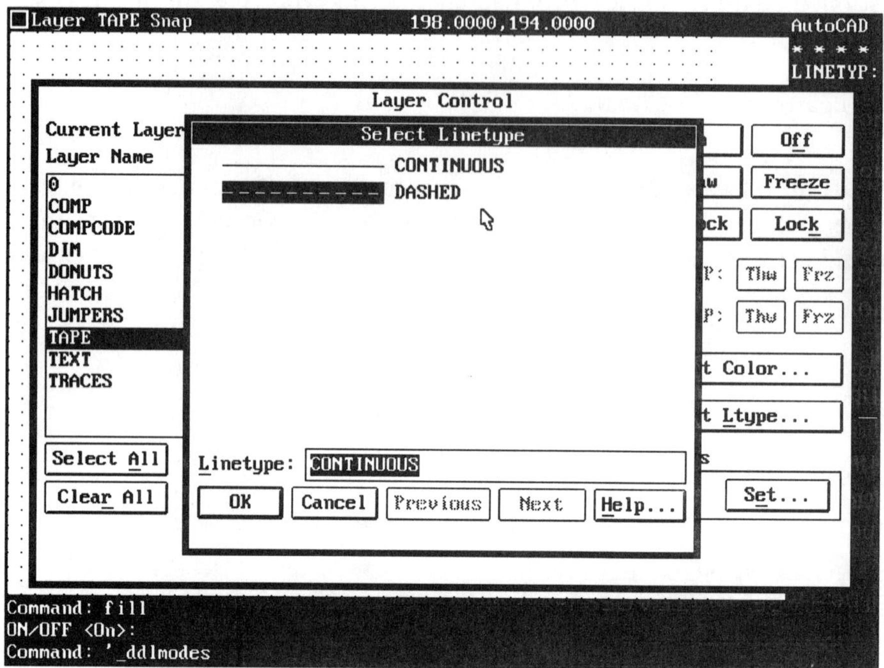

Figure 7-14. Select Linetype dialogue box.

Notice that the linetype assigned to the Tape layer is now Dashed in the Layer Control dialogue box. See Fig. 7-13.

```
Click on OK
```

The Trace lines on the Assembly View are dashed, but you may not see them because they are too close together. To provide better resolution for these lines, we'll increase the distance between the dashes by setting the LTSCALE to 8. At the command line type:

```
Command: LTSCALE (RETURN)
New scale factor <1.000>: 8 (RETURN)
```

After the drawing is regenerated, you will notice the distance between the dashes is farther apart, and the dashes are filled in. Compare Fig. 7-15 with Fig. 7-9, on page 257, which shows the Assembly View Trace Lines as solid lines.

We Are Now Ready To Printer Plot Our Drawing.

Make sure all your Layers are thawed and turned On.

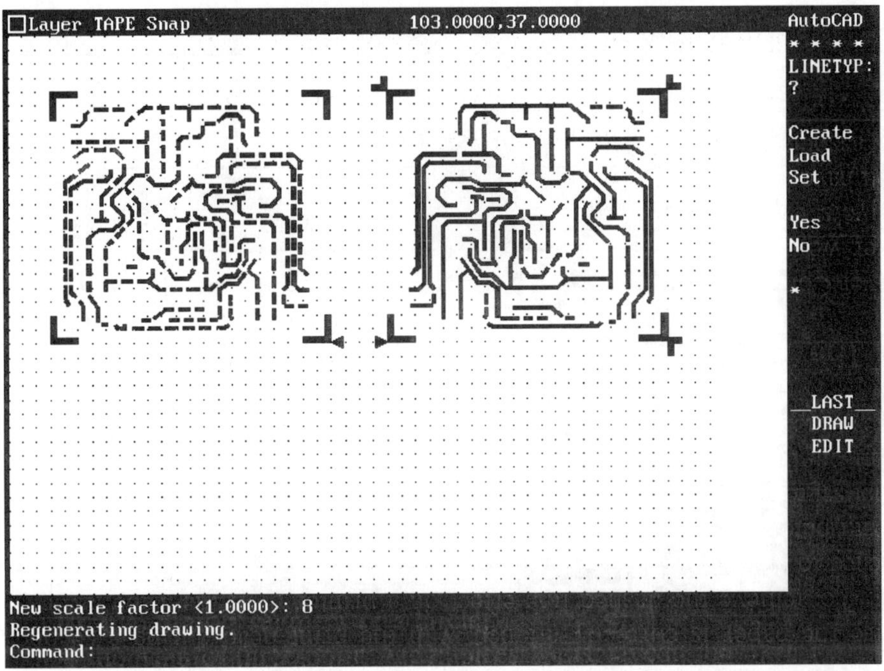

Figure 7-15. The Assembly View as Dashed lines.

```
Select PLOT/PRINT from the Files pull-down menu
From the Plot Configuration dialogue box, select your printer.
Click on the LIMITS Radio Button.
Click on MM
Click the Rotation and Origin button and Select 90 degrees, and OK.
Set the "Plotted MM = Drawing Units" at 1=1
```

Note: Make sure the Plot To File and Scale To Fit buttons are NOT selected.

```
Click the OK button of the Plot Configuration dialogue box.
```

Follow the instructions at the Command Line to start the plot:

```
Command: Plot effective plot area: 190 wide by 250 high.
Position paper in plotter.
Press RETURN to continue or S to Stop for hardware setup: (RETURN)
```

You will see that a regeneration is done and the (3150) raster data scan lines are calculated and sent to the printer. After several seconds, the printer generates the printed output, and your printer plot of the PCB-1 drawing is complete and looks similar to Fig.7-16.

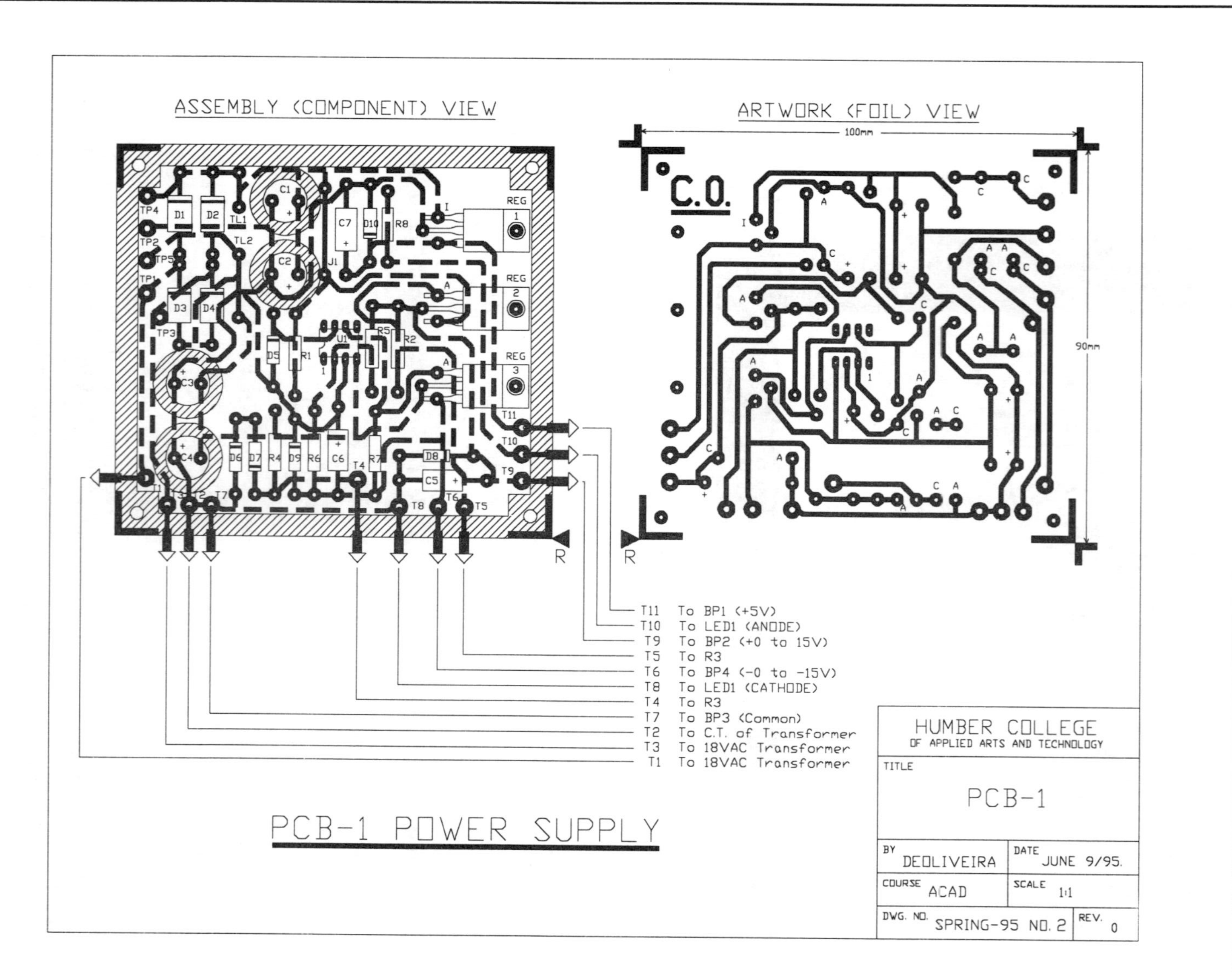

Figure 7-16 Printer plot with dashed lines in Assembly View.

chapter 8

Summary of AutoCAD Commands

DISCUSSION

In this chapter you will find, in alphabetical order, a summary of the AutoCAD Commands used in this text. It is laid out for quick reference, and the main points of each command are highlighted in the description area. Along with each command is an example of the typical screen display.

Beside each command is the AutoCAD version number, such as V11, V12, V13 for versions 11, 12 and 13, and the menu path to gain access to that command. When the menu path has been changed as a result of an AutoCAD version change, both paths are shown along with the AutoCAD version reference. An example of the SNAP command is shown below:

SNAP ***(V11, V12 AUTOCAD → SETTINGS → (Next) → SNAP: SNAP MENU)***
(V12 Pd SETTINGS → DRAWING AIDS... → DRAW AIDS Dialogue Box)
(V13 AUTOCAD → SERVICE → DRAWING AIDS...→

DRAWING AIDS Dialogue Box)
(V13 Pd OPTIONS → DRAWING AIDS... → DRAWING AIDS Dialogue Box)

As you can see, the Snap menu and Snap command for versions 11 and 12 are found in a submenu (Next) of the Settings section of the AutoCAD Root Menu. In version 13, Snap is part of the Drawing Aids dialogue box, which is a submenu of the Drawing Aids... submenu found under the Service section (submenu) of the AutoCAD Root Menu. That was quite a mouthful! Did you follow that?

Version 10 commands follow the same structure and path as version 11, thus they are not shown.

Frequently used AutoCAD commands can quickly be accessed, with the mouse, through pull-down menus that appear across the top of the screen when the cursor is placed in that area. The "***Pd***" abbreviation shown in the path, highlights that access to these commands is from a "***P***ull-***d***own" menu. Note Figure 1-2b on page 7, AutoCAD Screen and Pull-Down Menus.

When AutoCAD variables, pull-down menus, or commands are followed by an Ellipsis (...), three periods, it signifies there is a submenu, dialogue box or option associated with that item. For example; in version 12, the Snap dialogue box is directly accessed through the Settings, Drawing Aids**...** pull-down menu, as shown above. Note the ellipsis after Drawing Aids which indicates the access to the Snap Drawing Aids dialogue box.

You should remember that all keyboard commands should be followed by pressing the Return or Enter key, depending on how your keyboard is labelled. Most commands will allow you to use the Input Device (Mouse, Tablet, Joystick, and the like) as an alternative input response to the command. The right mouse button usually performs the same function as the Return key.

You may use the Space Bar or the Return key to repeat a command, such as when multiple selection of From/To points are required in the Line command.

With later versions of AutoCAD, "hot keys" were introduced. For example; pressing the E key will bring up the Erase menu, while pressing the Z key will display the Zoom options. These hot keys are also known as *alias* keys or Alias commands.

When instructed to **SELECT/PICK**, use the input device (mouse) to make your selection. When instructed to **CHOOSE/TYPE**, enter your choice from the keyboard. Don't forget to press the Return or Enter key after using keyboard commands.

APERTURE COMMAND. See OSNAP

ARC (V11, V12 AUTOCAD → DRAW → ARC)
(V12 Pd → DRAW → ARC → ARC MENU)
(V13 AUTOCAD → DRAW → ARC:... → ARC MENU)
(V13 Pd DRAW → ARC... → ARC MENU)

Description

Used to draw a segment of a circle.
A submenu of three-point Arc choices is displayed.
11 methods to draw an Arc can be chosen, one by specifying Arc and 10 as a three-point Arc by specifying: Start point, Center (radius), End point, Angle included, Direction, chord Length.
In all cases, the Arc is drawn counterclockwise except if the angle is specified as negative, in which case it is drawn clockwise.

Three-Point Arc Choices

	Version 13 Selection
3-Point: (Three points on Arc)	
S,C,E (Start, Center, End)	St,C,End
S,C,A (Start, Center, included Angle)	St,C,Ang
S,C,L (Start, Center, Length of chord)	St,C,Len
S,E,A (Start, End, included Angle)	St,E,Ang
S,E,R (Start, End, Radius)	St,E,Rad
S,E,D (Start, End, starting Direction)	St,E,Dir
C,S,E (Center, Start, End)	Ce,S,End
C,S,A (Center, Start, included Angle)	Ce,S,Ang
C,S,L (Center, Start, Length of chord)	Ce,S,Len
CONTIN (Connect an arc to a previously drawn line or arc)	

ARC Command Format

```
Command: ARC (RETURN)
Center/<Start point>: SELECT (point 1)
Center/End/<Second point>: SELECT (point 2)
Endpoint: SELECT (point 3)
```

A counterclockwise Arc is drawn between the points specified.

Three-Point Arc Command Format

```
Command: SELECT ARC FROM DRAW SCREEN MENU
Command: SELECT S,C,E: (FROM SCREEN MENU)
```

```
ArcCenter/<Start point>: SELECT (start point--1)
Center/End/<Second point>: SELECT (center point--2)
Angle/Length chord/<End point>: SELECT (end point--3)
```

A counterclockwise Arc is drawn between the points specified.

ARRAY (V11, V12 AUTOCAD → EDIT → ARRAY: → ARRAY) (V13 AUTOCAD → CONSTRCT → ARRAY: → ARRAY MENU) (V13 Pd CONSTRUCT → ARRAY: → ARRAY MENU)

Description

Used to create copies of objects in an array format.
May be placed in a Rectangular (columns and rows) or Polar (circular) arrangement.
Objects at final placement may be edited separately.
Distance between rows and/or columns are center-to-center distances.
Row distances may be different from column distances.
A window may be used to specify the row and/or column distances.

Rectangular Command Format

TIP: Requires an object on the screen for replication purposes.

```
Command: ARRAY (RETURN)
Select objects or Window or Last: SELECT OBJECT
Rectangular/Polar array (R/P): R (RETURN)
Number of rows (---) <1>: CHOOSE (RETURN)
Number of columns (111) <1>: CHOOSE (RETURN)
Unit cell distance between rows (---): CHOOSE (RETURN)
Unit cell distance between columns (111): CHOOSE (RETURN)
```

Polar (Circular) Command Format

```
Command: ARRAY (RETURN)
Select objects or Window or Last: SELECT
Rectangular/Polar array (R/P): P (RETURN)
Center point of array: SELECT
Number of items or -(degrees to fill): CHOOSE (RETURN)
Angle to fill (+ = CCW, - = CW)<360>: CHOOSE (RETURN)
Rotate objects as they are copied? <Y>: (RETURN)
```

In response to "Number of items," a positive reply will insert that number of items. A negative reply (-1 to -359) will fill that number of degrees with items at the specified angle.

BHATCH COMMAND. See HATCH COMMAND

BLOCK (V11, V12 AUTOCAD → BLOCKS → BLOCK: → BLOCK)
(V12, V13 Pd CONSTRUCT → BLOCK)
(V13 DRAW2 → INSERT → BLOCK)

Description

Used to capture a portion of the drawing for future use.
Note use of Wblock to save a Block to disk.
The Block is saved with the drawing as part of the drawing attributes.
An "insertion" or reference point is requested.
The Block name may be up to eight alphanumeric characters long.
Use Oops to recover entity after it is captured.

BLOCK Command Format

```
Command: BLOCK (RETURN)
Block name (or?): TYPE NAME (RETURN)
Insertion point: SELECT
Select object: CHOOSE or SELECT (RETURN)
```

The object becomes dotted. Pressing the Return key captures the object.

TIP: Use the Oops command to restore the blocked object back to the drawing.

BREAK (V11, V12 AUTOCAD → EDIT → BREAK: → BREAK MENU)
(V13 AUTOCAD → MODIFY → BREAK: → BREAK MENU)
V13 Pd MODIFY → BREAK: → BREAK MENU)

Description

Used to Break or remove part of a Line, Circle, Arc, Trace, or Polyline.
Object selection point becomes first break point.
May be used to trim a line.

BREAK Command Format

```
Command: BREAK (RETURN)
Select object: This selection point becomes the First BREAK point.
Enter second point (or F for first point): SELECT Second point.
```

The line between the two selected points is removed.

CHANGE (V11, V12 AUTOCAD → EDIT → CHANGE: → CHANGE) (V13 AUTOCAD → MODIFY → CHANGE: → CHANGE)

Description

Used to make changes to existing Lines, Circles, Text, Color, Linetypes, Thickness, and Layers.
Used to modify attribute definitions for Properties, Layer, and Elevation.
Object to be Changed may use "Select objects or Window or Last" format.

A Line Change: Given the line X, Y below, we'll connect it between X and Z.

```
Z                      Z1
X__________Y           X1__________Y1
```

```
Command: CHANGE (RETURN)
Select objects: SELECT the X end of the line
```

The line becomes highlighted. More lines may be selected if desired.

```
Select object: RETURN
Properties/<Change point>: SELECT the Z point.
```

The line is now drawn between X and Z. Let's try it again on X1 and Y1.

```
Command: CHANGE (RETURN)
Select object: SELECT Y1 END
Select object: RETURN
Properties/<Change point>: SELECT Z1 END
```

The line is now drawn between Y1 and Z1. The other end was changed!

TIP: The end taken as the "From Point" depends upon which end was closest to the original selection point.

Note: See Exercise 4-6C, Changing the Layer of an Inserted Block, and Exercise 7-3, "Preparing The PCB-1 Drawing For Plotting With Dashed Trace Lines In The Assembly Area." To Change the Layer of an object, type the following:

```
Command: CHANGE (RETURN)
Select object: SELECT the object for the layer change (RETURN)
Select object: (RETURN)
Properties/<Change point>: P  (RETURN)
Change what properties (Color/LAyer/LType/Thickness): LA (RETURN)
New layer <current>: CHOOSE/TYPE New layer name (RETURN)
```

The object now resides on the new layer.

CIRCLE (V11, V12 AUTOCAD → DRAW → CIRCLE)
(V13 AUTOCAD → DRAW → CIRCLE)
(V12, V13 Pd DRAW → CIRCLE: → CIRCLE MENU)

Description

There are six methods to draw a circle.

1. Center and Radius (Cen,Rad).
2. Center and Diameter (Cen,Dia).
3. Two-point Circles (2-Point).
4. Three-point Circles (3-Point).
5. Tangent, Tangent, Radius (tangent three-point circles).
6. Tangent circles.

Radius may be chosen by the keyboard numerical value, or by a screen displacement using the mouse.

Center and Radius Command Format (via Screen)

```
Command: CIRCLE (Select through Pd DRAW →CIRCLE)
Command: CEN,RAD (Select from screen menu)
3P/2P/TTR/<Center point>: SELECT (Center point on screen)
Diameter/<Radius>: (DRAG) (Drag with input device for desired circle
size)
```

Center and Diameter Command Format (via Screen)

```
Command: CIRCLE (Select through Pd DRAW →CIRCLE)
Command: CEN,DIA (Select from screen menu)
3P/2P/TTR/<Center point>: SELECT (Center point on screen)
Diameter/<Radius>: D (on keyboard)
Diameter): CHOOSE (number for diameter)
```

Two-Point Circle Command Format (via Keyboard)

```
Command: CIRCLE (Select through Pd DRAW → CIRCLE)
3P/2P/TTR/<Center point>: 2P (on keyboard)
First point on diameter: SELECT (on screen)
Second point on diameter: SELECT (on screen)
```

Three-Point Circle Command Format (via Keyboard)

```
Command: CIRCLE (Select through Pd DRAW →CIRCLE)
3P/2P/TTR/<Center point>: 3P (on keyboard)
First point: SELECT (on screen)
Second point: SELECT (on screen)
Third point: SELECT (on screen)
```

COLOR (V11, V12 AUTOCAD → SETTINGS → COLOR) (V13 AUTOCAD → DATA → COLOR: → COLOR MENU)

Description

Used to set the color of various entities on a layer.
Allows numerous colors on the same layer.
255 colors available, limited only by monitor.

COLOR Command Format

```
Command: COLOR (RETURN)
New entity color <default>: TYPE THE COLOR (RETURN)
```

If you type BYLAYER as the color, the default layer color is selected.

COPY (V11, V12 AUTOCAD → EDIT → COPY) (V13 AUTOCAD → CONSTRCT → COPY) (V13 Pd CONSTRUCT → COPY)

Description

Used to make a single or multiple copies of an object on the screen.
Objects are selected by pointing, or use of a Window or Last.
New location is selected by a displacement.

COPY Command Format

```
Command: COPY (RETURN)
Select objects or Window or Last: SELECT
<Base point or displacement>/multiple: SELECT
Second point or displacement: SELECT
```

This reproduces an exact copy of the original object.

TIP: It is preferable to select a point on the original object as a base point. This makes it easier to visualize the displacement of the object when copied.

COPY Command Format

```
Command: COPY (RETURN)
Select objects: SELECT
<Base point or displacement>/Multiple: M (RETURN)
Base point: SELECT OBJECT LOCATION
Second point of displacement: SELECT 1ST PT.
Second point of displacement: SELECT 2ND PT.
```

```
Second point of displacement: SELECT 3RD PT.
Second point of displacement: (RETURN)
```

The above commands duplicate the selected object three times at points 1, 2, and 3.

DDEDIT (V12 AUTOCAD → EDIT → MODIFY → EDIT TEXT Dialogue Box)
(V12 Pd MODIFY → ENTITY... → [Select Text] → MODIFY TEXT Dialogue Box)
(V13 AUTOCAD → MODIFY → DDEDIT:... → MODIFY TEXT Dialogue Box)

Description

Introduced in version 12, the Dynamic Dialogue Editor edits a single line of text or a single attribute using a dialogue box called the Edit Text dialogue box. See Fig.3-8b. When evoked from the pull-down menu through the Modify - Entity... menu, the Modify Text dialogue box appears, which allows many more attributes to be modified. See Fig. 3-8c.

DDEDIT Command Format

```
Command: DDEDIT (Return)
<Select a TEXT or ATTDEF object>/Undo: PICK text to modify.
```

After selecting the text to modify a dialogue box is displayed. Make your update and click on OK.

DDLMODES (V11 AUTOCAD → DDLMODES → MODIFY LAYER Dialogue Box)
(V12 AUTOCAD → LAYER: ... → LAYER CONTROL Dialogue Box)
(V12 Pd SETTINGS → LAYER CONTROL... → LAYER CONTROL Dialogue Box)
(V13 AUTOCAD → DATA → DDLMODES:)
(V13 Pd DATA → LAYERS... → LAYER CONTROL Dialogue Box)

Description

Since version 12 of AutoCAD, the Modify Layer dialogue box is replaced with the Layer Control dialogue box. It is evoked by either typing DDLMODES (Dynamic Dialogue Layer Modes) at the command prompt or by selecting

Layer... from the Data pull-down menu (version 13).
It is used to set up Layers.
A Layer is similar to a sheet of clear cellulose over the drawing.
Layer names should be short with no spaces.
Only one layer may be worked on at a time (current layer). See Status Line for active layer name.

DDLModes Options: New, Current, Select All, Clear All, Set Color, Set Ltype (linetype), Filters, Freeze, Thaw, Layer On/Off, Layer Lock/Unlock, Make. (See Fig. 4-9b and 4-9e).

Colors are selected by color or standard number:
Standard Colors are: # (1) red, (2) yellow, (3) green, (4) cyan, (5) blue, (6) magenta, (7) white, (8) Gray, (9) dark red
There are four Gray Shades: # (250), (251), (252), (253).
Full Color Palette contains 120 colors. (See Fig. 4-9d).
Colors are selected "By Layer" and/or "By Block".

DDLMODES Command Format

```
Command: DDLMODES (RETURN)
```

The Layer Control dialogue box appears. See Exercise 4-9a and Fig. 4-9b.

DIM (V11, V12 AUTOCAD → DIM:→ Pick Options)
(V12 Pd DRAW → DIMENSIONS → DIM: → Pick Options)
(V13 AUTOCAD → DRAWDIM:→ DIM MENU)
(V13 Pd DRAW → DIMENSIONING: → DRAWDIM:)

Description

In all AutoCAD versions up to and including version 12, the dimension mode is entered through the DIM command and is used to automatically dimension a drawing. Dimensioning methods may be Linear, Radial, Ordinate, Angular, and Leader. See options below.

DIM replaces the Command prompt in the dimension mode. To exxitthis mode, type EXIT or Ctrl-C. See Appendix E, Exercises 2a, 2b and 3.

Dimensioning in version 13 of AutoCAD has gone through a complete revision. The DIM prompt is no longer used. All dimensioning options are accessible from the command prompt and start with the DIM prefix. For example: Linear dimensioning is Dimlinear instead of Linear. This allows AutoCAD to make use of the various editing features that were locked out of the instruction set when the command prompt was DIM.

Although version 13 uses a new set of commands for dimensioning, the old set with the DIM prompt is still supported. Therefore, if you are comfortable with using the DIM and DIM1 commands, you can still make use of them.

DIM Command Format

```
Command: DIM or Select DIM from AUTOCAD Root Menu to allow option
menus to be displayed
DIM: (Note New prompt. Select options desired.)
```

To exit DIM mode, type:

```
DIM: EXIT or CTRL C (RETURN)
Command:
```

Note: The following are the dimension options available for the Linear, Radial and Ordinate methods:

Linear Options:	Radial Options:	Ordinate Options:
1. Horizontal	1. Diameter	1. Automatic
2. Vertical	2. Radius	2. X-Datum
3. Aligned	3. Center Mark	3. Y-Datum
4. Rotated		
5. Baseline		
6. Continue		

DIM1 (V11, V12 AUTOCAD → DIM: → DIM1 [Pick Options]) (V12 Pd DRAW → DIMENSIONS → DIM1 [Pick Options])

Description

DIM1 is a modification to the DIM command. It allows a quick way to execute a *single* dimension command. The command prompt is changed to DIM, but you do not have to EXIT to get back to the command prompt. This is done automatically upon completion of the command.

DIM1 Command Format

```
Command: DIM1 (Pick DIM1 from the AUTOCAD - DIM Root Menu to allow
the  DIM options to appear).
DIM: Pick dimension options.
Command:(The Command prompt reappears after the dimensioning is
complete.
```

DONUT (V11, V12 AUTOCAD → DRAW → DONUT)
(V13 AUTOCAD → DRAW → DONUT)
(V13 Pd DRAW → CIRCLE → DONUT)

Description

The Donut command creates either a solid circle (solid polyline) donut, the ones I call jelly filled donuts, or a ring-type donut with an open center, like a honey-dip donut. You can see the types of donuts I like.
Donuts are used to create land areas on printed circuit boards.
Solid donuts have an inside diameter of 0.
Donuts are repeatedly drawn until you press the Return or CTRL-C key.
Created by Donut or Doughnut command.

DONUT Command Format For A Solid Donut 3 Units Wide.

```
Command: DONUT (RETURN)
Inside diameter <default>: 0 (RETURN)
Outside diameter <default>: 3 (RETURN)
```

Using the mouse, click a number of solid donuts on the drawing. Press the Return key to terminate the command.

DONUT Command Format For A Ring Donut 3 Units Wide With A 1 Unit Center

```
Command: DONUT (RETURN)
Inside diameter <default>: 1 (RETURN)
Outside diameter <default>: 3 (RETURN)
```

DTEXT (V11, V12 AUTOCAD → DRAW → DTEXT)
(V13 AUTOCAD → DRAW2 → DTEXT:)
(V13 Pd DRAW → TEXT → TEXT/DYNAMIC TEXT)

Description

Dtext is similar to Text command, except that characters are displayed as the keys are pressed.
Justify "J" selection displays the justify options.

DTEXT Command Format

```
Command: DTEXT (RETURN)
Justify/Styles/<start point>: SELECT start point
Height<4.0000>: 2 (RETURN)
```

```
Rotation angle <0>: (RETURN)
Text: THIS IS DTEXT (RETURN)
```

The text "THIS IS DTEXT" appears as you type each letter.

END (V11 AUTOCAD → UTILITY → END)
(V12 AUTOCAD → UTILITY → (Next) → END)
(V12 Pd FILE → EXIT → AUTOCAD)
(V13 Not A Menu Command In Version 13, But Is Supported)

Description

Used to end a session of AutoCAD.
Saves a copy of the drawing and tags the extension as .DWG.
A new drawing name may be specified.
Creates a backup of the previous drawing, if you are editing, and tags the extension as .BAK.

END Command Format

```
Command: END (RETURN)
```

ERASE (V11, V12 AUTOCAD → EDIT → ERASE: → ERASE MENU)
(V13 AUTOCAD → MODIFY → ERASE: → ERASE MENU)
(V13 Pd MODIFY → ERASE: → ERASE MENU)

Description

Used to remove objects or lines from the drawing.
Entities are selected by Object, Window, or Last.
Object becomes dotted to highlight the selected item, and disappears when confirmed.
The Oops command redraws the selected item if a mistake is made.

ERASE Command Format

```
Command: ERASE (RETURN)
Select objects: CHOOSE (RETURN)
```

The selected object or objects become dotted when you pick your choice and disappear when the Return key is pressed to confirm the selection. See Appendix E, Exercise 1, A Practice Session Using Erase.

EXIT AUTOCAD (V12, V13 Pd FILES → EXIT AUTOCAD) (V13 AUTOCAD → FILE → EXIT)

Description

The Exit AutoCAD option under the Files pull-down menu will cause AutoCAD to exit. A Save Drawing Change dialogue box requester will appear if modifications were made to the drawing. If no changes were made, the exit is immediate. Used with versions 12 and 13.

EXIT Command Format

```
Command: From the FILES pull-down menu, Pick EXIT AUTOCAD.
```

AutoCAD ends.

EXIT DIM OPTION (DIM Command) (V11, V12 DIM: → Type EXIT) (V13 - Not A Command In Version 13, But Is Supported)

Description

The Exit option when used with the DIM (Dimension) Command will re-establish the Command Prompt back to "Command:" from the "DIM:" prompt. CTRL-C will also do the same thing.

TIP: Care should be used to assure you are at the DIM prompt when selecting Exit. When used at the Command prompt, "Exit" will exit AutoCAD completely.

EXIT (DIM Option) Command Format

```
DIM: EXIT or CTRL-C (Return)
Command: (The Command Prompt reappears)
```

EXPLODE (V11, V12 AUTOCAD → EDIT → EXPLODE) (V13 AUTOCAD → MODIFY → EXPLODE:) (V13 Pd MODIFY → EXPLODE)

Description

Used to disassemble entities once they have been inserted into a drawing.

EXPLODE Command Format

```
Command: EXPLODE (RETURN)
Select block reference, polyline or dimension:
```

TIP: Selection of the entity will cause it to be redrawn and exploded in the process. Also see Asterisk (*) option for the Insert command.

EXTEND (V11, V12 AUTOCAD → EDIT → (Next) → EXTEND) (V13 MODIFY → EXTEND: → EXTEND MENU) (V12, V13 Pd MODIFY → EXTEND)

Description

Extend is used to extend objects in a drawing to a given boundary line. You must first establish the boundary, and then select the object to be extended.
The extended line is drawn as an extension to an existing line, and thus is drawn straight, (not at an angle) to the boundary.

See Chapter 3, Exercise 3-6d for an in depth review of this command.

EXTEND Command Format

```
Command: EXTEND (RETURN)
Select boundary edge(s)...
Select objects: CLICK ON BOUNDARY LINE (RETURN)
Select objects to extend: CLICK ON OBJECT TO BE EXTENDED (RETURN)
Command:
```

Note the Return after each selection.

FILEDIA (V12, V13 Command Prompt: Type FILEDIA)

Description

The FILEDIA (Files Dialogue) command is a switch to allow the Standard File Dialogue Boxes to be active or not. When set to a 1, the dialogue boxes become active and visible. When set to a 0, they are switched off and are not visible. When the dialogue box is switched off, the Command line is used to input the data.

FILEDIA Command Format

```
Command: FILEDIA (Return)
New value for FILEDIA <1>: 0 (RETURN)
```

FILES (V10, V11 MAIN MENU, ITEM #6)
(V11, V12 AUTOCAD → UTILITY → FILES: → FILES MENU)
(V13 AUTOCAD → FILE → MANAGE → FILES: → FILES MENU)

Description

Used as a general file Utility.
Access gained via Main Menu, Root Menu, and Command line as FILES (Versions 10 and 11), or by typing FILES in versions 12 and 13.

FILES Command Format

```
Command: FILES (RETURN)
```

In versions 10 and 11, this evokes the File Utility Menu. To look at typical files on the A drive:

File Utility Menu

```
0. Exit File Utility Menu
1. List Drawing files
2. List user specified files
3. Delete files
4. Rename files
5. Copy files

Enter selection: 1 (RETURN)

Enter drive or directory: A: (RETURN)
```

The drawing files stored on the disk in drive A are displayed.

```
SCHEMA.DWG    SCHEMA1.DWG   TITLEBK.DWG
3 Files found
Press RETURN to continue: (RETURN)
```

We are now back at the File Utility Menu.

In versions 12 and 13, a File Utility Dialogue Box appears (Fig. 8-1).

FIGURE 8-1 File Utility dialogue box.

FILL (V11 AUTOCAD → DRAW → SOLID: → SOLID)
(V12, V12 AUTOCAD → DRAW → (Next) → SOLID: → SOLID → FILL)
(V13 AUTOCAD → OPTIONS → DISPLAY → FILL:)

Description

Used to fill solid objects, traces, and polylines, when turned ON.
May be turned OFF to save time when redrawing or regenerating a drawing.
In OFF mode, only the outline of the object is displayed, and subsequent polyline plots are drawn only as outlines.
To plot a drawing with solids and traces filled, regenerate the drawing before ending it. (See REGEN.)

FILL Command Format

```
Command: FILL (RETURN)
On/OFF <default>: CHOOSE (RETURN)
```

GRID (V11, V12 AUTOCAD → SETTINGS → GRID: → GRID)
(V11, V12 Pd SETTINGS → DRAWING AIDS... → DRAWING AIDS Dialogue Box)
(V13 AUTOCAD → SERVICE → DRAWING AIDS...→ DRAWING AIDS Dialogue Box)
(V13 Pd OPTIONS → DRAWING AIDS... → DRAWING AIDS Dialogue Box [Fig. 8-2})

Description

Establishes a Grid of dots over drawing area.
X and Y grid distances can be selected separately using Aspect "A" option.
May be toggled ON or OFF with the F7 function key.
At dense scale, may display "Grid too dense to display."

GRID Command Format (For Aspect)

```
Command: GRID (RETURN)
Grid spacing (X) or ON/OFF/Snap/Aspect<default>:  A (RETURN)
Horizontal spacing (X) <default>: CHOOSE (RETURN)
Vertical spacing (X)<default>: CHOOSE (RETURN)
```

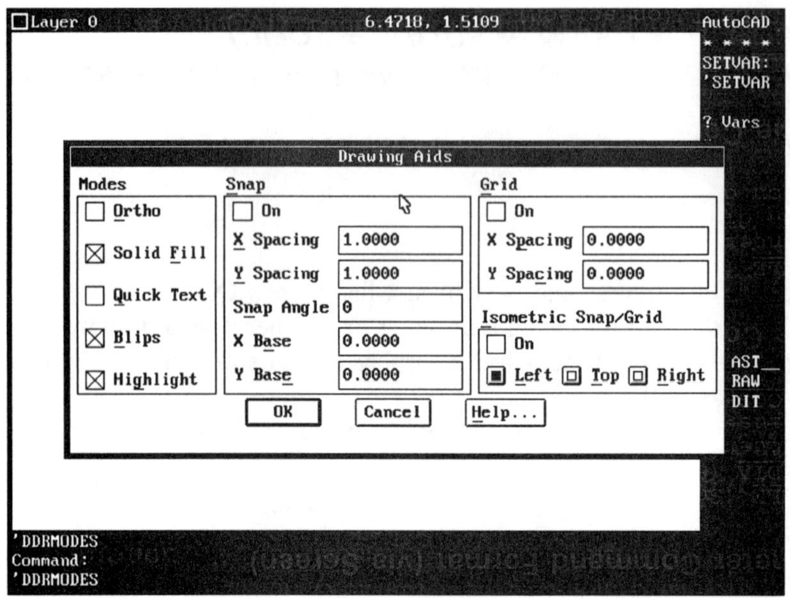

FIGURE 8-2 Versions 12 and 13 Drawing Aids dialogue box showing Grid/Snap.

GRIPS (V12, V13 COMMAND LINE OPTION → GRIPS)
(v12 Pd SETTINGS → GRIPS...→ GRIPS Dialogue Box)
(V13 Pd OPTIONS → GRIPS... → GRIPS Dialogue Box)

Description

Grips are small square areas on various portions of an entity that allow you to "grip" the entity and manipulate it, using such commands as Move, Stretch, Rotate, Scale, Mirror, Copy and Undo. Note Figure 8-3.
Grip boxes are usually at object snap locations.
Options include Grip Color, Grip Box size, Grip Block, and Grip Hot Points.
When Grip is active, the Grip Cursor contains a small square.
To capture an entity for Grip manipulation, either place a window around the entity using the mouse Grip Selection Box or just click on the desired entity.
Selecting the end point of a line will allow you to stretch or shrink the line.
Selecting the midpoint of a line will move the entire line instead of stretching it.
To see the Grip boxes, the DDGrip variable must be set to 1.
Grips are ideal to make modifications to printed circuit board trace lines as they allow you to move the polyline should it become too close to another line.

Note Exercise 7 in Appendix E for an in depth look at Grips.

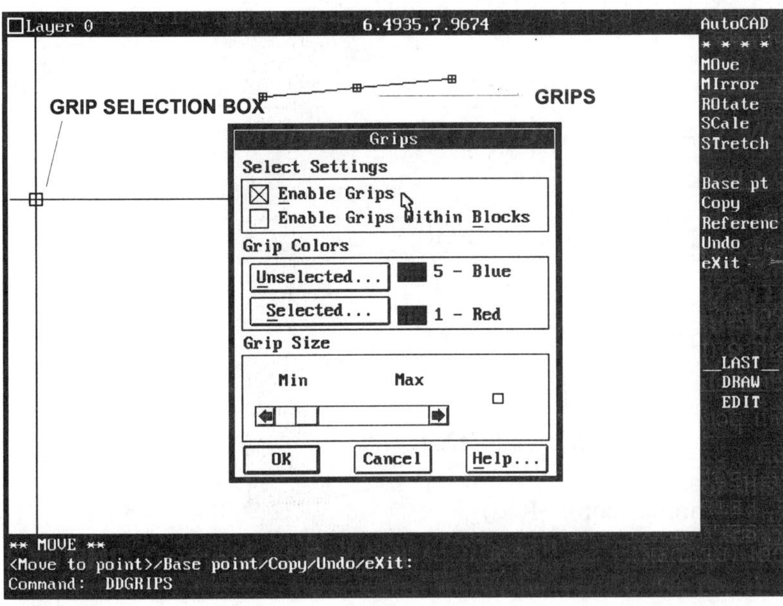

FIGURE 8-3 Grips dialogue box with Grip Selection Box and Grips.

GRIPS Command Format

```
Command: DDGRIPS (RETURN)
The GRIPS Dynamic Dialogue Box appears.
Enable GRIP editing by clicking on the Enable Grip Button, then click
on OK.
```

The cursor now has a small square around it, a Grip Selection Box. When this is displayed, objects can be selected in two ways. To use the first method, place the selection box at any point on the object and click the left mouse button. The object becomes highlighted and Grip boxes appear at the ends and in the middle of the object. The second method uses a window, developed by the Grip Selection Box, which is wrapped around the series of objects to be captured for inclusion as Grip entities.

To do this:

1. Place the Grip Selection Box outside the objects to be selected.
2. Hold the left mouse button down, and drag a window around the entities to be captured.
3. Click the left mouse button again to terminate the action.

When an entity has been captured (shown dotted), clicking on the entity again, either at the end or in the middle of it, will bring up the Grip options of:

STRETCH
<Stretch to point>/Base point/Copy/Undo/eXit:

MOVE
<Move to point>/Base point/Copy/Undo/eXit:

ROTATE
<Rotate angle>/Base point/Copy/Undo/eXit:

SCALE
<Scale factor>/Base point/Copy/Undo/eXit:

MIRROR
<Second point>/Base point/Copy/Undo/eXit:

The end clicked on turns another color (Red).

TIP: Pressing the space bar will cycle through these options.

Using the MOVE option, to move the end of a line, click the Grip Selection Box on the end Grip and drag the line where you want it.

This will bring you back to the command line.

HATCH (V11, V12 AUTOCAD → DRAW → HATCH)
(V12 Pd DRAW → HATCH/BHATCH... → BOUNDARY HATCH Dialogue Box)
(V13 AUTOCAD → CONSTRCT → BHATCH:)
(V13 Pd DRAW → HATCH: → HATCH... → HATCH Dialogue Box)

Description

Used to fill an area with a pattern representative of some material.
Sixty-seven patterns are available (version 13) plus user definable patterns. See Appendix H.
Hatch options are (N)ormal, (O)utermost, and (I)gnore.
Hatch area is treated as one entity.

A hatched pattern must be placed within the confines of an object. To demonstrate this, draw a circle as our object, then type the following:

HATCH Command Format

```
Command: HATCH (RETURN)
Pattern(? or name/U,style)<default>:  ANSI31,O (RETURN)
Scale for pattern <1.0000>: 10 (RETURN)
Angle for pattern <0>: (RETURN)
Select objects or Window or Last: W (RETURN)
```

Note the O after the ANSI31 to select the Outermost option.

Placing a window around the outside of the circle will cause it to become hatched on the inside area. See Exercise 5-2, Hatching PCB-1, for more detail.

BHATCH COMMAND (V13 AUTOCAD → CONSTRCT → BHATCH:)

Description

Bhatch is used to select specific boundary areas by using the cursor pick box. Allows selection of irregular shapes.

BHATCH Command Format (Boundary Hatch)

```
Command: SELECT DRAW HATCH or BHATCH
```

The Boundary Hatch dialogue box appears (Fig. 8-4).

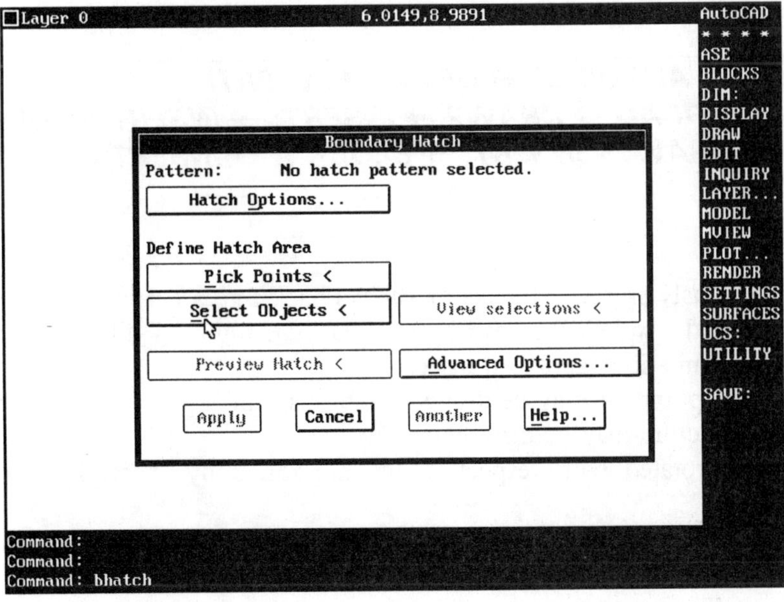

FIGURE 8-4 Boundary Hatch dialogue box.

Select a hatch pattern, then click on the Select Objects button.
The dialogue box leaves the screen.
Use the cursor "pick box" to select each line of the boundary area to be hatched.
Next, click on the Apply Button to complete the hatch.
Press Return to continue.

HELP or ? (V11, V12 → INQUIRY → HELP)
*(V11,V12 * * * * → HELP)*
(V13 AUTOCAD → HELP... → HELP MENU)
(V13 Pd HELP... → HELP MENU)

Description

Used to provide Help with AutoCAD commands.
Displays a list of Help commands available.

HELP Command Format

```
Command: HELP (RETURN)
Command name (Return for list): CHOOSE (RETURN)
```

If you reply with a command name, you will get a help message on that command; otherwise a list of valid commands is displayed. To cancel the Help display, enter CTRL-C.

INSERT (V11, V12 AUTOCAD → DRAW → INSERT)
(V13 AUTOCAD → DRAW2 → INSERT: → INSERT MENU)
(V13 Pd DRAW → INSERT: → BLOCK... → INSERT MENU)

Description

Used to place previously stored Wblocks or Blocks back into a drawing.
May be specially edited (Exploded) as part of drawing by using the (*) before the Block name; some limitations apply on insert.
Requires an "insert" point, or may be dragged into position.
The X and Y scale factors may be different ratios.
The block may be rotated with respect to the horizontal by specifying the Rotation Angle.
A reply with (?) to the Block name will list all blocks in the block storage area.

INSERT Command Format

```
Command: INSERT (RETURN)
```

```
Block name (or ? ): CHOOSE BLOCK NAME  (RETURN)
Insertion point: SELECT
X scale factor <1>/Corner/XYZ: CHOOSE (RETURN)
Y scale factor <default=X>: CHOOSE (RETURN)
Rotation angle: CHOOSE (RETURN)
```

Asterisk Option (*)

Using the asterisk (*) before the Block/Wblock name will break (Explode) the block into its basic parts. This will allow you to edit the entity after insertion.

```
Block name <or ?>: *TITLBK1
```

Also see the Explode command.

INSERTING BLOCKS through the DDINSERT Dialogue Box

To evoke the DDINSERT dialogue box, type the following:

```
Command: DDINSERT (RETURN)
```

The Insert dialogue box appears (Figure 8-5).

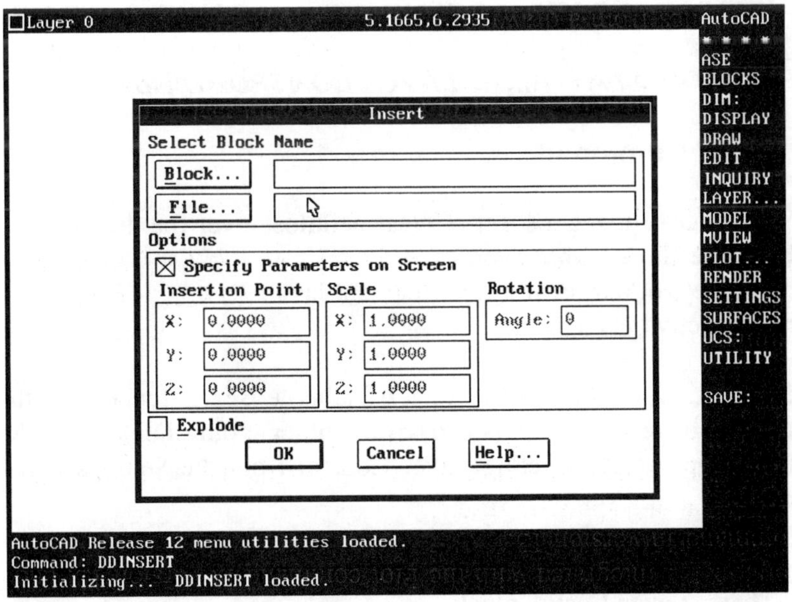

FIGURE 8-5 Insert dialogue box.

From this dialogue box, you can insert either a Block or a Wblock. Clicking on the *Block...* button will insert blocks and clicking on the *File...* button will insert selected drawing files (Wblocks).

Clicking the Block... button evokes the Define Blocks dialogue box, as well as gives a listing of blocks that are used with the drawing. From this list you can select a block for insertion by highlighting it and then clicking OK. Scale factors and rotation of the block can also be selected at the time of insertion through the Specify Parameters On Screen options button.

The above procedure is also followed when you insert a Wblock file using the Select Drawing File dialogue box.

TIP: When inserting Block and Wblock Drawing Files, they can be exploded by clicking the Explode button of the Insert dialogue box (Fig. 8-5).

LAYER ***(V11 AUTOCAD → LAYER: → LAYER MENU)***
(V11 AUTOCAD → DDLMODES → MODIFY LAYER Dialogue Box)
(V11 Pd SETTINGS → MODIFY LAYER... → MODIFY LAYER Dialogue Box)
(V12 AUTOCAD → LAYER:... → Layer Control Dialogue Box)
(V12 Pd SETTINGS → LAYER CONTROL... → Layer Control Dialogue Box)
(V13 AUTOCAD → DATA → DDLMODES... → LAYER CONTROL Dialogue Box)
(V13 Pd DATA → LAYER... → LAYER/CONTROL Dialogue Box)

Description For Layer Command

Using a Layer is similar to using a sheet of clear cellulose over the drawing.
Layer names should be short with no spaces.
Only one layer may be worked on at a time (current layer).
Note Status Line for active layer name.

From AutoCAD version 12 on, the Modify Layer Dialogue Box is replaced with the Layer Control Dialogue Box. This is evoked by either typing DDLMODES at the command prompt or by selecting Layer Control from the Settings pull-down menu.

DDLModes Options: ?, Make, Set, New, On, Off, Color, Ltype (linetype), Filters, Freeze, Thaw, Layer Lock/Unlock.

Colors are selected by color or standard number:
Standard Colors are: # (1) red, (2) yellow, (3) green, (4) cyan, (5) blue, (6) magenta, (7) white, (8) Gray, (9) dark red
There are four Gray Shades: # (250), (251), (252), (253).
Full Color Palette contains 255 colors, limited only by monitor.

Colors are selected "By Layer" and/or "By Block".

Since AutoCAD version 2.5, you can set a new color for each entity on a layer, in addition to the basic layer color. "Color" in this case is a command, not a layer option. Note the Color Command for more details.

LAYER Command Format

To Set a layer current, type the following:

```
Command: LAYER (RETURN)
?/Set/New/On/OFF/Color/Ltype/Freeze/Thaw: S (RETURN)
New current Layer <default>: TYPE NEW LAYER NAME (RETURN)
```

Note: See Exercise 4-5 in Chapter 4 for a detailed review of Layers.

LIMITS (V11, V12 AUTOCAD → SETTINGS → LIMITS: → LIMITS) (V13 AUTOCAD → DATA → LIMITS: → LIMITS MENU)

Description

Determines the outer framework of the drawing.
Should be part of the initialization procedure for the drawing.
Sets the boundaries which Grid and Zoom will display.
May be toggled ON and OFF via Limits command prompt.
Lower left corner default = 0,0, and may be set negative.
Upper right corner is usually set to outer edge of sheet size.
May be changed at any time.
Think of Limits as the sheet size you are working on.

LIMITS Command Format

```
Command: LIMITS (RETURN)
On/OFF/Lower left corner <default>: CHOOSE (RETURN)
Upper right corner <default>: CHOOSE (RETURN)
```

LINE (V11, V12 AUTOCAD → DRAW → LINE: → LINE MENU) (V13 AUTOCAD → DRAW → LINE: → LINE MENU) (V13 Pd DRAW → LINE:)

Description

Used to draw a line or lines.
Drawing aids used with Line are Close, Continue, Undo, and Cancel.

Placement of Lines is most easily done through input device.
Placement of Lines via the keyboard is done through cursor control keys.

LINE Command Format

```
Command: LINE (RETURN)
Line From point: SELECT
To point: SELECT
To point: SELECT
To point: (RETURN) Terminates line command.
```

For a detailed look at the Line command, review Exercise 3-3.

LINETYPE (V11, V12 AUTOCAD → SETTINGS → LINETYPE: → LINETYPE) (V13 AUTOCAD → DATA → DDLTYPE) (V13 Pd OPTIONS → LINETYPE: → LINETYPE MENU)

Description

Used to select various line styles.
Linetype options: ?, Load, Create, Set.
A. *?* provides a list of linetypes available.
B. *Load* loads the selected linetype into memory.
C. *Create* allows you to create a custom linetype.
D. *Set* sets a line to a specified linetype.

*Note:*To get back to the linetype as specified for the layer, set the linetype to the BYLAYER option.

The linetype scale is adjusted by LTScale. See LTScale command.
To find the default linetypes, search the ACAD File. For custom linetypes, specify the file under which you saved the new linetype.

LINETYPE Command Format

```
Command: LINETYPE (RETURN)
?/Create/Load/Set: L (RETURN)
Name of linetype to load: DASHED (RETURN)
File to search: ACAD (RETURN)
```

Now you can specify the Dashed line as a linetype on the Layer Control dialogue box menu. See Appendix A-2 for linetype.

LTSCALE (V11, V12 AUTOCAD → SETTINGS → (Next) → LTSCALE)
(V13 AUTOCAD → OPTIONS: → DISPLAY → LTSCALE)

Description

Used to change the scale of a linetype. This increases the space between dashed lines.

LTSCALE Command Format

```
Command: LTSCALE (RETURN)
New scale factor <default>: TYPE NEW FACTOR (RETURN)
```

Entering numbers greater than one increases the line segments. Entering numbers less than one shortens the line segment.

MIRROR (V11, V12 AUTOCAD → EDIT → [Next] → MIRROR)
(V13 AUTOCAD → CONSTRCT → MIRROR: → MIRROR)
(V13 Pd CONSTRUCT → MIRROR)

Description

Used to make a mirror image of an object.
The mirror image is made about a mirror line.
The mirror line must be vertical or horizontal.
Original object may be deleted or kept.
Objects to be mirrored may be selected or Window or Last.

MIRROR Command Format

```
Command: MIRROR (RETURN)
Select objects or Window or Last: SELECT
First point of mirror line: SELECT
Second point: SELECT
Delete old objects? <N>: Y or N (RETURN)
```

See Exercise 5-7, in Chapter 5 for a detailed view of the mirror command.

MOVE (V11, V12 AUTOCAD → EDIT → [Next] → MOVE) (V13 AUTOCAD → MODIFY → MOVE: → MOVE) (V13 Pd MODIFY → MOVE:)

Description

Allows you to move objects to another area of the drawing.
Objects are selected by pointing, or by a Window or Last.
Distance moved is given by a displacement.

MOVE Command Format

```
Command: MOVE (RETURN)
Select objects : SELECT (PICK OBJECT)
Base point of displacement: SELECT
Second point of displacement: SELECT
```

TIP: It is preferable to select a point on the object as your Base point. This makes it easier to visualize where the object is being moved to.

MTEXT (V13 DRAW2 → MTEXT:)

Description

Mtext is used to create multiple lines of text (paragraphs) as compared to the Text and Dtext commands which create only a single line of text.
Mtext uses an external text editor to establish the text such as the one found with DOS Shell. After the text is formatted by the editor, leaving the editor brings you back to the Mtext command where it is inserted into the AutoCAD drawing.
The Mtext command prompts you to pick two points on the screen to form a rectangle where the text will be located after it is formatted by the editor.

MTEXT Command Format

```
Command: MTEXT (RETURN)
Attach/Rotate/Style/Height/Direction/ <Insertion point>: SELECT 1ST
POINT
Attach/Rotate/Styles/Height/Direction/Width/2Point/<Other corner>:
SELECT 2ND CORNER
```

DOS Shell Editor now appears. See Fig. 3-9d, on page 81.
Format your text, save the file and exit the editor.
Upon exiting the editor, AutoCAD places the text in the rectangle you specified earlier.

NEW (V12, V13 Pd FILES → NEW)
(V12 AUTOCAD → UTILITY → (Next) → NEW)
(V13 AUTOCAD → FILE → NEW)

Description

AutoCAD versions 12 and 13 use a Windows approach to evoke the drawing, thus making the start-up procedure look similar to the standard format used by most Windows software developers. The standard Main Menu that is familiar to AutoCAD users no longer exists from version 12 on.

When ACAD is initiated, or the ACADR12/ACADR13 batch file command is typed, AutoCAD immediately displays the Drawing Editor Screen. To start a New drawing, pick New from the Files pull-down menu, and type the drawing name into the Create New Drawing dialogue box. See Fig 3-2. Typing New at the command line will also bring up the Create New Drawing dialogue box.

NEW Command Format

```
Command: New (RETURN)
The Create New Drawing Dialogue Box appears. See Fig.3-2
Type in the drawing name and Pick OK.
```

OOPS (V11, V12 AUTOCAD → EDIT → ERASE → OOPS)
(V12 Pd MODIFY → ERASE → OOPS)
(V13 AUTOCAD → MODIFY → OOPS:)
(V13 Pd MODIFY → OOPS:)

Description

Oops restores entities that were erased from the drawing.
You should use it immediately after erasing an entity to be sure of the recovery of the entity. You *may* be able to Oops back an entity after one or more AutoCAD commands have been issued followed the Erase command, but there is no assurance this will always work.

OOPS Command Format

```
Command: OOPS (RETURN)
```

OPEN (V12 AUTOCAD → UTILITY → (Next) → OPEN)
(V12, V13 Pd FILE → OPEN)
(V13 AUTOCAD → FILE → OPEN)

Description

The Open command is used to Open an existing AutoCAD file either by means of a dialogue box or from the command line.
For the dialogue box to appear, the FILEDIA variable must be set to 1. (See FILEDIA.)
Open performs the same function as "Edit an Existing Drawing", Item #2 of the Main Menu of AutoCAD versions 10 and 11.

OPEN Command Format

```
Command: OPEN (RETURN)
Enter name of drawing: CHOOSE/TYPE IN NAME (RETURN)
```

OSNAP (V11 AUTOCAD → SETTINGS → [Next] → OSNAP → OSNAP)
*(V11, V12 * * * * → OSNAP MENU)*
*(V13 * * * * DDOSNAP → RUNNING OBJECT SNAP BOX)*
(V13 AUTOCAD → ASSIST → DDOSNAP → OSNAP MENU)
(V13 Pd OPTIONS RUNNING OBJECT OSNAP... OSNAP MENU)

Description

Object Snap is similar to Snap, except that it locks onto objects that may not be on a specific grid point, such as the outer part of a circle.
OSNAP uses a small window or aperture at the cursor cross-hairs to capture the object in question.
OSNAP options may be turned On or Off.
The Aperture Command sets the size of the capture window. See the command below.
OSNAP allows you to capture an object by: CENter, ENDpoint, INSert, INTersec, MIDpoint, NEArest, NODe, PERpend, QUAdrant, QUICK, or TANgent.
Use OSNAP in conjunction with the draw commands such as Line, Circle, and the like.
See Exercise 4-4b, Starting Our Schematic Drawing.

OSNAP Command Format

A line X is drawn on a page with no reference to grid marks. A second line Y will now be joined to the end of line X by using the OSNAP "ENDpoint" option.

```
Command: LINE (RETURN)
From point: END (RETURN)
```

OSNAP is evoked by typing END while in the Line command; an aperture window (Fig. 8-6) appears at the end of the cursor cross-hairs. Place the aperture window on the end of line X and press the input device select button. The new Y line snaps to the end of line X. It has been captured byOSNAP! Now to terminate the line.

```
To point: SELECT A TERMINATION POINT
To point: (RETURN)
Command:
```

APERTURE Command Format

```
Command: APERTURE (RETURN)
Object snap target size (1 - 50 pixels)<default>: CHOOSE (RETURN)
```

Choose an aperture size that is workable with the limits and grid size of your drawing.
When the drawing is re-entered, the last aperture set is the new default value.

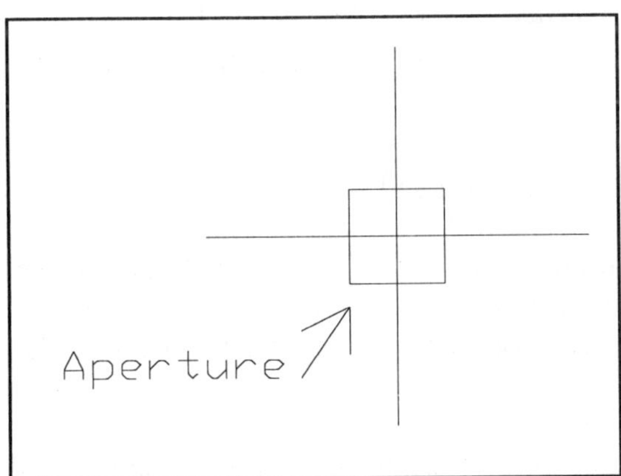

Figure 8-6 OSNAP aperture window.

PAN (V11, V12 AUTOCAD → DISPLAY → PAN:)
(V13 AUTOCAD → VIEW → PAN: → PAN MENU)
(V13 Pd VIEW → PAN: → PAN MENU)

Description

Pan is used while in the Zoom mode to move to a different part of the drawing, by dragging the zoom window across the drawing screen.
A displacement is required either through the keyboard in X, Y units or via the input device.

PAN Command Format

```
Command: PAN (RETURN)
Displacement: CHOOSE or SELECT
Second point:  SELECT
```

PRINT (V13 - See PLOT)

PLINE (V11, V12 AUTOCAD → DRAW → PLINE)
(V13, AUTOCAD → DRAW → PLINE)
(V13 Pd DRAW → POLYLINE)

Description

Polyline or Pline is used to draw variable-width, solid lines.
Options include Arc, Close, Halfwidth, Length, Undo, and Width.
The start width may be different from the end width.
Fill On/Off will make the Pline a solid or an outline.
The width of the line is drawn about its midpoint.

PLINE Command Format

To set the Pline line width to two units, type the following:

```
Command: PLINE (RETURN)
From point: SELECT A POINT
Current line width <0.0000>Arc/Close/Halfwidth/Length/Undo/Width/
 <endpoint of line>: W (RETURN)
Starting width <0.0000>: 2 (RETURN)
End width <2.0000>: (RETURN)
```

See Exercises 5-5, 5-5A, and 5-5B, for more details on Pline.

PLOT (V11 Through MAIN MENU, item #3)
(V11, V12 AUTOCAD → PLOT...)
(V12 Pd FILES → PLOT)
(V13 AUTOCAD → FILE → PRINT... → Plot Configuration Dialogue Box)
(V13 Pd FILE → PRINT... → Plot Configuration Dialogue Box)

Description (Versions 10 and 11)

In versions 10 and 11, used to start a "pen" plot.
Options for plot: Display, Extents, Limits, View, or Window.
Plot displays the default setup values of the plotter, and then asks if there are changes.
Be sure to turn Fill on and regenerate drawing if objects are to be plotted as solids.

Description (Versions 12 and 13)

In versions 12 and 13, Plot will evoke the Plot Configuration dialogue box for pen or printer plots.
In version 13, the Plot command is supported, but shows up in File Menu as the Print command.

PLOT/PRINT Command Format

```
Command: PLOT/PRINT (RETURN)
```

AutoCAD clears the screen and displays the plot options, or the Plot Configuration Dialogue Box.

```
Specify the part of the drawing to be plotted by entering:
Display, Extents, Limits, View, or Window <D>: L (RETURN)
```

AutoCAD now displays the default setup and checks for changes. If there are none, you are asked to insert the paper and set up the plotter. Pressing the Return key then starts the plotting process. From the dialogue box, click on OK when you have selected the appropriate options. See Chapter 7 - Plotting.

PRPLOT Also accessed through MAIN MENU, item #4.
(V10, V11 AUTOCAD → PLOT: → PLOT MENU)
(V12, V13 AUTOCAD. Not available; use PLOT/PRINT)

Description

Used to start a "printer" plot.
Note description under pen plot.
Not used after version 11. Integrated with the Plot command in version 12, and with the Print command in version 13.

PRPLOT Command Format

```
Command: PRPLOT (RETURN)
```

A menu of items similar to that of "plot" appears on the screen. For a detailed description of Plot and PRPlot, see Chapter 7.

PURGE (V12 AUTOCAD → UTILITY → PURGE: → PURGE MENU)
(V13 AUTOCAD → DATA → PURGE: → PURGE MENU)

Description

Used to Purge object files from a drawing.
Options include purging of Blocks, Layers, Dimension Styles, Linetypes, Shapes, Styles, All objects.
Must be the first command used after opening the drawing.
You cannot Purge the layer O, the Continuous Linetype, or the Standard text style.

PURGE Command Format

```
Command: PURGE (RETURN)
Purge unused Blocks/Dimstyles/LAyers/LTypes/SHapes/STyles/All: CHOOSE
(RETURN)
```

AutoCAD next steps through the selected area for the various items to Purge.

QSAVE (V12, V13 AUTOCAD → UTILITY → [Next] → QSAVE)

Description

Qsave (Quick Save) is similar to the Save command, except the dialogue box is not displayed. The drawing is immediately saved to disk under the name given at start up.

QSAVE Command Format

```
Command: QSAVE (RETURN)
```

QTEXT (V11, V12 AUTOCAD → SETTINGS → [Next] → QTEXT:) (V13 AUTOCAD → OPTIONS: → DISPLAY → QTEXT:) (V13 Pd SETTINGS → DRAWING MODES... → DRAWING AIDS Dialogue Box)

Description

Used to speed up drawing regeneration by replacing text in a rectangular box.
Toggled ON and OFF.
Requires a REGEN to view text after being turned ON.
Qtext can be set from the Drawing Aids dialogue box. Also selected from the DDRModes.

QTEXT Command Format

```
Command: QTEXT (RETURN)
On/OFF <default>: CHOOSE (RETURN)
```

QUIT (V11 AUTOCAD → UTILITY → QUIT) (V12 AUTOCAD → UTILITY → (Next) → QUIT) (V13 See EXIT)

Description

Used to exit a drawing when nothing is to be saved. The drawing remains as it was when you entered it.
All changes will be disregarded.
In versions 12 and 13, Quit is supported at the command line but is replaced by Exit in the file pull-down menu.

QUIT Command Format

```
Command: QUIT (RETURN)
Really want to discard all changes to drawing?: Y (RETURN)
```

In versions 12 and 13, the Drawing Modification dialogue box appears.

REDO. See UNDO

REDRAW (V11, V12 AUTOCAD → DISPLAY → REDRAW)
(V12 AUTOCAD → VIEW → REDRAW)
(V13 Pd VIEW → REDRAW VIEW)

Description

Redraws the screen and its entities.
Used to clean up the display after editing.
Does *not* regenerate the display; merely "redraws" it.
Turning the Grid On and Off (F7) takes similar action to Redraw command.

REDRAW Command Format

```
Command: REDRAW (RETURN)
```

REGEN (V11, V12 AUTOCAD → DISPLAY → REGEN)
(V13 AUTOCAD → TOOLS → SHADE → REGEN)

Description

Regenerates the drawing Data Base.
Recalculates line end points, hatching.
CTRL-C will cancel a Regeneration.
Useful to bring drawing up to date.

REGEN Command Format

```
Command: REGEN (RETURN)
```

SAVE (V11, V12 AUTOCAD → SAVE)
(V12 Pd FILES)
(V13 AUTOCAD → FILE → SAVE)

Description

Used to "back up" your drawing without exiting the drawing editor.
Allows drawing to be saved under a different file name from that presently being worked on.

Include disk drive suffix if the drawing is being saved under a new name.

SAVE Command Format

```
Command: SAVE (RETURN)
File name <default>: A:NAME (RETURN)
```

After the drawing is saved, the Command line appears.

SAVEAS (V12 AUTOCAD → UTILITY → [Next] → SAVEAS:) (V12 Pd FILE → SAVEAS:) (V13 AUTOCAD → FILE → SAVEAS: → SAVEAS Dialogue Box)

Description

Saveas allows you to establish a new drawing file name or change the existing drawing file name. If the new file name already exists, AutoCAD warns you and asks for confirmation to overwrite the old file.
The Save Drawing As dialogue box appears.

SAVEAS Command Format

```
Command: SAVEAS (RETURN)
SAVE DRAWING AS Dialogue Box Appears: TYPE in name and drive to save
file to, then click OK
```

SAVETIME (V12, V13 AUTOCAD → OPTIONS → SAVETIME) (V13 Pd OPTIONS → AUTO SAVE TIME)

Description

Used to set the time interval which causes AutoCAD to automatically save the drawing.
When FILEDIA is set to 1, the SAVE TIME dialogue box will appear.
If time is set to 0, the command is disabled.
Time is reset after a Save, Saveas, or Qsave.

SAVETIME Command Format

```
Command: SAVETIME (RETURN)
The Savetime Dialogue Box appears.
Click on the desired interval, Two, Five, Ten or Disable, or type in
desired time.
Click on OK to exit.
```

SHELL (V11, V12 AUTOCAD → UTILITY → EXTERNAL COMMANDS → SHELL) (V13 Pd TOOLS → EXTERNAL → SHELL)

Description

Shell lets you execute a DOS command while still in AutoCAD.
A short form of Shell is the SH command.
To return to AutoCAD from DOS, type EXIT.
Do not evoke CHKDSK/F through Shell, because AutoCAD keeps *open files* and CHKDSK will close them, getting you into one real mess.
Requires the DOS COMMAND.COM file in the default drive.
You may require a Redraw to put your drawing back in order after a Shell command is issued.

SHELL Command Format (Single DOS Command)

To get a Directory of the A disk, type:

```
Command: SHELL (RETURN)
DOS command: DIR A: (RETURN)
```

A listing of the files in the A drive are displayed, then you are immediately returned to AutoCAD.

SHELL Command Format (Multiple DOS Commands)

To use multiple DOS commands, press the Return key after the "DOS command" prompt.

```
Command: SHELL (RETURN)
DOS command: (RETURN)
C:>>
```

TIP: The DOS prompt is followed with two ">>" identifiers to signify you are in a DOS Shell operation.

At the C:>> prompt you can execute multiple DOS commands. To get back to AutoCAD, type Exit.

With large AutoCAD files, there may not be enough computer RAM memory to evoke the SHELL command. In this case you will get the message:

```
Insufficient memory for SHELL command
```

And the AutoCAD Command line reappears.

TRAP: Be careful not to erase any files which AutoCAD may have open when you look at your AutoCAD directory. AutoCAD stores a number of temporary files on your work disk which you will see when the directory command is issued. These files usually have the AC$, $AC or $A extension. If these files are removed while you are still using AutoCAD, it will crash either when you exit AutoCAD or when you exit DOS Shell. Again, you will be in a real mess.

SNAP ***(V11, V12 AUTOCAD → SETTINGS → (Next) → SNAP: → SNAP MENU)***
(V12 Pd SETTINGS → DRAWING AIDS... → DRAWING AIDS Dialogue Box)
(V13 AUTOCAD → SERVICE → DRAWING AIDS...→ DRAWING AIDS Dialogue Box)
(V13 Pd OPTIONS → DRAWING AIDS... → DRAWING AIDS Dialogue Box)

Description

Establishes an invisible grid over the drawing to allow the screen cursor to align itself to known reference points.
Snap grid is independent of the grid pattern established by the Grid command.
DDOSNAP invokes the Running Object Snap dialogue box, see page 133.
Aspect (A) option allows independent X and Y snap locations.
May be toggled ON and OFF with F9 Function key.
When FILEDIA is set (1), the Drawing Aids dialogue box appears for easy snap selection.

SNAP Command Format (Using Aspect)

```
Command: SNAP (RETURN)
Snap spacing or On/OFF/Aspect/Rotate/Style:  A (RETURN)
Horizontal spacing <default>: CHOOSE (RETURN)
Vertical spacing <default>: CHOOSE (RETURN)
```

Comments

Snap command On/Off toggles snap on and off.
Snap spacing is a numeric input that establishes the placement resolution.
Aspect allows different X and Y resolution values.
Rotate establishes a baseline and rotation angle from which the grid is drawn.
Style selects a grid display of “orthographic/standard” (90 - 90 degree pattern)

or "isometric" (30-90-150 degree pattern).
See Figure 8-2, the Drawing Aids dialogue box, which shows the Snap and Grid controls.

SPELLING (V13 AUTOCAD → TOOLS → SPELLING) (V13 Pd TOOLS → SPELLING)

Description

New in version 13.
Used to check the spelling of text.
Similar to spelling check programs found in Windows applications.
Has several foreign language directories.
Will check spelling of text created by Text, Dtext and Mtext commands.

SPELLING Command Format

```
Command: SPELLING (RETURN)
Select objects: Pick desired text to verify spelling.
```

When the spell checker finds a word it does not recognize, the Check Spelling dialogue box appears. Spell checker suggests alternate words, allowing you to click on your desired replacement and the Change Button to replace it. If many occurrences of the misspelled word are in your text, the ALL option allows you to change every occurrence at one time.

Options include:
Ignore - Skip the word.
Ignore All - Skip all occurrences.
Add - Add word to the dictionary.
Lookup - Look for another word.
Change Directory - Look in another Spell Check directory.
Context - Displays the phrase in which the word was originally found.

STATUS (V11, V12 AUTOCAD → INQUIRY → STATUS: → STATUS MENU) (V13 AUTOCAD → DATA → STATUS: → STATUS MENU)

Description

Used to display the various default values presently set up in AutoCAD.

STATUS Command Format

```
Command: STATUS (RETURN)
```

Text, similar to the following, is now displayed:

```
    93 entities in A:ASSEM1C

Model space limits are X:    0.0000  Y: 800.0000 (Off) (World)
                       Y:    0.0000  Y: 500.0000
Model space uses       X:   90.0000  Y: 710.0000
                       Y:  350.0000  Y: 447.0000
Display shows          X:    0.0000  Y: 800.0000
                       Y:    0.0000  Y: 617.9757

Insertion base is      X:    0.0000  Y: 0.0000   Z: 0.0000
Snap resolution is     X:    1.0000  Y: 1.0000
Grid spacing is        X:    1.0000  Y: 1.0000

Current Space:       Model space

Current layer:       TEXT1
Current Color:       BYLAYER  - 7 (white)
Current Linetype:    BYLAYER  - CONTINUOUS

Current elevation:   0.0000     thickness: 0.0000
Fill on   Grid on    Ortho off Qtext off Snap on   Tablet off
Object snap modes:   Endpoint, Midpoint

Free disk: 23748592 bytes
Virtual memory allocated to program: 1826K

Amount of program in physical memory/Total (virtual) program
  size: 29%
Total conventional memory: 316K   Total extension memory: 376K
Swap file size: 388KB
Page faults: 374  Swap Writes: 137  Swap reclaims: 99
```

STRETCH (V12 AUTOCAD → EDIT → [Next] → STRETCH) *(V13 MODIFY → STRETCH)* *(V12, V13 Pd MODIFY → STRETCH)*

Description

Selected objects may be moved using the Stretch command. Entities attached to the object being moved will tag along for the ride, similar to stretching a rubber band (the rubber-band effect).

Object selection requires the "Crossing" option C, or the Window option.

Lines, Arcs, Solids, Traces, and Polylines may be stretched.

When Line, Arc and Polyline segments are chosen using the "Crossing" option, the endpoints contained within the window are moved, while those outside the window remain fixed.

Objects entirely enclosed by a Window are moved similar to the Move command.

If a window is not used in object selection, AutoCAD displays the following message: *YOU MUST SELECT A WINDOW TO STRETCH.*

STRETCH Command Format

```
Command: STRETCH (RETURN)
Select objects to stretch by window or polyline...
Select object: C (RETURN)
First corner: WRAP A WINDOW AROUND OBJECT TO STRETCH
Second corner: COMPLETE WINDOW SELECTION (RETURN)
Base point: SELECT
New point: SELECT
Command:
```

Note: For more detailed information on the Stretch command, refer to Appendix - E, Exercise 7.

STYLES (V12 AUTOCAD → DRAW → TEXT/DTEXT → STYLE: SELECT → TEXT Dialogue Box)
(V12 AUTOCAD → SETTINGS → [Next] → STYLE: → SELECT TEXT Dialogue Box)
(V12 Pd DRAW → TEXT → SET STYLE... → SELECT TEXT FONT Dialogue Box)
(V13 AUTOCAD → DRAW → TEXT → SET STYLE: → SELECT STYLE Dialogue Box)
(V13 Pd DRAW → TEXT STYLE: → SELECT STYLE Dialogue Box)

Description

Used to modify the text font style.
Base font is Standard.
The "?" option will display the text styles available.
Note exercise 4A in Appendix E for a practice exercise on creating and using the Style command.
Leaving Height option at 0.0000 will have AutoCAD request a text height when text is inserted into the drawing.
PostScript and True Type fonts are available.

STYLE Command Format to set STANDARD as current text style

```
Command: STYLE (RETURN)
Text style name (or ?) <STANDARD>: STANDARD (RETURN)
The Select Font File Dialogue Box appears. Select TXT.SHT and click
OK
```

The dialogue box disappears, and command line says:

```
Existing Style Height <0.0000>: (RETURN)
Width Factor <1.0000>: (RETURN)
Obliquing Angle <0>: (RETURN)
Backwards? <N>: (RETURN)
Upside-Down? <N>: (RETURN)
Vertical? <N>: (RETURN)
Standard is now the current text style.
Regenerating drawing
Command:
```

TEXT (V11, V12 AUTOCAD → DRAW → [Next] → TEXT: → TEXT) TEXT/ DTEXT (V11 AUTOCAD → DRAW → DTEXT → DTEXT/ TEXT) (V12, V13 Pd DRAW → TEXT: → TEXT/DYNAMIC TEXT)

Description

Used to place text on the drawing.
True Type and PostScript text fonts are available (version 13).
Text options: Justify/Styles. When Justify is selected, the following justify options become available: Aligned, Fit, Centered, Middle, Right, TL, TC, TR, ML, MC, MR, BL, BC, BR, where T = Top, M = Middle, B = Bottom, L = Left, C = Center, R = Right.

TEXT Command Format

To place text, 2 units high, to the right of a selected point, type the following:

```
Command: TEXT (RETURN)
Justify/Style/<Start point>: Pick a point
Height <4.0000>: 2 (RETURN)
Rotation angle <0>: (RETURN)
Text: THIS IS TEXT (RETURN)
```

The text "THIS IS TEXT" appears to the right of the point selected.

Note: The basic difference between Text and Dtext (dynamic text) is as follows: Using Dtext, text appears in your drawing as you type each letter, whereas in the Text mode, the text appears at the command line as you type it and is transferred to the drawing after pressing the Return key. See Exercise 3-4b.

To evoke the Select Text (Styles) dialogue box in versions 12 and 13, select: DRAW → TEXT → Set Styles... *Note:* Appendix A1, Figs. A1 and A2.

TILEMODE (V12 AUTOCAD → MVIEW → TILEMODE)
(V13 AUTOCAD → VIEW → TILEMODE)

Description

Tilemode, when ON (set to 1), allows you to work only in Model Space and the standard UCS Icon appears. Model Space is your standard work environment. It is the space used by this tutorial, and how AutoCAD appears when you first turn it on.

When Tilemode is OFF (set to 0), you can work in Paper Space. A triangular UCS Icon appears in this mode. Think of Paper Space as allowing you to place a glass window or viewport over your drawing. This viewport can be placed anywhere around this model you have created on the computer screen. It can be placed underneath the screen, viewed from the top of the screen, or from any point of view. You can zoom in close, or far away. Think of it as a roving camera. You can have several of these windows open at a time.

Getting Out Of Paper Space

You may accidently find yourself in the Paper Space environment, having clicked on one of the various pull-down viewport commands, and find that you can no longer work on your drawing as you knew it.

When this happens switch the Paper Space Mode off by setting Model Space On or to 1, as follows:

TILEMODE Command Format

```
Command: TILEMODE (RETURN)
New value for TILEMODE <0>: 1 (RETURN)
```

TRACE (V12 AUTOCAD → RAW → [Next] → RACE:)
(V13 AUTOCAD → DRAW2 → TRACE)

Description

Trace lines may be as wide as you want.
Fill ON/OFF will make them solid or outlined.
Trace lines have several disadvantages when compared to Polylines. These are:
a) You cannot Curve, Close, Continue, or Undo a Trace or Trace segment.

b) They cannot be made a Dashed Linestyle, a useful item when developing printed circuit boards.

The Trace linestyle is *not* recommend for PCB trace lines.

TRACE Command Format

```
Command: TRACE (RETURN)
Trace width <default>: CHOOSE (RETURN)
From point: PICK
To point: PICK
To point: PICK
To point: (RETURN) To terminate the command.
Command:
```

TRIM (V11, V12 AUTOCAD → EDIT → [Next] → TRIM)
(V13 AUTOCAD → MODIFY → TRIM: → TRIM MENU)
(V13 Pd MODIFY → TRIM:)

Description

Used to remove or trim unwanted portions of an entity.

A line or "cutting edge" is first specified to establish the trim point.

Trimming is performed as a square edge; thus polylines with thickness may have the trimmed end protrude beyond the cutting edge.

Some entities cannot be used as a cutting edge. A "NO EDGES SELECTED" message identifies these entities.

TRIM Command Format

```
Command: TRIM (RETURN)
Select cutting edge(s)...
Select objects: SELECT LINE TO USE AS A CUTTING EDGE AND PRESS
(RETURN)
<Select object to trim>/Project/Edge/Undo: SELECT OBJECT (RETURN)
```

Note: To terminate each selection, press the Return key.

UCSICON (V11, V12 AUTOCAD → SETTINGS → [Next] → UCSICON)
(V13 AUTOCAD → VIEW → UCS: → UCSICON)
(V13 AUTOCAD → OPTIONS → UCSICON:)

Description

AutoCAD can display a drawing in any of three axes (X, Y, and Z), and the

UCSIcons (User Coordinate System Icon) allow you to visualize these axes or viewpoint orientations of your drawing.
This perspective is mostly related to three-dimensional (3D) drawing models.
Two UCS Icons are used by AutoCAD; a right angle icon representing Model Space and a triangular icon for Paper Space. See Figure 1-2c.
The drawings associated with this tutorial are two-dimensional (2D), thus the UCSIcons are not needed and are turned OFF.

UCSICON Command Format

```
Command: UCSICON (RETURN)
ON/OFF/All/Noorgin/ORgin/<ON>: OFF (RETURN)
```

UNDO (V11, V12 AUTOCAD → EDIT → [Next] → UNDO)
(V13 AUTOCAD → ASSIST → UNDO: → UNDO)
(V13 Pd ASSIST → UNDO)

Description

Used to undo a single command or multiple commands in one operation.
The *U* key is a *Hot* key or Alias key for Undo.
Caution should be used with Undo, as you can Undo all commands performed in the present drawing session with the "Back" option.

Undo works in conjunction with the following commands: DXFOUT, End, Files, Help, ID, IGESOUT, List, Mslide, Quit, Resume, Rscript, Save, Script, Status, Wblock.

TIP: The Redo command acts as an antidote to Undo, but it must be used immediately after Undo.

UNDO Command Format

```
Command: UNDO (RETURN)
Auto/Back/Control/End/Group/Mark/<number>:
```

Options to Note

Mark - Places an identity mark at the instruction being performed as a reference mark that can later be used with the Back option to "undo" to this mark.

Back - Causes AutoCAD to undo all commands until it finds the previously established "reference mark." If no reference mark was established, AutoCAD displays:

```
This will undo everything. OK? <Y>:
```

TRAP Be careful, a Yes will undo all commands since you last entered the drawing editor, if you are at the command prompt.

Number - A number response will undo that number of commands as a single entry. A single Undo will undo the last command entered. You may step backward one step at a time.

UNITS (V11, V12 AUTOCAD → SETTINGS → [Next] → UNITS: → UNITS MENU) (V13 AUTOCAD → DATA → UNITS: → UNITS MENU)

Description

Establishes the system in which measurements are recorded.
Five choices are available:

1. Scientific
2. Decimal
3. Engincering
4. Architectural
5. Fractional

Should be part of the initialization procedure for AutoCAD drawings.

UNITS Command Format

```
Command: UNITS (RETURN)
1. Scientific
2. Decimal
3. Engineering
4. Architectural
5. Fractional

Enter choice, 1 to 5 <default>: CHOOSE (RETURN)
(Metric = 2)
```

If 2 was chosen, you are asked:

```
Number of digits to right of decimal point (0 to 8)<4>: CHOOSE (RETURN)

System of angle measurement      Examples
1. Decimal degrees               45.000
2. Degrees/minutes/seconds       45d 0' 0"
3. Grads                         50.0000g
4. Radians                       0.7854r
5. Surveyor's Units              N45d 0' 0" E
```

```
   Enter choice, 1 to 5 <1>: CHOOSE

Number of fractional places for display of angles (0 to 8)<0>: CHOOSE
(RETURN)

Direction for angle 0:
  East 3 o'clock    = 0
  North 12 o'clock  = 90
  West 9 o'clock    = 180
  South 6 o'clock   = 270

Enter direction for angle 0 <0>: (RETURN)
Do you want angles measured clockwise? <N>: (RETURN)
```

WBLOCK (V11, V12 AUTOCAD → BLOCKS → WBLOCK: → WBLOCK)

Description

Captures a Block on a drawing and saves it to the disk as a separate drawing. Provides a simple way to extract one part of a drawing and use it in another drawing. Like creating a rubber stamp.
Used with Insert, the Asterisk (*) option, and Explode.

How to use WBLOCK

1. First create a Block and store it in the drawing block area.
2. Next, use the Wblock command to save the stored Block to disk under a separate file name.

WBLOCK Command Format

The following procedure uses Wblock to save the title block (TITLBK1), drawn in Exercise 3-4, under the file name WTITLBK1.

Step 1. Save the Block.

```
Command: BLOCK (RETURN)
Block name (or ?): TITLBK1 (RETURN)
Insertion base point: SELECT LOWER RIGHT CORNER OF THE TITLE BLOCK
Select objects or Window or Last: W (RETURN)
First point: SELECT A CORNER (wrap a window around the title block)
Second point: SELECT DIAGONAL CORNER
```

The title block drawing becomes dotted; press Return to confirm the selection. The title block disappears. Now type Oops to recover the title block.

Step 2. Next, we'll save the blocked title block (TITLBK1) to the A drive under the file name WTITLBK1, using the Wblock command. Note that a W is placed in front of the Wblock name to identify it as a Wblock when you are looking at that directory. Note Exercise 3-5a.

```
Command: WBLOCK (RETURN)
File name: A:\WTITLBK1 (RETURN)
Block name: TITLBK1 (RETURN)
```

Refer to Exercises 3-5 and 4-3b for more details on Blocks and Wblocks.

ZOOM (V11, V12 AUTOCAD → DISPLAY → ZOOM)
(V13 AUTOCAD → VIEW → ZOOM: → ZOOM Dialogue Box)
(V13 Pd VIEW → ZOOM: → ZOOM Dialogue Box)

Description

Selected from keyboard or menu.
Allows you to enlarge or reduce a portion of the drawing.
Magnification Scale of overall drawing as well as eight zoom choices available: All, Center, Dynamic, Extents, Left, Previous, Vmax Window <scale (X/XP)>:.
Hot key or alias key is "Z".

ZOOM Command Format

```
Command: ZOOM (RETURN)
All/Center/Dynamic/Extents/Left/Previous/Vmax/Window/<Scale(X/XP)>:
CHOOSE/SELECT (RETURN)
```

A response to: "Scale" with a number, increases or decreases the entire drawing by that scale factor. If the Zoom factor is followed by an X, the zoom is computed relative to the current display.

Zoom scale factors can only be positive numbers. For reductions in zoom size, use a decimal number. For example, a reduction in size of 50 percent is a zoom scale of .5.

(A)ll causes the complete drawing to be displayed, including entities outside the limits (orphans). Drawing may be regenerated in process.

(C)enter allows you to choose an entity and place it in the center of the screen.

(D)ynamic presents a special view selection screen containing information about current and possible view screen selections. Each view (Drawing extents - white, Current view - green, View box/Pan mode - white) are labelled and noted as to their respective color.

(E)xtents displays only that area of the drawing which contains entities; an area with no entities is not displayed.

(L)eft corner is the same as Center, except that the entity is placed lower left.

(P)revious allows you to return to the last zoom used.

(V)max, (Maximum Virtual Screen) zooms to the maximum virtual screen size that does not force a regeneration of the drawing.

(W)indow allows you to place a window around the area to be zoomed.

appendix A

AutoCAD Fonts, Linetypes, and Character Sets

APPENDIX A1 AutoCAD Text Fonts Styles

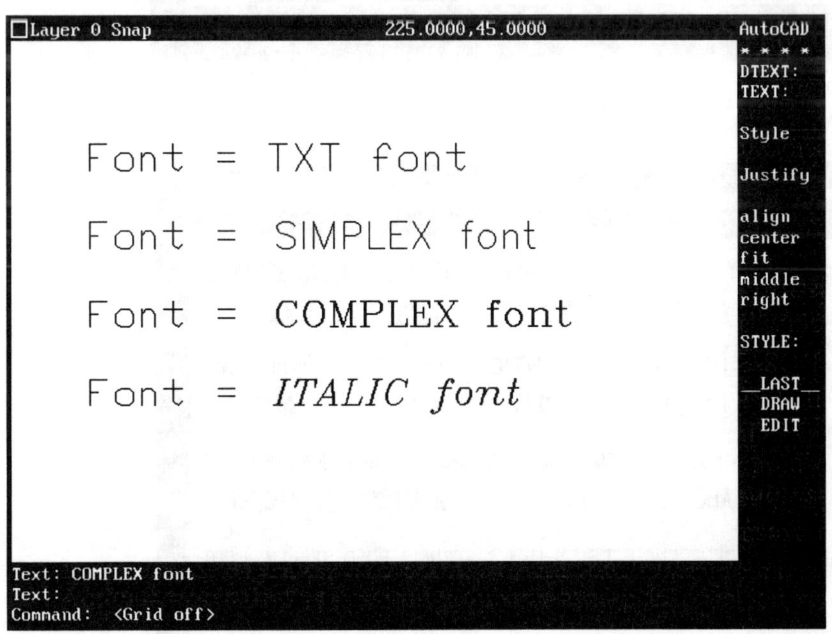

Figure A-1 AutoCAD text fonts styles.

Appendix A1 AutoCAD Text Fonts Styles cont.

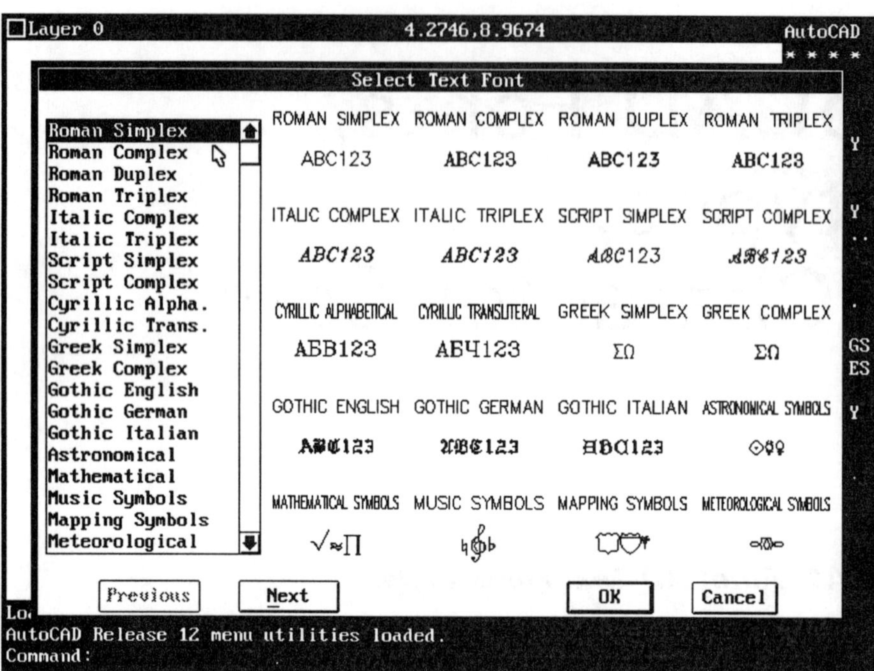

Figure A-2 Text Fonts (Part 1)

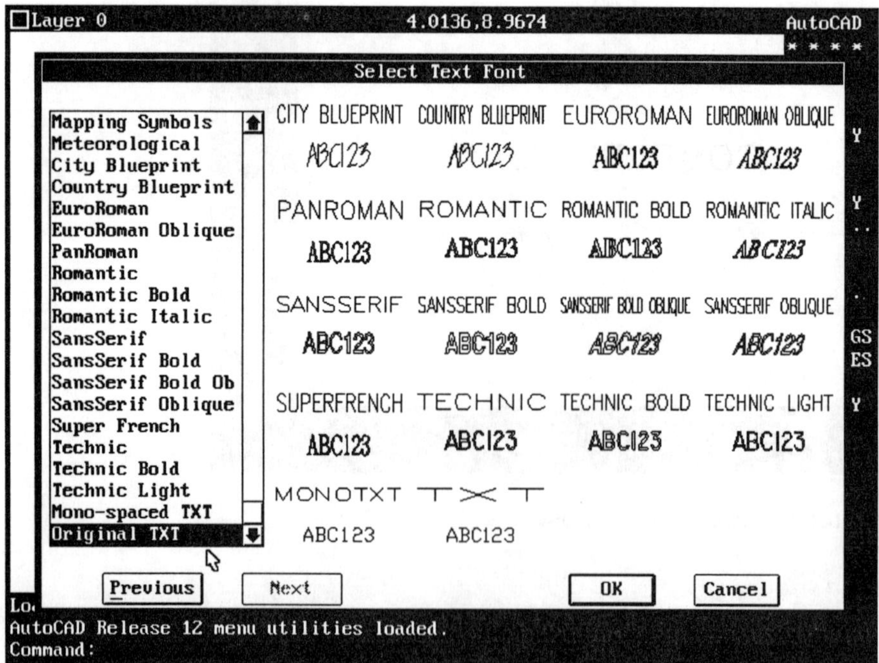

Figure A-3 Text Fonts (Part 2)

APPENDIX A2 LINETYPES

BORDER
BORDER2
BORDERX2
CENTER
CENTER2

CENTERX2
DASHDOT
DASHDOT2
DASHDORX2
DASHED

DASHED2
DASHEDX2
DIVIDE
DIVIDE2
DIVIDEX2

DOT
DOT2
DOTX2
HIDDEN
HIDDEN2

HIDDENX2
PHANTOM
PHANTOM2
PHANTOMX2

appendix B

Function Key Designation

FUNCTION KEYS

AutoCAD makes use of the standard Function Keys found on the Touch-Type Keyboard. These keys allow quick access to several commonly used AutoCAD commands to give greater flexibility when using the system.

Some keys act as a toggle switch and allow you to turn a function on or off at the mere touch of a key. For example: To turn ON the Grid, simply press the F7 key once. To turn it OFF, press the F7 key again. You can verify the operation and status of various keys by observing the Mode Area of the screen (upper left corner).

Key	*AutoCAD Function*
F1	Switch the screen from the draw mode to the status mode
F2	Not used
F3	Not used
F4	Not used
F5	Not used
F6	Toggle Coordinates ON & OFF
F7	Toggle Grid ON & OFF
F8	Toggle Ortho ON & OFF
F9	Toggle Snap ON & OFF
F10	Toggle Tablet ON & OFF

The template for the Enhanced keyboard is shown in Fig. B-1.

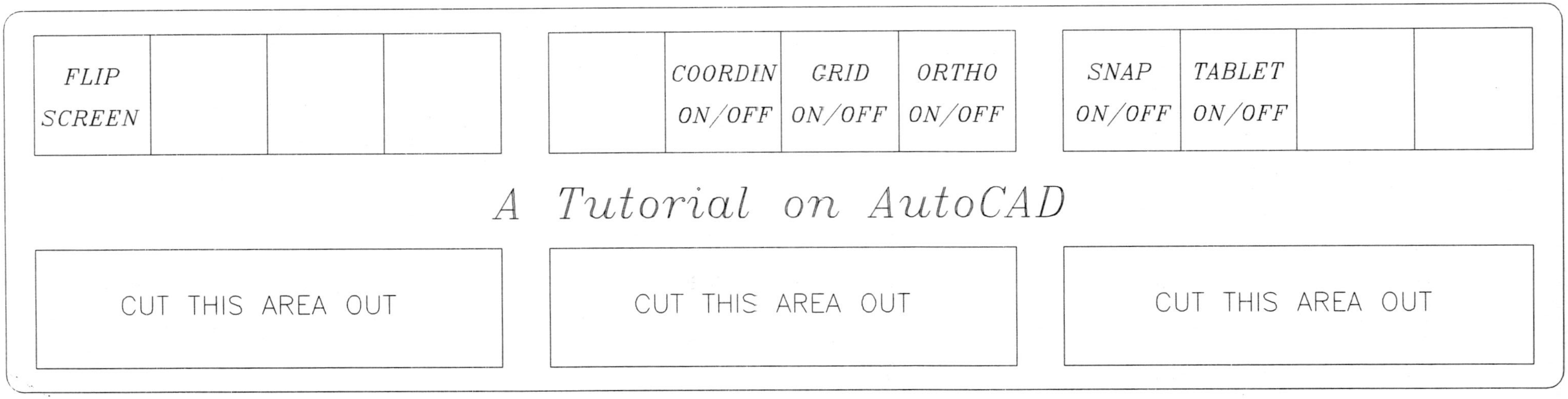

FLIP SCREEN					COORDIN ON/OFF	GRID ON/OFF	ORTHO ON/OFF	SNAP ON/OFF	TABLET ON/OFF		

A Tutorial on AutoCAD

CUT THIS AREA OUT	CUT THIS AREA OUT	CUT THIS AREA OUT

Figure B-1 Function key template for 101 enhanced keyboard.

appendix C

Major Parts List

COMPCODE	*COMPVAL (in ohms)*

RESISTORS: All resistors are 1/2 watt, 20%, unless otherwise specified.

R1	820
R2, R7	120
R3	1K - Variable resistor (KMU1521) PEC
R4	1K
R5, R6	10K
R8	150

CAPACITORS:	**VALUE**	**VOLTAGE RATING**
C1, C2, C3, C4	1000uf	25V
C5, C6, C7	10uf	25V

SEMICONDUCTORS:

U-1	LM741CN OP-AMP or LF735 OP-AMP	
REG-1	LM7805CT	Voltage Regulator, fixed +5V
REG-2	LM317T	Voltage Regulator, Positive Adjustable
REG-3	LM337T	Voltage Regulator, Negative Adjustable
D1 - D4	1N5402	Diode, 3A @ 100V
D5	1N752A	Zener Diode 5V6 zener
D6 - D10	1N4001	Diode, 1A @ 100V
LED-1	521-9216	LED, Red. Dialight

MISCELLANEOUS PARTS:

Qty.	*Part #*	*Compcode*	*Manufacturer*	*Description*
1	1454L	Case	Hammond	Instrument case, 4x6x5 inches
1	17237	Cord	Belden	AC Power Card
1	41-230-0	SW-1	Mode	SPST toggle switch
1	55-910-0	F1	Mode	Fuse, 1Amp.
1	55-802-0		Mode	Fuse holder
1	166K35	TR-1	Hammond	Transformer, 35V ct @ 1Amp
1	111-0102-001	BP-1	Johnson	Binding post, Red
1	111-0107-001	BP-2	Johnson	Binding post, Yellow
1	111-0103-001	BP-3	Johnson	Binding post, Black
1	111-0110-001	BP-4	Johnson	Binding post, Blue
1	111-0104-001	BP-5	Johnson	Binding post, Green
12	PLT1M		Panduit	Cable ties
1	633		Keystone	Right angle mounting bracket
1	46-169		Cinch	#8 solder lug
1	54-228-0		Mode	Plastic knob
1			Sheet aluminum 1 3/4 x 3 inches	
1			Printed Circuit Board ~ 100mm x 110mm	
1	H01B		Lenlite	LED Holder
1	5P		Lenline	Strain relief

HARDWARE:

7	54-522-0	Mode	#4-40 machine screw
7	54-527-0	Mode	#4 flat washer
7	54-528-0	Mode	#4 lock washer
7	54-526-0	Mode	#4-40 hex nuts
2	54-541-0	Mode	#8-32 machine screw
2	54-547-0	Mode	#8 flat washer
2	54-548-0	Mode	#8 lock washer
2	54-546-0	Mode	#8-32 hex nuts
3	3049	Keystone	shoulder washer
3	4672	Keystone	mica insulators

WIRE:

5ft, Belden 22 AWG wire
8530 - 1000 - 2 red
8530 - 1000 - 4 yellow
8530 - 1000 - 6 blue
8530 - 1000 - 10 black
8530 - 1000 - 5 green
8530 - 1000 - 9 white

appendix D

Manufacturers' Data Sheets

Discussion

In the following data sheets you will find electrical and mechanical information on the components used in this project. The data sheets are supplied strictly as a reference. They are here for you to become familiar with the typical layout of data sheets used in a parts catalog. The prices are by no means current, but have been included for relative comparison purposes between various types of components.

Many thanks to Mr. Collin Webb, President of Electro Sonic Inc. of Toronto, distributors of electronic and electrical components, for allowing me to use their parts catalog for the data sheets used in the first and second edition of this text.

Electro Sonic Inc. reserve the right to make changes in the parts specifications at any time and without notice, and no responsibility is assumed by their use.

INDEX TO COMPONENTS IN DATA SHEETS

Positive & Negative Voltage Regulators

OSITIVE & NEGATIVE ADJUSTABLE VOLTAGE REGULATORS

IO Max Amps	POSITIVE Cat. No.	NEGATIVE Cat. No.	Manufacturer	Output Voltage Range		Input Voltage Range		Regulation		Temp Range	Package	Net Price Each	
				Min.	Max.	Min.	Max.	Line % V Max.	Load % V max.			1-99	100-Up
.5	—	**LM137H**	National Semi.	-1.2	-37	-5.0	-40	.04	0.5	0 to +150°C	TO-39	**19.28**	**14.11**
.5	—	**LM337MP**	National Semi.	-1.2	-37	-5.0	-40	.04	1.0	0 to +125°C	TO-202	**1.89**	**1.26**
.5	—	**LM337MT**	Motorola	-1.2	-37	-5.0	-40	.04	1.0	0 to +125°C	TO-220	**1.67**	**1.67**
.5	—	**LM337H**	National Semi.	-1.2	-37	-5.0	-40	.04	1.0	0 to +125°C	TO-39	**6.44**	**4.71**
.5	—	**LM337H**	Linear Tech.	-1.2	-37	-5.0	-40	.04	1.0	0 to +125°C	TO-39	**4.44**	**4.44**
.5	—	**LM337HVH**	National Semi	-1.2	-47	-5.0	-50	.04	1.0	0 to +125°C	TO-39	**10.55**	**7.76**
1.0	**LM2941T**	—	National Semi.	5.0	20	7.0	31	—	—	-40 to +125°C	5L-TO220	**2.57**	**1.71**
1.0	**LM78GCP**	—	National Semi.	5.0	30	7.5	40	1.0	1.0	-65 to +150°C	TO-202	**3.05**	**2.08**
1.5	**LM117K STEEL**	—	National Semi.	1.2	37	5.0	40	.04	0.5	-55 to +150°C	TO-3	**10.81**	**7.89**
1.5	**LM117HVK STEEL**	—	National Semi.	1.2	57	3.0	60	.02	0.3	-55 to +150°C	TO-3	**38.43**	**28.09**
1.5	**LM117HVK**	—	Linear Tech.	1.2	57	3.0	60	.02	0.3	-55 to +150°C	TO-3	**25.24**	**19.69**
1.5	**LM317K STEEL**	—	National Semi	1.2	37	5.0	40	.04	0.5	0 to +125°C	TO-3	**4.94**	**3.60**
1.5	**LM317K**	—	Linear Tech.	1.2	37	5.0	40	.04	0.5	0 to +125°C	TO-3	**3.81**	**2.91**
1.5	**LM317KC**	—	National Semi	1.2	37	5.0	40	.04	0.5	0 to +125°C	TO-3 ALUMINUM	**3.35**	**2.43**
1.5	**LM317T**	—	Motorola	1.2	37	5.0	40	.04	0.5	0 to +125°C	TO-220	**.94**	**.94**
1.5	**LM317T**	—	National Semi	1.2	37	5.0	40	.04	0.5	0 to +125°C	TO-220	**1.46**	**1.00**
1.5	**LM317T**	—	Linear Tech.	1.2	37	5.0	40	.04	0.5	0 to +125°C	TO-220	**2.01**	**1.59**
1.5	**LM317HVK STEEL**	—	National Semi	1.2	57	3.0	60	.02	0.3	0 to +125°C	TO-39	**8.02**	**5.83**
1.5	**LM317HVK**	—	Linear Tech.	1.2	57	3.0	60	.02	0.3	0 to +125°C	TO-39	**7.63**	**5.89**
1.5	**LT317AH**	—	Linear Tech.	1.2	37	3.0	40	.01	0.1	0 to +125°C	TO-39	**4.51**	**3.81**
1.5	**LT317AT**	—	Linear Tech.	1.2	37	3.0	40	.01	0.1	0 to +125°C	TO-220	**2.63**	**2.08**
1.5	—	**LM137HVK**	Linear Tech.	-1.2	-37	-5.0	-40	.01	0.3	-55 to +150°C	TO-3	**39.18**	**32.41**
1.5	—	**LM337K STEEL**	National Semi	-1.2	-37	-5.0	-40	.04	1.0	-55 to +150°C	TO-3	**7.41**	**5.41**
1.5	—	**LM337K**	Linear Tech.	-1.2	-37	-5.0	-40	.04	1.0	-55 to +150°C	TO-3	**6.20**	**5.20**
1.5	—	**LM337T**	Motorola	-1.2	-37	-5.0	-40	.04	1.0	0 to +125°C	TO-220	**1.65**	**1.65**
1.5	—	**LM337T**	National Semi	-1.2	-37	-5.0	-40	.04	1.0	0 to +125°C	TO-220	**2.18**	**1.51**
1.5	—	**LM337T**	Linear Tech.	-1.2	-37	-5.0	-40	.02	0.3	0 to +125°C	TO-220	**2.49**	**1.85**
1.5	—	**LT337AT**	Linear Tech.	-1.2	-37	-5.0	-40	.01	0.2	0 to +125°C	TO-220	**3.06**	**2.42**
1.5	—	**LM337HVK STEEL**	National Semi	-1.2	-47	-5.0	-50	.02	0.3	0 to +125°C	TO-3	**13.34**	**10.69**
1.5	**LT1086CT**	—	Linear Tech.	1.5	30	1.5	30	0.2	0.3	0 to +125°C	TO-220	**3.47**	**2.70**

TO-5 TO-220AB TO-202AB (NON-ISOLATED) TO-3 TO-39 TO-92

Output Voltage Typical (V)	Output Current IO Max. (Amps)	POSITIVE Regulator Part No.	NEGATIVE Regulator Part No.	Manufacturer	Output Voltage		Input Voltage		Line Reg. Max. (mV)	Load Reg. Max. (mV)	Package	Net Price Each	
					Min.	Max.	Min.	Max.				1-24	25-Up
5	.1	**MC78L05CP**	—	Motorola	4.5	5.5	6.7	30	200	60	TO-92	**$.38**	**$.38**
5	.1	—	**MC79L05CP**	Motorola	-4.5	-5.5	6.7	30	200	60	TO-92	**.38**	**.38**
5	.1	**LM2931T-5.0**	—	National Semi.	4.5	5.5	5.5	33	30	50	TO-220	**1.23**	**1.23**
5	.1	**LM2931M-5.0**	—	National Semi.	4.5	5.5	5.5	33	30	50	SO-8	**.93**	**.75**
5	.1	**LM2931AT-5.0**	—	Motorola	4.75	5.25	5.5	33	30	50	TO-220	**2.05**	**2.05**
5	.1	**LM2931AT-5.0**	—	National Semi	4.75	5.25	5.5	33	30	50	TO-220	**2.12**	**1.59**
5	.1	**LM2931AM-5.0**	—	National Semi	4.75	5.25	5.5	33	30	50	SO-8	**1.20**	**.96**
5	.1	**LM2931AZ-5.0**	—	Motorola	4.75	5.25	5.5	33	30	50	TO-92	**.79**	**.79**
5	.1	**LM2931AZ-5.0**	—	National Semi	4.75	5.25	5.5	33	30	50	TO-92	**1.06**	**.79**
5	.1	**MC78L05ACP**	—	Motorola	4.75	5.25	6.7	30	150	60	TO-92	**.39**	**.39**
5	.1	**LM78L05ACH**	—	National Semi	4.75	5.25	6.7	30	75	60	TO-39	**1.69**	**1.69**
5	.1	**LM78L05ACZ**	—	National Semi	4.75	5.25	6.7	30	75	60	TO-92	**.60**	**.45**
5	.1	—	**MC79L05ACP**	Motorola	-4.75	-5.25	6.7	30	150	60	TO-92	**.39**	**.39**
5	.1	—	**LM79L05ACZ**	National Semi	-4.75	-5.25	7.0	35	60	50	TO-92	**.67**	**.50**
5	.1	—	**LM320LZ-5.0**	National Semi	-4.75	-5.25	7.5	20	50	50	TO-92	**.68**	**.68**
5	.1	**LM140LAH-5.0**	—	National Semi	4.8	5.2	7.2	20	30	40	TO-39	**4.22**	**3.37**
5	1.5	**MC7805BT**	—	Motorola	4.5	5.5	8.0	35	100	100	TO-220	**1.17**	**1.17**
5	.15	**LM330T-5.0**	—	National Semi.	4.8	5.2	5.6	26	60	50	TO-220	**1.38**	**1.38**
5	.15	**LM2930T-5.0**	—	National Semi.	4.5	5.5	5.0	30	30	80	TO-220	**1.07**	**.85**
5	.2	—	**LM120H-5.0**	National Semi.	-4.9	-5.1	7.0	25	25	50	TO-39	**12.72**	**10.20**
5	.2	**LM109H**	—	National Semi.	4.7	5.3	7.0	35	50	50	TO-5	**12.65**	**12.65**
5	.2	—	**LM320H-5.0**	National Semi.	-4.75	-5.25	6.7	30	100	100	TO-39	**7.69**	**7.69**
5	.2	**LM309H**	—	National Semi.	4.8	5.2	7.0	35	50	50	TO-5	**4.77**	**4.77**
5	.25	**LM342P-5.0**	—	National Semi.	4.75	5.25	7.5	20	55	50	TO-202	**1.07**	**.99**
5	.5	**MC78M05CT**	—	Motorola	4.6	5.4	7.0	35	100	100	TO-220	**.73**	**.73**
5	.5	—	**MC79M05CT**	Motorola	-4.8	-5.2	7.0	35	100	100	TO-220	**.73**	**.73**
5	.5	—	**LM320MP-5.0**	National Semi.	-4.75	-5.25	7.5	20	50	50	TO-202	**4.00**	**4.00**
5	.5	**LM341P-5.0**	—	National Semi.	4.75	5.25	7.5	20	100	100	TO-202	**1.11**	**1.11**
5	1.0	**LM2940T-5.0**	—	National Semi.	4.75	5.25	6.25	26	50	50	TO-220	**2.57**	**1.93**
5	1.0	**LM2940CT-5.0**	—	National Semi.	4.75	5.25	6.25	26	50	50	TO-220	**2.12**	**1.59**
5	.1	**LM78L05ACM**	—	National Semi.	4.75	5.25	6.7	30	75	60	SO-8	**.67**	**.67**
5	1.0	—	**LM2990T-5.0**	National Semi.	-4.75	-5.25	-26	+0.3	40	40	TO-220	**3.07**	**2.86**
5	1.5	**LM109K STEEL**	—	National Semi.	4.7	5.3	7.0	35	50	100	TO-3S	**18.03**	**18.03**
5	1.5	—	**LM320T-5.0**	National Semi.	-4.8	-5.2	7.5	25	40	100	TO-220	**3.30**	**3.30**
5	1.5	**LM309K**	—	National Semi.	-4.8	-5.2	7.0	35	50	100	TO-3	**2.94**	**2.35**
5	1.5	**MC7805CT**	—	Motorola	4.75	5.25	7.0	35	100	100	TO-220	**.73**	**.73**
5	1.5	**MC7805ACT**	—	Motorola	4.9	5.1	7.5	35	50	100	TO-220	**1.25**	**1.25**
5	1.5	**LM340K-5.0**	—	National Semi.	4.75	5.25	7.0	20	50	50	TO-3	**4.80**	**3.85**
5	1.5	**LM340T-5.0**	—	Motorola	4.75	5.25	7.0	20	50	50	TO-220	**.76**	**.76**
5	1.5	**LM340AT-5.0**	—	National Semi.	4.75	5.25	7.0	20	50	50	TO-220	**2.27**	**1.70**
5	1.5	—	**MC7905CT**	Motorola	-4.75	-5.25	7.0	35	100	100	TO-220	**.96**	**.64**
5	1.5	—	**LM7905CT**	National Semi.	-4.75	-5.25	7.0	20	50	100	TO-220	**1.30**	**.98**
5	1.5	—	**LM320K-5.0**	National Semi.	-4.75	-5.25	7.0	20	50	50	TO-3	**6.86**	**6.86**

Page D1

Operational Amplifiers

Part No.	Manufacturer	Input Offset Voltage (mV) Max.	Input Offset Current (nA) Max.	Input Bias Current (nA) Max.	Max. Supply Voltage	Voltage Gain Min. (V/mV)	Slew (V/µs)	Package	Description	Net Price Each 1-99	100
AUTOMOTIVE TEMPERATURE RANGE (–40°C TO +85°C)					**SINGLE OP-AMP**				**INTERNALLY COMPENSATE**		
LM604IJ	National Semi.	5.0	12	100	+18	25	1.5	18L-Cer-DIP	4-Channel Mux-Amp	**8.23**	5
LM6181IN	National Semi.	7.0	—	10µA	±18	±10	2000	8L-DIP	Current Feedback	**5.11**	5
LTC1052CN	Linear Tech.	5µV	.3	.3	18	1000	4	14L-DIP	Precision, Low Noise	**8.32**	6
LTC1052CN8	Linear Tech.	5µV	.03	.03	18	1000	4	8L-DIP	Precision, Low Noise	**7.07**	5
LTC1150CN8	Linear Tech.	5µV	.06	.06	36	10K	3	8L-DIP	Precision	**6.86**	5
LTC7652CH	Linear Tech.	5µV	.03	.03	18	1000	4	8L-TO5	Precision	**9.01**	6
MILITARY TEMPERATURE RANGE (–55°C TO +125°C)					**SINGLE OP-AMP**				**INTERNALLY COMPENSATE**		
LM107H	National Semi.	2.0	10	75	±22	160	0.3	8L-TO5	General Purpose	**2.84**	2
LM143H	National Semi.	5.0	3.0	20	±40	180K	2.5	8L-TO5	High Voltage	**35.98**	26
LM741H	National Semi.	5.0	200	500	±22	200	0.5	8L-TO5	General Purpose	**3.74**	2
LM741CH	National Semi.	6.0	200	500	±18	200	0.5	8L-TO5	General Purpose	**2.41**	1
LM741CJ	National Semi.	6.0	200	500	±18	200	0.5	8L-Cer-DIP	General Purpose	**1.32**	1
LM741CN	National Semi.	6.0	200	500	±18	200	0.5	8L-DIP	General Purpose	**.85**	
LM741EN	National Semi.	3.0	30	80	±22	50	0.7	8L-DIP	General Purpose	**1.54**	1
LT1010CH	Linear Tech.	—	—	250µA	±22	1.0	75	3L-TO39	Fast ±150mA Power Buffer	**6.38**	4
LT1010CT	Linear Tech.	—	—	250µA	±22	1.0	75	5L-TO220	Fast ±150mA Power Buffer	**5.06**	3
LT1012CH	Linear Tech.	50	150	±150pA	±20	2000	0.2	8L-TO5	Universal Precision	**6.79**	5
LT1012CN8	Linear Tech.	50	150	±150pA	±20	2000	0.2	8L-DIP	Universal Precision	**5.27**	3
LT1012MH	Linear Tech.	35	100	±100	±20	2000	0.2	8L-TO5	Low VOS Low Power	**11.79**	9
LT1028MJ8	Linear Tech.	80µV	130	±240	±22	5000	11	8L-Cer-DIP	Low Noise, Precision	**21.91**	16
MC1536U	Motorola	5.0	3.0	20	±40	500K	2.0	8L-Cer-DIP	High Voltage	**7.89**	7
MC1741U	Motorola	5.0	200	500	±22	200	0.5	8L-Cer-DIP	General Purpose	**1.84**	1

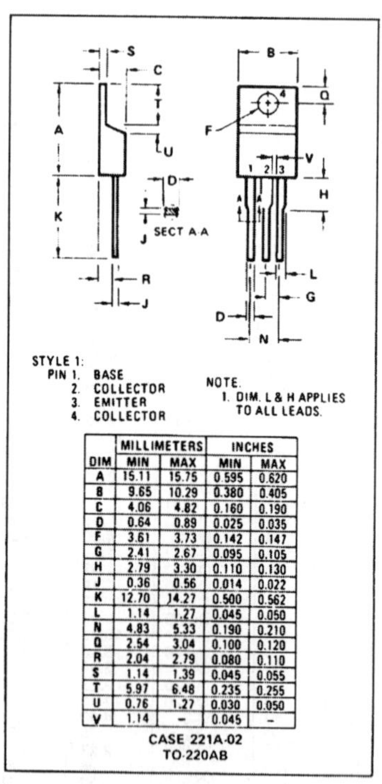

STYLE 1:
PIN 1. BASE
2. COLLECTOR
3. EMITTER
4. COLLECTOR

NOTE.
1. DIM. L & H APPLIES TO ALL LEADS.

DIM	MILLIMETERS MIN	MILLIMETERS MAX	INCHES MIN	INCHES MAX
A	15.11	15.75	0.595	0.620
B	9.65	10.29	0.380	0.405
C	4.06	4.82	0.160	0.190
D	0.64	0.89	0.025	0.035
F	3.61	3.73	0.142	0.147
G	2.41	2.67	0.095	0.105
H	2.79	3.30	0.110	0.130
J	0.36	0.56	0.014	0.022
K	12.70	14.27	0.500	0.562
L	1.14	1.27	0.045	0.050
N	4.83	5.33	0.190	0.210
Q	2.54	3.04	0.100	0.120
R	2.04	2.79	0.080	0.110
S	1.14	1.39	0.045	0.055
T	5.97	6.48	0.235	0.255
U	0.76	1.27	0.030	0.050
V	1.14	–	0.045	–

CASE 221A-02
TO-220AB

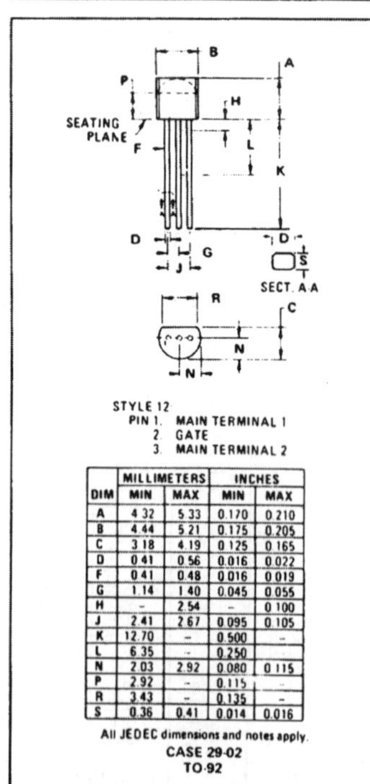

STYLE 12
PIN 1. MAIN TERMINAL 1
2. GATE
3. MAIN TERMINAL 2

DIM	MILLIMETERS MIN	MILLIMETERS MAX	INCHES MIN	INCHES MAX
A	4.32	5.33	0.170	0.210
B	4.44	5.21	0.175	0.205
C	3.18	4.19	0.125	0.165
D	0.41	0.56	0.016	0.022
F	0.41	0.48	0.016	0.019
G	1.14	1.40	0.045	0.055
H	–	2.54	–	0.100
J	2.41	2.67	0.095	0.105
K	12.70	–	0.500	–
L	6.35	–	0.250	
N	2.03	2.92	0.080	0.115
P	2.92	–	0.115	–
R	3.43	–	0.135	–
S	0.36	0.41	0.014	0.016

All JEDEC dimensions and notes apply.
CASE 29-02
TO-92

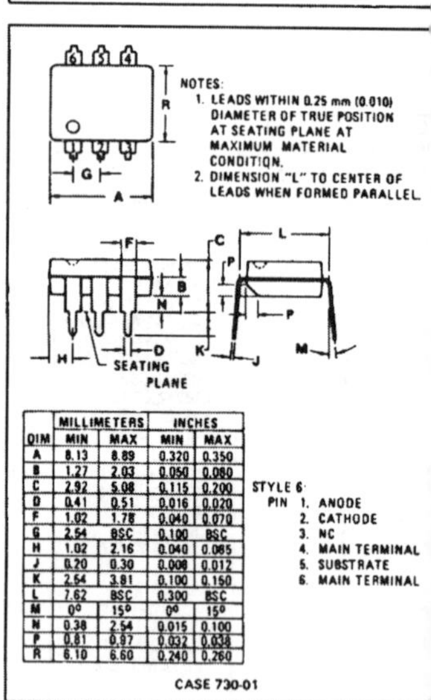

NOTES:
1. LEADS WITHIN 0.25 mm (0.010) DIAMETER OF TRUE POSITION AT SEATING PLANE AT MAXIMUM MATERIAL CONDITION.
2. DIMENSION "L" TO CENTER OF LEADS WHEN FORMED PARALLEL.

DIM	MILLIMETERS MIN	MILLIMETERS MAX	INCHES MIN	INCHES MAX
A	8.13	8.89	0.320	0.350
B	1.27	2.03	0.050	0.080
C	2.92	5.08	0.115	0.200
D	0.41	0.51	0.016	0.020
F	1.02	1.78	0.040	0.070
G	2.54	BSC	0.100	BSC
H	1.02	2.16	0.040	0.085
J	0.20	0.30	0.008	0.012
K	2.54	3.81	0.100	0.150
L	7.62	BSC	0.300	BSC
M	0°	15°	0°	15°
N	0.38	2.54	0.015	0.100
P	0.81	0.97	0.032	0.038
R	6.10	6.60	0.240	0.260

STYLE 6:
PIN 1. ANODE
2. CATHODE
3. NC
4. MAIN TERMINAL
5. SUBSTRATE
6. MAIN TERMINAL

CASE 730-01

Silicon Rectifiers

GENERAL PURPOSE SILICON RECTIFIERS

DO-41

Type Number	Manufacturer	PRV Volts	Case	Net Price Each 1-99	Net Price Each 100-Up
1 AMP 75°C GLASS PASSIVATED					
1N4001GP	—	50	DO-41	$.15	$.10
1N4002GP	—	100		.15	.11
1N4003GP	—	200		.16	.11
1N4004GP	—	400		.19	.13
1N4005GP	—	600		.19	.13
1N4006GP	—	800		.21	.14
1N4007GP	—	1000		.25	.17

DO-35 DO-7

Type Number	Manufacturer	PRV Volts	Case	Net Price Each 1-99	Net Price Each 100-Up
1 AMP 75°C MINIATURE EPOXY					
1N4001	—	50	DO-41	.07	.05
1N4002	—	100		.08	.05
1N4003	—	200		.08	.05
1N4004	—	400		.08	.06
1N4005	—	600		.08	.06
1N4006	—	800		.09	.06
1N4007	—	1000		.09	.06

Zener Diodes

Type Number	Zener Volts	Test Current mA.	Max. Imp. @Izt Ohms	Net Price Each 1-5999	Net Price Each 6000-Up
500 MILLIWATT ZENER DIODES "DO-35" CASE STYLE, VOLT. TOL. 5%					
1N4370A	2.4	20	30	$.20	$.15
1N4371A	2.7	20	30	.20	.15
1N4372A	3.0	20	29	.20	.15
1N746A	3.3	20	28	.10	.08
1N747A	3.6	20	24	.10	.08
1N748A	3.9	20	23	.10	.08
1N749A	4.3	20	22	.10	.08
1N750A	4.7	20	19	.10	.08
1N751A	5.1	20	17	.11	.08
1N752A	5.6	20	11	.10	.08
1N753A	6.2	20	7	.10	.08
1N754A	6.8	20	5	.10	.08
1N957B	6.8	18.[illegible]	4.5	.21	.15
1N755A	7.5	20	6	.10	.08
1N958B	7.5	16.[illegible]	5.5	.20	.15
1N756A	8.2	20	8	.10	.08
1N959B	8.2	15	6.5	.20	.15
1N757A	9.1	20	10	.10	.08
1N960B	9.1	14	7.5	.20	.15
1N758A	10	20	17	.10	.08
1N961B	10	12.5	8.5	.10	.08
1N962B	11	11.5	9.5	.20	.15
1N759A	12	20	30	.10	.08
1N963B	12	10.5	11.5	.10	.08
1N964B	13	9.5	13	.11	.08
1N965B	15	8.5	16	.10	.08
1N966B	16	7.8	17	.10	.08
1N967B	18	7	21	.10	.08
1N968B	20	6.2	25	.20	.15
1N969B	22	5.6	29	.19	.15
1N970B	24	5.2	33	.19	.16
1N971B	27	4.6	41	.19	.16
1N972B	30	4.2	49	.19	.16
1N973B	33	3.8	58	.19	.16
1N974B	36	3.4	70	.19	.16
1N975B	39	3.2	80	.86	.72
1N976B	43	3	93	.72	.60
1N977B	47	2.7	105	1.01	.85
1N978B	51	2.5	125	.70	.53
1N979B	56	2.2	150	1.01	.84
1N980B	62	2	185	.92	.42
1N981B	68	1.8	230	.92	.42
1N982B	75	1.7	270	.86	.72
1N983B	82	1.5	330	1.04	.86
1N984B	91	1.4	400	.86	.72

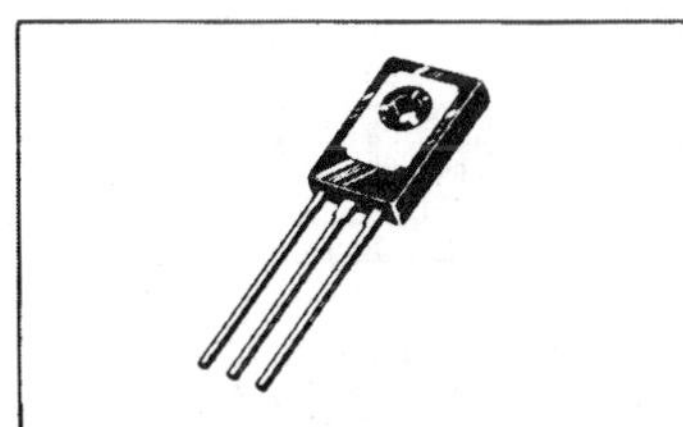

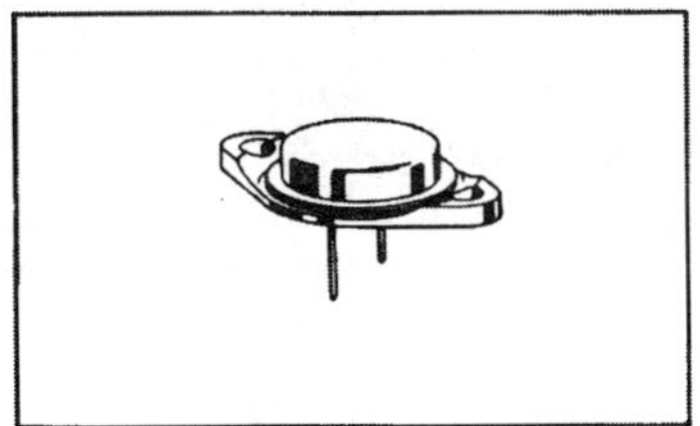

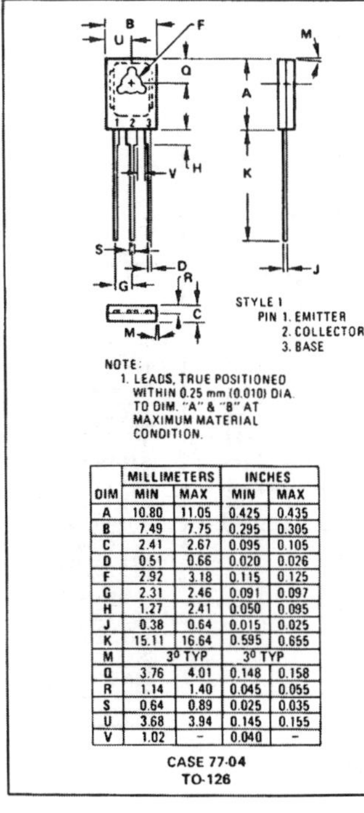

NOTE:
1. LEADS, TRUE POSITIONED WITHIN 0.25 mm (0.010) DIA. TO DIM. "A" & "B" AT MAXIMUM MATERIAL CONDITION.

DIM	MILLIMETERS MIN	MILLIMETERS MAX	INCHES MIN	INCHES MAX
A	10.80	11.05	0.425	0.435
B	7.49	7.75	0.295	0.305
C	2.41	2.67	0.095	0.105
D	0.51	0.66	0.020	0.026
F	2.92	3.18	0.115	0.125
G	2.31	2.46	0.091	0.097
H	1.27	2.41	0.050	0.095
J	0.38	0.64	0.015	0.025
K	15.11	16.64	0.595	0.655
M	3° TYP		3° TYP	
Q	3.76	4.01	0.148	0.158
R	1.14	1.40	0.045	0.055
S	0.64	0.89	0.025	0.035
U	3.68	3.94	0.145	0.155
V	1.02	–	0.040	–

CASE 77-04
TO-126

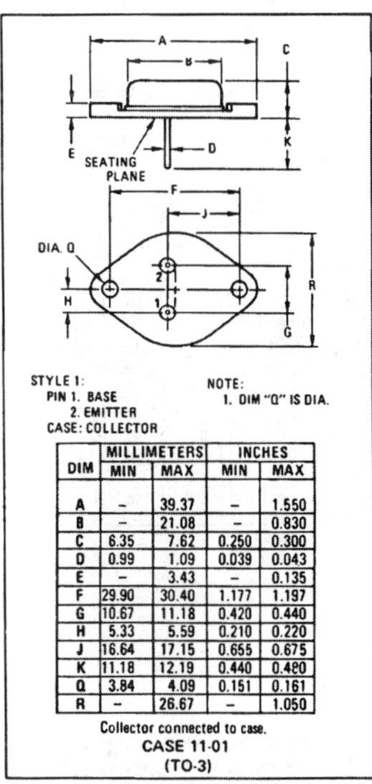

DIM	MILLIMETERS MIN	MILLIMETERS MAX	INCHES MIN	INCHES MAX
A	–	39.37	–	1.550
B	–	21.08	–	0.830
C	6.35	7.62	0.250	0.300
D	0.99	1.09	0.039	0.043
E	–	3.43	–	0.135
F	29.90	30.40	1.177	1.197
G	10.67	11.18	0.420	0.440
H	5.33	5.59	0.210	0.220
J	16.64	17.15	0.655	0.675
K	11.18	12.19	0.440	0.480
Q	3.84	4.09	0.151	0.161
R	–	26.67	–	1.050

Collector connected to case.
CASE 11-01
(TO-3)

Siemens Opto Devices

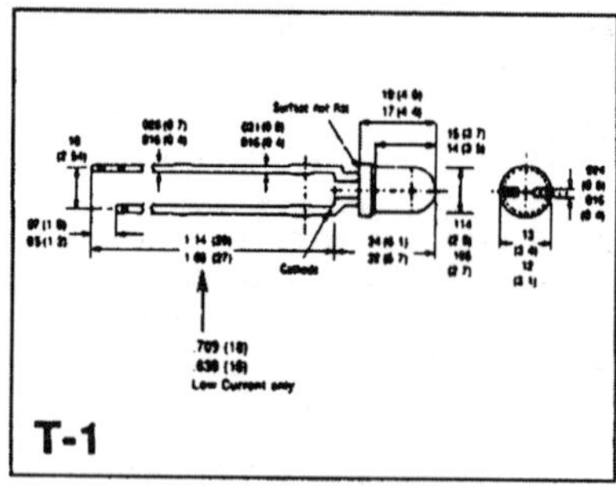

T-1

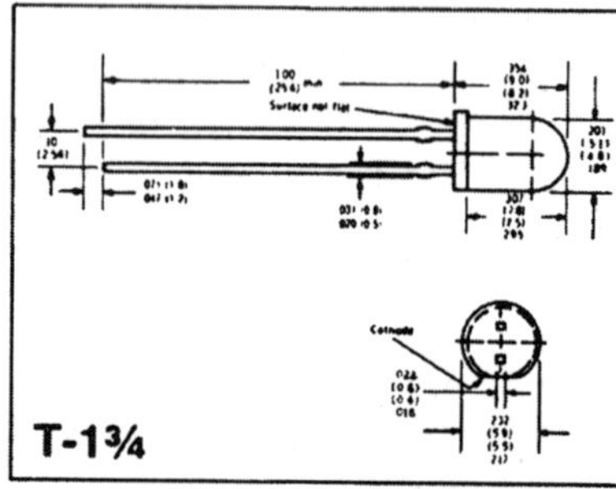

T-1¾

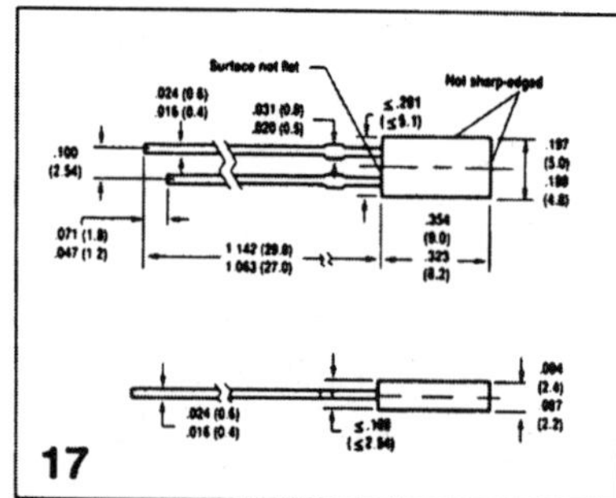

17

SIEMENS LED INDICATOR LAMPS

Cat. Number	Color	Lens	View Angle	LV (mcd) Min.	LV (mcd) mA	Fwd. I (mA)	Fig.	Net Price Each 1-99	Net Price Each 100-Up
3MM DIA. LEDS									
LR3360-G	Red	Diff.	70°	1.6	10	45	T-1	**$.30**	**$.30**
LR3360-FJ	Red	Diff.	70°	1.0	10	45	T-1	**.19**	**.25**
LS3360-HL	Super Red	Diff.	70°	2.5	10	45	T-1	**.24**	**.24**
LY3360-HL	Yellow	Diff.	70°	2.5	10	45	T-1	**.24**	**.24**
LG3360-GK	Green	Diff.	70°	1.6	10	45	T-1	**.24**	**.24**
5MM DIA. LEDS									
LR5360-FJ	Red	Diff.	50°	1.0	10	45	T-1¾	**.25**	**.25**
LS5360-KN	Super Red	Diff.	50°	6.3	10	45	T-1¾	**.27**	**.27**
LY5360-HL	Yellow	Diff.	50°	2.5	10	45	T-1¾	**.24**	**.24**
LG5360-GK	Green	Diff.	50°	1.6	10	45	T-1¾	**.24**	**.24**
5MM DIA. SYMBOL LEDS									
LSB480-EH	Super Red	P. Diff.*	100°	0.63	10	45	17	**.29**	**.29**
LGB480-EH	Green	P. Diff.*	100°	0.63	10	45	17	**.29**	**.29**

P. Diff.* = Partly Diffused

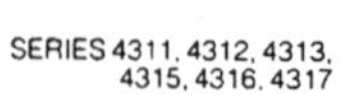

SERIES 4311, 4312, 4313, 4315, 4316, 4317

SERIES 4321, 4323, 4322, 4325, 4326, 4327

SERIES 4331, 4332, 4333, 4335, 4336, 4337

SERIES 4351 THRU 4357

SERIES 4341 THRU 4347

SERIES 4371, 4372, 4375, 4376, 4377 LENSES; 4370 SPACER

IDI PANEL MOUNT LENSES

Provide quick and economical panel mounting, plus electrostatic discharge protection for T-1 and T-1¾ LEDs. Constructed from polycarbonate material. The 4331-4337 series snap into .170/.174 (4.32/4.42) dia. hole; the 4351-4357 series snap into .187/.192 (4.75/4.87) dia. hole; 4311-4317, 4321-4322 and 4371-4377 into .248/.252 (6.30/6.40) dia. hole and 4341-4347 into .354/.360 (9.00/9.14) dia. hole. Series 4311-4317, 4321-4327 and 4371-4370 feature molded in Fresnel rings that spread light evenly, providing wide viewing angle viewing (up to 180°) with a bright attractive appearance. Series 4341-4347 and 4371-4377 can be used with narrow beam LEDs. The 4370 spacer is required with the 4371-4377 series lenses where the panel thickness is less than .19 (.48). T-1 LEDs use 4330, 4350 Series lenses; T-1¾ LEDs use 4310, 4320 and 4370 Series, and 4340 series fit T-1¾ narrow beam LEDs as well as Incandescent and Neon Lamps.

Lens Color	For T-1 (3mm) LEDs Cat. No.	Net Price	Cat. No.	Net Price	For T-1¾ (5mm) LEDs Cat. No.	Net Price	Cat. No.	Net Price
PANEL MOUNT LENSES								
Red	**4331**	**$.39**	**4351**	**$.45**	**4311**	**$.39**	**4321**	**$.45**
Clear	—		**4352**	**.45**	**4312**	**.39**	**4322**	**.45**
Amber	—		**4353**	**.45**	—		—	
White	—		**4354**	**.45**	—		—	
Green	**4335**	**.39**	**4355**	**.45**	**4315**	**.39**	**4325**	**.45**
Blue	—		**4356**	**.45**	—		—	
Yellow	**4337**	**.39**	**4357**	**.45**	**4317**	**.39**	**4327**	**.45**

Lens Color	For T-1¾ (5mm) LEDs Cat. No.	Net Price	Cat. No.	Net Price
PANEL MOUNT LENSES				
Red	**4341**	**$.45**	**4371**	**$.45**
Clear	**4342**	**.45**	**4372**	**.45**
Amber	**4343**	**.45**	—	
White	**4344**	**.45**	—	
Green	**4345**	**.45**	**4375**	**.45**
Blue	**4346**	**.45**	—	
Yellow	**4347**	**.45**	**4377**	**.45**
Spacer	—		**4370**	**.13**

DIALIGHT SERIES 521
DIODE-LITE SOLID STATE LAMPS

High luminance light sources offer longer lamp life, lower current at low cost. For applications requiring direct mount through a panel. IC compatible; vibration and shock resistant; wide viewing angle. Solid state reliability. Leads are flexible enough for bending and rigid enough for direct mounting to a PC board.

Cat. No.	Color	Rated Volts	Current mA	Lumen mcd	Illus.	Net Price Each 1-99	Net Price Each 100-Up
T-1-¾ BUILT-IN INTEGRAL CURRENT LIMIT RESISTOR							
521-9183	Red Diff.	5	15	2.0	K	**$.93**	**$.78**
T-1 WITH DIFFUSED LENS							
521-9216	Red Diff.	1.7	20	2.0	P	**.67**	**.56**
521-9210	Green Diff.	2.4	20	2.0	P	**.67**	**.56**
521-9211	Yellow Diff.	2.2	10	2.5	P	**.67**	**.56**
T-1-¾ HIGH EFFICIENCY WITH DIFFUSED LENS							
521-9246	Red Diff.	2.2	10	4.0	O	**.71**	**.58**
521-9247	Red Non-Diff.	2.2	10	24.0	O	**.71**	**.58**
521-9248	Yellow Diff.	2.2	10	3.0	O	**.71**	**.58**
521-9249	Yellow Non-Diff.	2.2	10	16.0	O	**.71**	**.58**
521-9250	Green Diff.	2.4	20	3.0	O	**.71**	**.58**
521-9251	Green Non-Diff.	2.3	20	16.0	O	**.71**	**.58**
SNAP-IN MOUNTING HARDWARE FOR 521 SERIES REAR MOUNTED LEDS							
515-0004	Use with T-1-¾ Standard Profile LEDs; K, O, etc.					**.59**	**.44**
515-0005	Use with T-1-¾ Low Profile LEDs					**.88**	**.59**
515-0006	Use with T-1 LEDs; P, etc.					**.88**	**.59**

Page D4

Wakefield Engineering

AKEFIELD PRE-DRILLED HEAT SINK DIMENSIONS

Cat. No.	Width in. (mm)	Depth in. (mm)	Height in. (mm)
301-M	2.00(50,80)	0.75(19,05)	2.00(50,80)
301-N	2.00(50,80)	0.75(19,05)	2.00(50,80)
401-F	4.70(119,38)	1.50(38,10)	1.25(31,75)
403-F	4.70(119,38)	3.00(76,20)	1.50(38,10)
413-F	4.75(120,65)	3.00(76,20)	1.88(47,75)
421-F	4.75(120,65)	3.00(76,20)	2.63(66,80)
476-W	5.00(127,00)	6.00(152,40)	5.00(127,00)
601-E	2.00(50,80)	1.25(31,75)	.562(14,27)
601-F	2.00(50,80)	1.25(31,75)	.562(14,27)
695-1-B	.62(15,75) I.D.	1.33(33,78) O.D.	.50(12,70)

AKEFIELD PRE-DRILLED HEAT SINKS

ıt. .	Material	Finish	Net Price Each 1-24	25-49	50-Up
-32 TAPPED, .625" DEEP		**[D0-203AA (D0-4)]**			
1-M	Aluminum	Black Anodized	**$11.34**	**$8.58**	**$6.91**
28 TAPPED, .625" DEEP		**[D0-203AB (D0-5)/T0-208AA (T0-48)/(T0-65)]**			
1-N	Aluminum	Black Anodized	**12.22**	**9.08**	**7.22**
00" DIA. HOLE (D0-4)					
1-E	Aluminum	Black Anodized	**7.06**	**5.38**	**4.52**
5-1-B	Aluminum	Black Anodized	**.83**	**.83**	**.83**
70" DIA. HOLE (D0-5)					
1-F	Aluminum	Black Anodized	**12.96**	**9.61**	**8.36**
3-F	Aluminum	Black Anodized	**11.91**	**9.21**	**7.79**
3-F	Aluminum	Black Anodized	**17.58**	**13.59**	**11.49**
1-F	Aluminum	Black Anodized	**14.07**	**10.88**	**8.89**
1-F	Aluminum	Black Anodized	**6.97**	**5.31**	**4.46**
55" DIA. HOLE (T0-93)					
6-W	Aluminum	Black Anodized	**42.96**	**35.79**	**31.81**

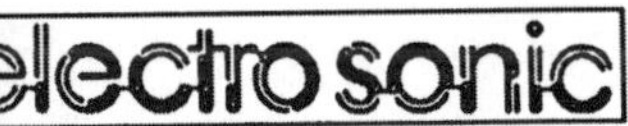

Page D5

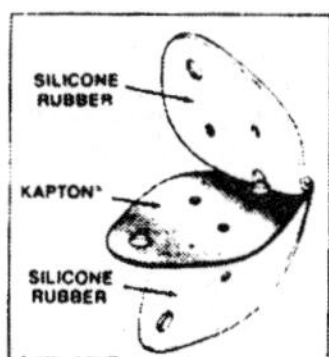

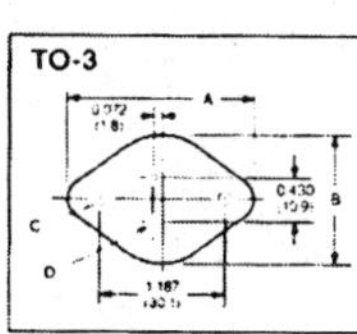

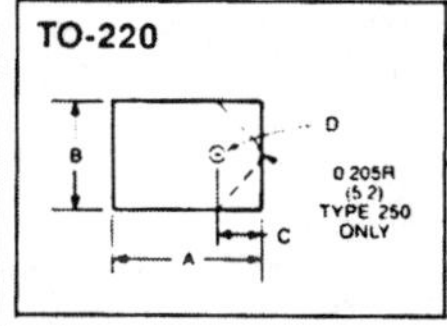

WAKEFIELD 175 SERIES GREASELESS, THERMALLY CONDUCTIVE KAPTON™ REINFORCED INSULATORS

A highly effective, thermally conductive, greaseless way to interface transistors to heat sinks. New improved Kapton™ film gives a high dielectric, physically tough insulator that withstands high voltages. No messy thermal grease is required. These insulators are the most effective products of their type. Available in standard shapes to fit TO-3 and TO-220 semiconductors.

Cat. No.	Dimensions Inches (mm) A	B	C	D
FITS TO-3 CASE				
1756310P	1.593(40.5)	1.100(27.9)	0.156(4.0)	0.062(1.6)
1756320P	1.650(41.9)	1.140(29.0)	0.122(3.1)	0.062(1.6)
1756330P	1.650(41.9)	1.140(29.0)	0.140(3.6)	0.093(2.4)
FITS TO-220 CASE				
1756210P	0.687(17.4)	0.562(14.3)	0.218(5.5)	0.125(3.2)
1756220P	0.710(18.0)	0.500(12.7)	0.160(4.1)	0.141(3.6)
1756230P	0.750(19.1)	0.500(12.7)	0.187(4.7)	0.125(3.2)
1756240P	0.750(19.1)	0.500(12.7)	0.187(4.7)	0.147(3.7)
1756250P	0.865(22.0)	0.650(16.5)	0.250(5.2)	0.140(3.6)

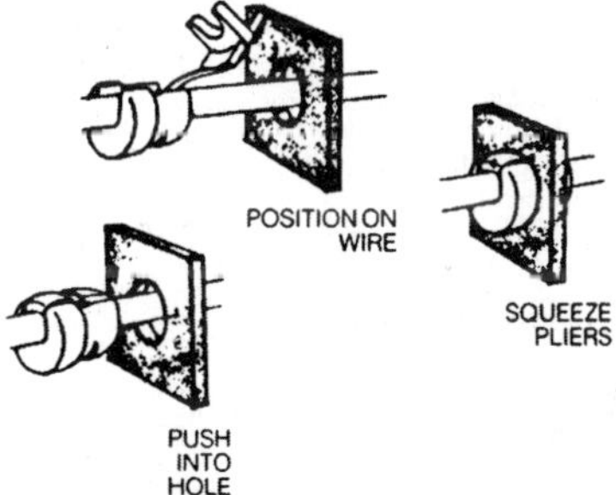

HEYCO STRAIN RELIEFS

These precision nylon bushings securely anchor power supply cords to the chassis. Snaps into chassis and absorbs all pull, push or strain.

No.	Wire Type	Hole Size	Net Price Per 100 1-99	100-Up
3P	POSJ, POT	7/16"	**$15.50**	**$13.00**
5P	SV, HPD, HPN	1/2"	**16.50**	**15.10**
6P	SJ	5/8"	**19.00**	**16.50**

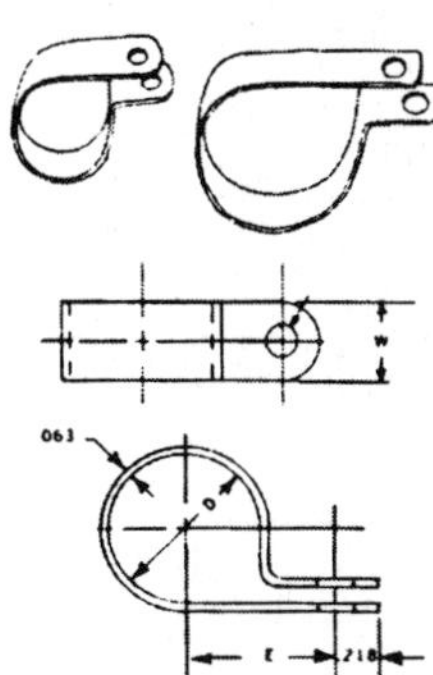

Cat. No. Dia.	E	W Width	Net Price Per 1000 1-99	100-Up
CLN-1/8	.328	3/8	**$70.95**	**$56.76**
CLN-3/16	.390	3/8	**70.95**	**56.76**
CLN-1/4	.421	3/8	**70.94**	**56.76**
CLN-5/16	.453	3/8	**56.31**	**45.05**
CLN-3/8	.483	3/8	**59.69**	**47.75**
CLN-7/16	.531	3/8	**60.81**	**48.65**
CLN-1/2	.562	3/8	**63.06**	**50.45**
CLN-9/16	.593	1/2	**102.48**	**81.98**
CLN-5/8	.625	1/2	**108.11**	**86.49**
CLN-11/16	.656	1/2	**115.99**	**92.79**
CLN-3/4	.765	1/2	**122.75**	**98.20**
CLN-7/8	.812	1/2	**134.01**	**107.21**
CLN-1	.906	1/2	**138.51**	**110.81**
CLN-1-1/8	.970	5/8	**183.56**	**146.85**
CLN-1-3/16	1.000	5/8	**186.93**	**149.54**
CLN-1-1/4	1.000	5/8	**193.69**	**154.96**
CLN-1-3/8	1.125	5/8	**159.46**	**159.46**
CLN-1-1/2	1.125	5/8	**207.21**	**165.77**

LENLINE NYLON CABLE CLAMPS

A durable clamp that is tougher, resists acids better and has a higher operating temperature. Material and flammability rating 94V-2. Color natural.

P.E.C. MIL Style Potentiometers

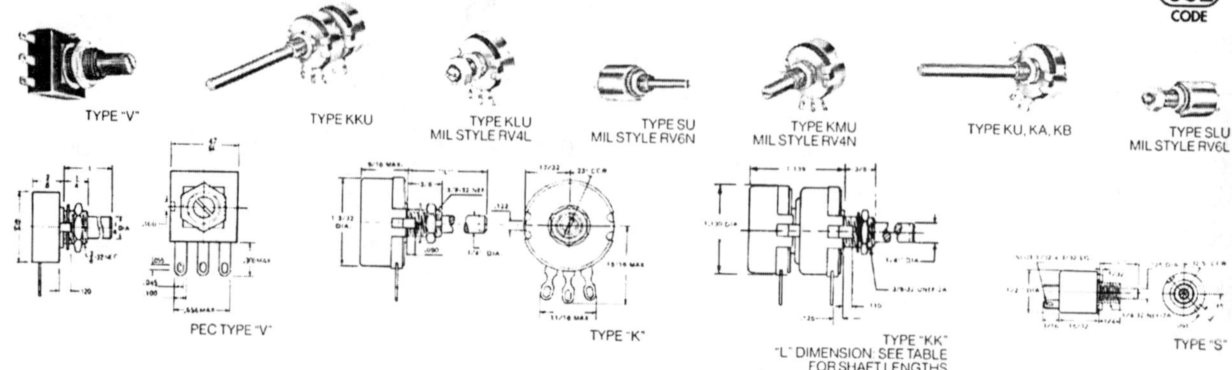

P.E.C. TYPE V 2-WATT CERMET POTENTIOMETERS

Electrical and environmental performance exceeds MIL-R-94/RV4—rotational life 500,000 cycles. Non-metallic housing, shaft, and bushing. Dielectric strength 2500V RMS at sea level. Operating temperature -50° to +110°C. Mechanical angle 274° ± 3°. Housing ¾" square. Bushing ⅜" NEF x ⅜" long. Slotted shaft ¼" dia. x ⅞" long. Solder lug terminals. Linear taper.

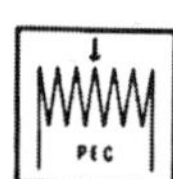

PRECISION ELECTRONIC COMPONENTS
MOLDED COMPOSITION POTENTIOMETERS, ½ AND 2 WATTS

PEC potentiometers feature a hot-molded resistance element integrated with a low resistance collector, gold-plated terminals and an insulating base, to form a thermally stable, reliable unit. The bridging contact brush is a single block of corrosion-free carbon. The rotational life is far in excess of film-type units. Noise level is extremely low. There is no electrolytic metal transfer to the resistance track from the carbon brush and terminals. MIL Style RV4 potentiometers have ⅜" bushing and ¼" diameter shaft; operating voltage is 500V. Miniature RV6 potentiometers have ¼" bushing and ⅛" diameter shaft; operating voltage is 350V. All units are dust and splash proof, fungus and corrosion reistant and meet the latest requirements of MIL-R-94. Units are stamped with MIL and commercial designations. Made in Canada.

P.E.C. POTENTIOMETERS

	CERMET	MOLDED COMPOSITION												
		MIL STYLE RV4, 2 WATT (1 ¹⁄₃₂" DIA.)								MIL STYLE RV6, ½ WATT (½" DIA.)			
	2 Watt Single Unit ⅞" Slotted Shaft	Single Unit ⅞" Slotted Shaft		Single Unit 2" Unslotted Shaft			Locking Type ⅝" Slotted Shaft		Dual Unit 2" Shaft	Miniature Locking ⅝" Slotted Shaft		Miniature ⅞" Slotted Shaft	
Resistance Ohms	Type V Linear	Type KMU Linear	MIL RV4N AYSD-	Type KU Linear	Type KA CW Log	Type KB CCW Log	Type KLU Linear	MIL RV4L-AYSA-	Type KKU Linear	Type SLU Linear	MIL RV6L-AYSA-	Type SU Linear	MIL RV6N-AYSD-
50	—	KMU5001	500A	KU5001	—	—	KLU5001	500A	—	—	—	—	—
100	V6U1011	KMU1011	101A	KU1011	—	—	KLU1011	101A	—	SLU1011	101A	SU1011	101A
150	—	KMU1511	151A	KU1511	—	—	KLU1511	151A	—	—	—	—	—
250	V6U2511	KMU2511	251A	KU2511	—	—	KLU2511	251A	—	SLU2511	251A	SU2511	251A
350	—	KMU3511	351A	KU3511	—	—	KLU3511	351A	—	—	—	—	—
500	V6U5011	KMU5011	501A	KU5011	—	—	KLU5011	501A	—	SLU5011	501A	SU5011	501A
750	—	KMU7511	751A	KU7511	—	—	KLU7511	751A	—	—	—	—	—
1000	V6U1021	KMU1021	102A	KU1021	—	—	KLU1021	102A	—	SLU1021	102A	SU1021	102A
1500	—	KMU1521	152A	KU1521	—	—	KLU1521	152A	—	—	—	—	—
2500	V6U2521	KMU2521	252A	KU2521	—	—	KLU2521	252A	—	SLU2521	252A	SU2521	252A
3500	—	KMU3521	352A	KU3521	—	—	KLU3521	352A	—	—	—	—	—
5000	V6U5021	KMU5021	502A	KU5021	—	—	KLU5021	502A	—	SLU5021	502A	SU5021	502A
7500	—	KMU7521	752A	KU7521	—	—	KLU7521	752A	—	—	—	—	—
10000	V6U1031	KMU1031	103A	KU1031	—	KB1031	KLU1031	103A	KKU1031	SLU1031	103A	SU1031	103A
15000	—	KMU1531	153A	KU1531	—	—	KLU1531	153A	—	—	—	—	—
25000	V6U2531	KMU2531	253A	KU2531	—	KB2531	KLU2531	253A	KKU2531	SLU2531	253A	SU2531	253A
35000	—	KMU3531	353A	KU3531	—	—	KLU3531	353A	—	—	—	—	—
50000	V6U5031	KMU5031	503A	KU5031	—	KB5031	KLU5031	503A	KKU5031	SLU5031	503A	SU5031	503A
75000	—	KMU7531	753A	KU7531	—	—	KLU7531	753A	—	—	—	—	—
.1 meg	V6U1041	KMU1041	104A	KU1041	KA1041	—	KLU1041	104A	KKU1041	SLU1041	104A	SU1041	104A
.15 meg	—	KMU1541	154A	KU1541	—	—	KLU1541	154A	—	SLU1541	154A	SU1541	154A
.25 meg	V6U2541	KMU2541	254A	KU2541	KA2541	—	KLU2541	254A	KKU2541	SLU2541	254A	SU2541	254A
.35 meg	—	KMU3541	354A	KU3541	—	—	KLU3541	354A	—	—	—	—	—
.5 meg	V6U5041	KMU5041	504A	KU5041	KA5041	—	KLU5041	504A	KKU5041	SLU5041	504A	SU5041	504A
.75 meg*	—	KMU7542	754A	KU7542	—	—	KLU7542	754B	—	—	—	—	—
1.0 meg*	V6U1052	KMU1052	105B	KU1052	KA1052	—	KLU1052	105B	KKU1052	SLU1052	105B	SU1052	105B
1.5 meg*	—	KMU1552	155B	KU1552	—	—	KLU1552	155B	—	—	—	—	—
2.0 meg*	—	KMU2052	205B	KU2052	—	—	KLU2052	205B	—	SLU2052	205B	SU2052	205B
2.5 meg*	—	KMU2552	255B	KU2552	KA2552	—	KLU2552	255B	—	SLU2552	255B	SU2552	255B
3.5 meg*	—	KMU3552	355B	KU3552	—	—	KLU3552	355B	—	—	—	—	—
5.0 meg*	—	KMU5052	505B	KU5052	—	—	KLU5052	505B	—	SLU5052	505B	SU5052	505B
Net Each													
1-9	$5.65		$6.23	$6.75	$6.83	$6.83		$7.13	$15.20		$7.03		$6.75
10-24	5.65		5.60	6.08	6.14	6.14		6.41	13.68		6.32		6.08
25-49	4.52		5.04	5.47	5.53	5.53		5.77	12.31		5.69		5.47
50-99	4.52		4.54	4.92	4.98	4.98		5.19	11.08		5.12		4.92
100-249	3.39		3.86	4.18	4.23	4.23		4.41	9.42		4.35		4.18

Resistance Tolerance ±10%, except* ±20%. **NOTE:** Prices vary with resistance value. Prices shown are for most popular values.

Page D6

Philips Resistors

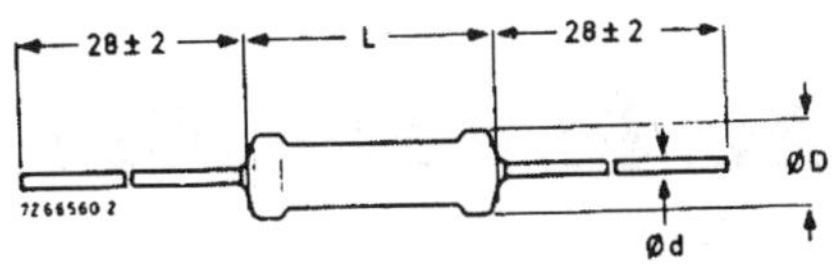

PHILIPS CARBON FILM AND METAL FILM RESISTORS (CONT'D.)

Cat. No.	Wattage	Tolerance	Resistance Range	Decade Chart	Dimensions (mm) L	ϕD	ϕd	Working Voltage	Net Price Per 1000 1-99	100-999	1000-Up
CARBON FILM											
CR25*	¼	5%	1 ohm-10 meg	A	6.5	2.5	0.6	250	**$35.63**	**21.38**	**10.33**
METAL FILM											
SFR16SJ*	½	5%	1 ohm-9.1 ohms	A	3.7	1.9	0.5	200	**62.43**	**37.46**	**24.97**
SFR16SJ*	½	5%	10 ohms-1 meg	A	3.7	1.9	0.5	200	**56.64**	**33.99**	**22.66**
SFR16SJ*	½	5%	1.1 meg-2.4 meg	A	3.7	1.9	0.5	200	**56.64**	**33.99**	**22.66**
SFR25-0*	—	—	Zero Ohm Jumper	—	6.5	2.5	0.6	—	**68.52**	**41.11**	**27.41**
SFR25HJ*	½	5%	1 ohm-9.1 ohms	A	6.5	2.5	0.6	250	**91.38**	**54.83**	**36.55**
SFR25HJ*	½	5%	10 ohms-1 meg	A	6.5	2.5	0.6	250	**76.14**	**45.68**	**30.45**
SFR25HJ*	½	5%	1.1 meg-10 meg	A	6.5	2.5	0.6	250	**91.38**	**54.83**	**36.55**
PRECISION METAL FILM											
MRS25F*	0.6	1%	1 ohm-9.76 ohms	B	6.5	2.5	0.6	350	**184.25**	**110.55**	**49.41**
MRS25F*	0.6	1%	10 ohms-1 meg	B	6.5	2.5	0.6	350	**105.07**	**63.04**	**25.83**
MRS25F*	0.6	1%	1.02 meg-10 meg	B	6.5	2.5	0.6	350	**184.25**	**110.55**	**49.41**
POWER METAL FILM											
PR02J	2.0	5%	1.0 ohms-9.1 ohms	A	10.0	3.9	0.8	500	**418.75**	**251.25**	**167.50**
PR02J	2.0	5%	10 ohms-1 meg	A	10.0	3.9	0.8	500	**251.25**	**150.75**	**100.50**
PR02J	2.0	5%	1.1 meg-33 meg	A	10.0	3.9	0.8	500	**251.25**	**150.75**	**100.50**
HIGH-VOLTAGE METAL FILM											
VR37J	0.5	1%	1 meg-33 meg	A	9.0×3.7×0.7			700	**1256.25**	**753.75**	**502.50**

*Also available in reels of 5000.

Cat. No.	Wattage	Tolerance	Resistance Range	Decade Chart	Dimensions (mm)	Working Voltage	Net Price Per 1000 1-999	1000-Up
SURFACE MOUNT CHIP—SIZE 1206								
RC01-0	—	—	Zero Ohm Jumper	—	3.2×1.6×0.6	—	**$39.59**	**$23.75**
RC01J	¼	5%	1 ohm-9.1 ohms	A	0.2×1.6×0.6	200	**64.71**	**38.83**
RC01J	¼	5%	10 ohms-1 meg	A	3.2×1.6×0.6	200	**39.59**	**23.75**
RC01J	¼	5%	1.1 meg-10 meg	A	3.2×1.6×0.6	200	**64.71**	**33.83**
RC02H	⅛	1%	100 ohms-1 meg	A	3.2×1.6×0.6	200	66.23	**39.74**

IRC MOLDED COMPOSITION RESISTORS

1/4 W

1/2 W

- Superior Moisture resistance, high frequency characteristics and solderability
- Molded-in leads giving greater lead strength than slug types
- Molded jackets for ease in use with automated cutting and forming or insertion

Toler-ance	Watts @70°C	Type	Dimensions Body Length	Body Diameter	Lead Length	Lead Diameter	Maximum Voltage	Resistance Range in Ohms	Net Price per 100 1-99	100-249	250-999	1000-Up
5%	¼	**IBT4**	0.250"	0.090"	1.5"	.025"	250	10–910	**$45.76**	**$34.32**	**$22.88**	**$17.16**
								1K–1M	**41.60**	**31.20**	**20.80**	**15.60**
	½	**IBT2**	0.390"	0.140"	1.5"	.031"	350	10–910	**46.54**	**34.94**	**23.30**	**17.47**
								1K–1M	**43.68**	**32.76**	**21.84**	**16.38**
								1M1–20M	**45.76**	**34.32**	**22.88**	**17.16**

DECADE CHART "A"
STANDARD 5% VALUES (Ω): NOTE *10% VALUES

1.0	13	91	620	4300	30K	200K	1.8M*
1.2	15	100	680	4700	33K	220K	2.2M*
1.5	16	110	750	5100	36K	240K	2.7M*
1.8	18	120	820	5600	39K	270K	3.3M*
2.2	20	130	910	6200	43K	300K	3.9M*
2.7	22	150	1000	6800	47K	330K	4.7M*
3.3	24	160	1100	7500	51K	360K	5.6M*
3.9	27	180	1200	8200	56K	390K	6.8M*
4.3	30	200	1300	9100	62K	430K	8.2M*
4.7	33	220	1500	10K	68K	470K	10M*
5.1	36	240	1600	11K	75K	510K	12M
5.6	39	270	1800	12K	82K	560K	15M
6.2	43	300	2000	13K	91K	620K	18M
6.8	47	330	2200	15K	100K	680K	22M
7.5	51	360	2400	16K	110K	750K	27M
8.2	56	390	2700	18K	120K	820K	33M
9.1	62	430	3000	20K	130K	910K	
10	68	470	3300	22K	150K	1.0M	
11	75	510	3600	24K	160K	1.2M*	
12	82	560	3900	27K	180K	1.5M*	

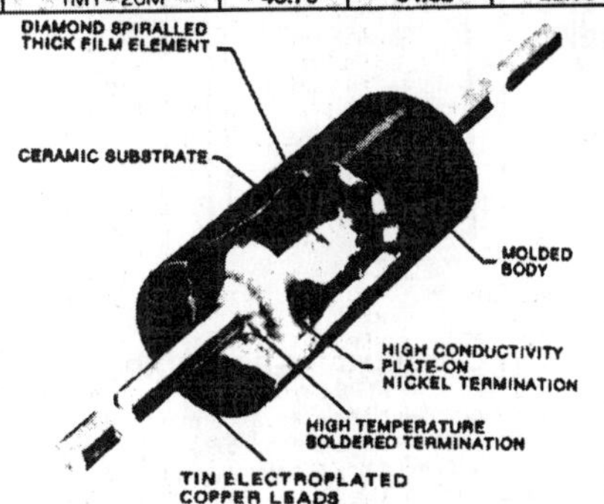

DIMENSIONS INCHES (MM)
LENGTH: .515" ± .010 (13.08 ± .254)
DIAMETER: .255" ± .008 (6.48 ± .203)
LEADS: 1.5" × .032" ± .002 (38.10 × .813 ± .051)

IRC TYPE G5-3 THICK FILM METAL GLAZE POWER RESISTORS

The GS-3 RG Metal Glaze Resistor reflects the high power handling capacity of both Metal Glaze and RG type construction.

Page D7

Philips Ceramic Capacitors

GENERAL PURPOSE CERAMIC DISC CAPACITORS
TYPE DD

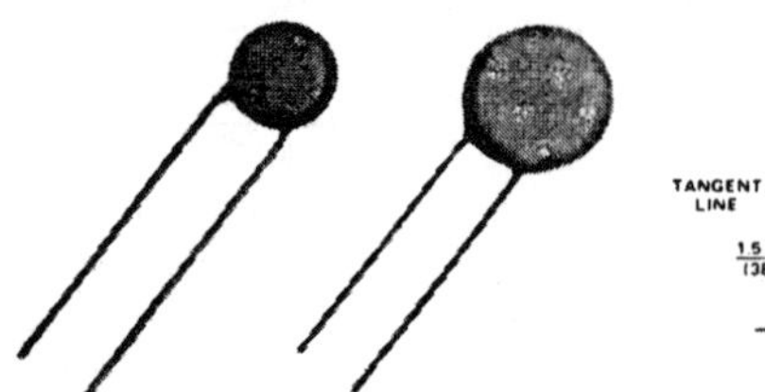

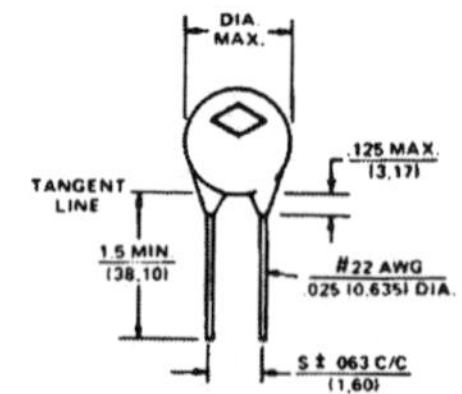

General Purpose capacitors provide a wide choice of 1000 Volt DC rated units for general purpose applications. These high dielectric ceramic discs offer small physical size, tighter tolerances and a rugged humidity resistant coating. Capacitors are primarily designed for non-critical coupling, bypass and filter applications found in all types of entertainment, industrial and medical equipment. Operating temperature: -55°C to +85°C. Rated at 1000 VDCW, except DD6-103 thru DD-104 which are 600 VDCW. Dielectric strength: 200% of rated working voltage. Insulation resistance: 10,000 megohms minimum. Tested at 500 volts DC maximum. Leads: tinned wire, minimum length 1½". Maximum thickness: .156", except DD-303, 503 and 104 which are .250". Encapsulation: Phenolic coated, wax impregnated.

Cap.	Cat. No.	Tol.	EIA T.C. Code	Max. Dia.	Lead Space	Net Price Each 1-99	Net Price Each 100-Up
DD-SERIES; 1000 VOLT DCW.							
3.3 pF	**DD-3R3**	±0.5 pf	S2L	.290"	.250"	**$.32**	**$.13**
5	**DD-050**	±10%	S21	.290"	.250"	**.32**	**.13**
6	**DD-060**	±10%	S2L	.290"	.250"	**.32**	**.13**
6.8	**DD-6R8**	±10%	S2L	.290"	.250"	**.32**	**.13**
8	**DD-080**	±10%	S2L	.290"	.250"	**.32**	**.13**
10	**DD-100**	±10%	S2L	.290"	.250"	**.32**	**.13**
12	**DD-120**	±10%	S2L	.290"	.250"	**.32**	**.13**
15	**DD-150**	±10%	S2L	.290"	.250"	**.32**	**.13**
18	**DD-180**	±10%	S2L	.290"	.250"	**.32**	**.13**
20	**DD-200**	±10%	S2L	.290"	.250"	**.32**	**.13**
22	**DD-220**	±10%	S2L	.290"	.250"	**.32**	**.13**
27	**DD-270**	±10%	S2L	.290"	.250"	**.32**	**.13**
30	**DD-300**	±10%	S2L	.290"	.250"	**.32**	**.13**
33	**DD-330**	±10%	S2L	.290"	.250"	**.32**	**.13**
39	**DD-390**	±10%	S2L	.290"	.250"	**.32**	**.13**
47	**DD-470**	±10%	S2L	.290"	.250"	**.32**	**.13**
50	**DD-500**	±10%	S3N	.290"	.250"	**.32**	**.13**
56	**DD-560**	±10%	S3N	.305"	.250"	**.32**	**.13**
68	**DD-680**	±10%	S3N	.305"	.250"	**.32**	**.13**
75	**DD-750**	±10%	S3N	.305"	.250"	**.32**	**.13**
82	**DD-820**	±10%	S3N	.305"	.250"	**.32**	**.13**
100	**DD-101**	±10%	S3N	.305"	.250"	**.23**	**.09**
120	**DD-121**	±10%	S3N	.305"	.250"	**.32**	**.13**
150	**DD-151**	±10%	S3N	.305"	.250"	**.32**	**.13**
180	**DD-181**	±10%	S3N	.305"	.250"	**.32**	**.13**
200	**DD-201**	±10%	S3N	.305"	.250"	**.32**	**.13**
220	**DD-221**	±10%	Z5F	.305"	.250"	**.32**	**.13**
240	**DD-241**	±10%	Z5F	.305"	.250"	**.32**	**.13**
250	**DD-251**	±10%	Z5F	.305"	.250"	**.32**	**.13**
270	**DD-271**	±10%	Z5F	.305"	.250"	**.32**	**.13**
300	**DD-301**	±10%	Z5F	.350"	.250"	**.32**	**.13**
330	**DD-331**	±10%	Z5F	.305"	.250"	**.32**	**.13**
390	**DD-391**	±10%	Z5F	.305"	.250"	**.32**	**.13**

Cap.	Cat. No.	Tol.	EIA T.C. Code	Max. Dia.	Lead Space	Net Price Each 1-99	Net Price Each 100-Up
DD-SERIES; 1000 VOLT DCW.							
470 pF	**DD-471**	±10%	Z5F	.305"	.250"	**$.32**	**$.13**
500	**DD-501**	±10%	Z5F	.305"	.250"	**.32**	**.13**
510	**DD-511**	±10%	Z5F	.305"	.250"	**.32**	**.13**
560	**DD-561**	±10%	Z5F	.305"	.250"	**.32**	**.13**
680	**DD-681**	±10%	Z5R	.305"	.250"	**.32**	**.13**
750	**DD-751**	±10%	Z5R	.305"	.250"	**.32**	**.13**
820	**DD-821**	±10%	Z5F	.305"	.250"	**.32**	**.13**
.001 mF	**DD-102**	±10%	Z5F	.405"	.250"	**.31**	**.12**
.001	**DD-102G**	GMV	Z5U	.305"	.250"	**.38**	**.15**
.0012	**DD-122**	±10%	Z5R	.405"	.250"	**.35**	**.14**
.0015	**DD-152**	±10%	Z5R	.405"	.250"	**.35**	**.14**
.0018	**DD-182**	±10%	Z5R	.405"	.250"	**.35**	**.14**
.002	**DD-202**	±20%	Z5U	.405"	.250"	**.35**	**.14**
.0022	**DD-222**	±20%	Z5U	.405"	.250"	**.35**	**.14**
.0025	**DD-252**	±20%	Z5U	.405"	.250"	**.35**	**.14**
.0027	**DD-272**	±20%	Z5U	.405"	.250"	**.35**	**.14**
.003	**DD-302**	±20%	Z5U	.405"	.250"	**.35**	**.14**
.0033	**DD-332**	±20%	Z5U	.405"	.250"	**.35**	**.14**
.0039	**DD-392**	±20%	Z5U	.405"	.250"	**.46**	**.18**
.0047	**DD-472**	±20%	Z5U	.500"	.250"	**.44**	**.19**
.005	**DD-5022**	±20%	Z5U	.500"	.250"	**.53**	**.21**
.005	**DD-502**	GMV	Z5U	.605"	.375"	**.58**	**.23**
.0056	**DD-562**	±20%	Z5U	.500"	.250"	**.49**	**.19**
.0068	**DD-682**	±20%	Z5U	.605"	.375"	**.58**	**.23**
.0082	**DD-822**	±20%	Z5U	.605"	.375"	**.58**	**.23**
.01	**DD-1032**	±20%	Z5U	.705"	.375"	**1.05**	**.53**
.01	**DD-103**	GMV	Z5U	.705"	.375"	**.89**	**.36**
.015	**DD-153**	+80-20%	Z5U	.690"	.375"	**.79**	**.32**
.02	**DD-203**	+80-20%	Z5U	.690"	.375"	**.82**	**.33**
.03	**DD-303**	+80-20%	Z5U	.718"	.375"	**3.26**	**1.30**
.05	**DD-503**	+80-20%	Z5U	.875"	.375"	**4.19**	**1.67**
.1	**DD-104**	+80-20%	Z5U	.975"	.375"	**5.46**	**2.19**

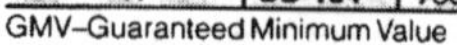

GMV–Guaranteed Minimum Value

EPOXY COATED GENERAL PURPOSE
TYPE CY, CZ

General Purpose capacitors offer high capacitance, miniature size units for general purpose applications. These miniature monolithic capacitors feature radial leads for convenient printed circuit board installation and a rugged epoxy coating for environmental protection. The CY and CZ-Series capacitors are ideally suited for a wide range of non-critical general purpose bypass, coupling and filtering applications. Operating temperature: -55°C to +85°C. Dielectric strength: 200% of rated working voltage. Insulation resistance: 100,000 megohms microfarads whichever is less. Tested at rated working voltage and +25°C. Leads: Tinned wire. Encapsulation: Epoxy coated.

Cap.	Cat. No.	Cap. Tol.	Size Code	Net Price Each 1-99	Net Price Each 100-999	Net Price Each 1000-Up
CY-SERIES; 50 VOLT DCW. EIA TEMPERATURE CHARACTERISTIC Y5V						
.001 µF	**CY15C102M**	±20%	15	**$.41**	**$.16**	**$.12**
.0022	**CY15C222M**	±20%	15	**.54**	**.21**	**.16**
.0033	**CY15C332M**	±20%	15	**1.13**	**1.13**	**1.13**
.0068	**CY15C682M**	±20%	15	**.54**	**.21**	**.16**
.01	**CY15C103M**	±20%	15	**.43**	**.17**	**.12**
.01	**CY15C103P**	+100-0%	15	**.41**	**.16**	**.12**
.022	**CY15C223M**	±20%	15	**.55**	**.22**	**.16**
.1	**CY20C104M**	±20%	20	**.41**	**.16**	**.12**
.1	**CY20C104P**	+100-0%	20	**.39**	**.16**	**.11**
.15	**CY20C154M**	±20%	20	**.59**	**.24**	**.17**
1.0	**CY30C105M**	±20%	30	**1.46**	**.58**	**.42**
1.0	**CY30C105P**	+100-0%	30	**1.44**	**.58**	**.42**
CY-SERIES; 100 VOLT DCW. EIA TEMPERATURE CHARACTERISTIC Y5V						
.01	**CY15A103P**	+100-0%	15	**.43**	**.17**	**.12**
.047	**CY20A473M**	±20%	20	**.84**	**.34**	**.24**
.1	**CY20A104M**	±20%	20	**.80**	**.32**	**.23**
.1	**CY20A104P**	+100-0%	20	**.76**	**.30**	**.22**
.47	**CY30A474M**	±20%	20	**1.86**	**.74**	**.54**
CZ-SERIES; 50 VOLT DCW., EIA TEMPERATURE CHARACTERISTIC Z5U						
.01	**CZ15C103M**	±20%	15	**.37**	**.15**	**.11**
.022	**CZ15C223M**	±20%	15	**.38**	**.15**	**.11**
.047	**CZ20C473M**	±20%	20	**.46**	**.18**	**.13**
.1	**CZ20C104M**	±20%	20	**.36**	**.14**	**.10**
.1	**CZ20C104M-244***	±20%	20	**.36**	**.14**	**.10**
CZ-SERIES; 100 VOLTS DCW. EIA TEMPERATURE CHARACTERISTIC Z5U						
.01	**CZ15A103M**	±20%	15	**.39**	**.15**	**.11**
.1	**CZ20A104M**	±20%	20	**.34**	**.15**	**.11**

*.200" Lead Spacing

Page D8

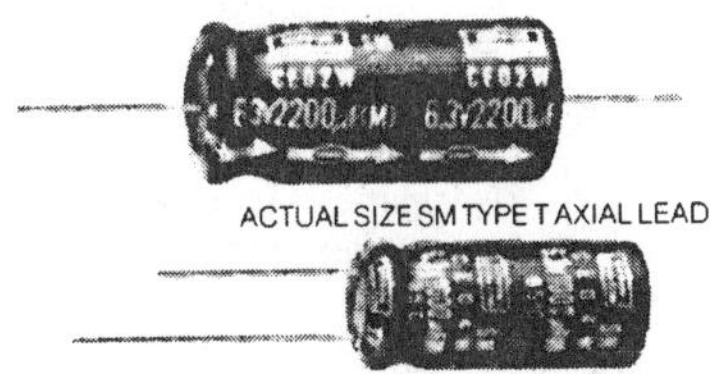

ACTUAL SIZE SM TYPE T AXIAL LEAD

ACTUAL SIZE SM TYPE VB RADIAL LEAD

UNITED CHEMI-CON SM SERIES
MINIATURE ALUMINUM ELECTROLYTIC CAPACITORS
SMALL SIZE, HIGH PERFORMANCE, GENERAL PURPOSE

M Series aluminum electrolytic capacitors are far smaller than conventional eneral-purpose capacitors because of a newly developed high-gain foil. This ew high performance standard series features improved characteristics made ossible by reducing dispersion of leakage current and capacitance. Operating emperature range −40°C to +85°C for 6.3V to 100V; −25° to +85°C for 160 to 50V. Leakage current 0.01CV(µA) or 3µA, whichever is greater, after 2 minutes or 6.3 to 100V; 0.03CV + 15(µA), CV≤1000, or 0.02CV + 25(µA), CV 1000, after 5 ninutes for 160V to 450V. Standard capacitance tolerance ±20%.

ap. F	Part Number	Nom. D×L (mm)	Max. Leakage µA*	Net Price Per 1000		
				1-99	100-499	500-Up
0 VDC, T/AXIAL LEAD, 85°C						
22	**SME10T22RM5X12LL**	5×12.5	3.0	**$320.05**	**$213.37**	**$160.02**
33	**SME10T33RM5X12LL**	5×12.5	3.3	**320.05**	**213.37**	**160.02**
47	**SME10T47RM5X12LL**	5×12.5	4.7	**320.05**	**213.37**	**160.02**
100	**SME10T101M6X16LL**	6.3×16	10.0	**388.63**	**259.09**	**194.32**
220	**SME10T221M6X16LL**	6.3×16	22.0	**388.63**	**259.09**	**194.32**
330	**SME10T331M8X16LL**	8×16	33.0	**502.94**	**335.29**	**251.47**
470	**SME10T471M8X16LL**	8×16	47.0	**525.79**	**350.53**	**262.90**
1000	**SME10T102M10X20LL**	10×20	100.0	**777.27**	**518.18**	**388.63**
2200	**SME10T222M12X25LL**	12.5×25	220.0	**1005.88**	**670.58**	**502.94**
3300	**SME10T332M12X30LL**	12.5×30	330.0	**1188.76**	**792.51**	**594.38**
6 VDC, T/AXIAL LEAD, 85°C						
22	**SME16T22RM5X12LL**	5×12.5	3.5	**320.05**	**213.37**	**160.02**
33	**SME16T33RM5X12LL**	5×12.5	5.3	**320.05**	**213.37**	**160.02**
47	**SME16T47RM6X12LL**	6.3×12.5	7.5	**365.77**	**243.85**	**182.88**
100	**SME16T101M6X16LL**	6.3×16	16.0	**388.63**	**259.09**	**194.32**
220	**SME16T221M8X16LL**	8×16	35.2	**502.94**	**335.29**	**251.47**
330	**SME16T331M8X16LL**	8×16	52.8	**539.01**	**269.50**	**269.50**
470	**SME16T471M8X20LL**	8×20	75.2	**571.52**	**381.01**	**285.76**
1000	**SME16T102M10X25LL**	10×25	160.0	**822.99**	**548.66**	**411.49**
2200	**SME16T222M12X30LL**	12.5×30	352.0	**1234.48**	**822.99**	**617.24**
3300	**SME16T332M16X25LL**	16×25	528.0	**1600.26**	**1066.84**	**800.13**
10000	**SME16T103M22X40LL**	22×40	1600.0	**4983.68**	**3322.45**	**2491.84**
25 VDC, T/AXIAL LEAD, 85°C						
10	**SME25T10RM5X12LL**	5×12.5	3.0	**320.05**	**213.37**	**160.02**
22	**SME25T22RM5X12LL**	5×12.5	5.5	**320.05**	**213.37**	**160.02**
33	**SME25T33RM6X12LL**	6.3×12.5	8.3	**365.77**	**243.85**	**182.88**
47	**SME25T47RM6X16LL**	6.3×16	11.8	**388.63**	**259.09**	**194.32**
100	**SME25T101M6X16LL**	6.3×16	25.0	**388.63**	**259.09**	**194.32**
220	**SME25T221M8X16LL**	8×16	55.0	**525.79**	**350.53**	**262.90**
330	**SME25T331M8X20LL**	8×20	82.5	**571.52**	**392.44**	**285.76**
470	**SME25T471M10X20LL**	10×20	117.5	**822.99**	**565.12**	**411.49**
1000	**SME25T102M12X25LL**	12.5×25	250.0	**1005.88**	**670.58**	**502.94**
2200	**SME25T222M16X25LL**	16×25	550.0	**1554.54**	**1067.45**	**777.27**
3300	**SME25T332M16X30LL**	16×25	825.0	**1783.15**	**1224.43**	**891.57**
4700	**SME25T472M18X40LL**	18×40	1175.0	**2743.30**	**1828.87**	**1371.65**
35 VDC, T/AXIAL LEAD, 85°C						
4.7	**SME35T4R7M5X12LL**	5×12.5	3.0	**320.05**	**213.37**	**160.02**
10	**SME35T10RM5X12LL**	5×12.5	3.5	**320.05**	**213.37**	**160.02**
22	**SME35T22RM6X12LL**	6.3×12.5	7.7	**365.77**	**243.85**	**182.88**
33	**SME35T33RM6X16LL**	6.3×16	11.6	**388.63**	**259.09**	**194.32**
47	**SME35T47RM6X16LL**	6.3×16	16.5	**388.63**	**259.09**	**194.32**
100	**SME35T101M8X16LL**	8×16	35.0	**502.94**	**335.29**	**251.47**
220	**SME35T221M8X20LL**	8×20	77.0	**571.52**	**381.01**	**285.76**
330	**SME35T331M10X20LL**	10×20	115.5	**800.13**	**533.42**	**400.06**
470	**SME35T471M10X25LL**	10×25	164.5	**822.99**	**548.66**	**411.49**
1000	**SME35T102M12X30LL**	12.5×30	350.0	**1143.04**	**762.03**	**571.52**
2200	**SME35T222M16X30LL**	16×30	770.0	**1874.59**	**1249.73**	**1007.22**
4700	**SME35T472M22X40LL**	22×40	1645.0	**5120.84**	**3413.90**	**2560.42**

Page D9

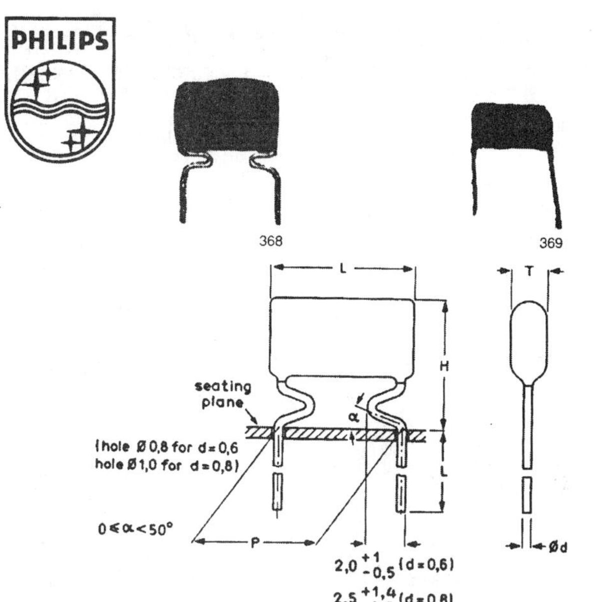

CAPACITORS

PHILIPS METALLIZED POLYESTER FILM CAPACITORS SERIES 368

Capacity (µF)	Dimensions (mm)				Crimped Lead Cat. No.	Net Price Per 100		
	T	L	H	P		1-99	100-999	1000-Up
100 VDC WORKING, METALLIZED POLYESTER FILM CAPACITORS (PETP ±10%)								
0.068	4	12.5	12	10.16	**368-25683**	**$16.14**	**$12.10**	**$10.89**
0.10	4	12.5	12	10.16	**368-25104**	**17.66**	**13.25**	**11.92**
0.15	4.5	12.5	12.5	10.16	**368-25154**	**19.79**	**14.84**	**13.36**
0.22	5.5	12.5	13.5	10.16	**368-25224**	**21.32**	**15.99**	**14.39**
0.33	5	17.5	14	15.24	**368-25334**	**38.37**	**28.78**	**25.90**
0.47	6	17.5	15	15.24	**368-25474**	**38.37**	**28.78**	**25.90**
0.68	7	17.5	16	15.24	**368-25684**	**46.29**	**34.72**	**31.24**
1.0	8.5	17.5	17.5	15.24	**368-25105**	**52.07**	**39.06**	**35.15**
1.5	6.5	26	18.5	22.86	**368-25155**	**99.28**	**74.46**	**67.01**
2.2	8	26	20	22.86	**368-25225**	**99.28**	**74.46**	**67.01**
3.3	9.5	26	21.5	22.86	**368-25335**	**182.42**	**136.81**	**123.13**
4.7	11	30	23	27.94	**368-25475**	**239.97**	**179.98**	**161.98**
6.8	11.5	30.0	23.5	27.94	**368-25685**	**279.87**	**279.87**	**279.87**
250 VDC WORKING, METALLIZED POLYESTER FILM CAPACITORS (PETP ±10%)								
0.033	4	12.5	12	10.16	**368-45333**	**16.14**	**12.10**	**10.89**
0.047	4	12.5	12	10.16	**368-45473**	**16.14**	**12.10**	**10.89**
0.068	4.5	12.5	12.5	10.16	**368-45683**	**17.66**	**13.25**	**11.92**
0.10	5	12.5	13	10.16	**368-45104**	**19.18**	**14.39**	**12.95**
0.15	5	17.5	14	15.24	**368-45154**	**27.10**	**20.33**	**18.29**
0.22	6	17.5	15	15.24	**368-45224**	**27.10**	**20.33**	**18.29**
0.33	7	17.5	16	15.24	**368-45334**	**39.89**	**29.92**	**26.93**
0.47	5.5	26	17.5	22.86	**368-45474**	**44.77**	**33.57**	**30.22**
0.68	6.5	26	18.5	22.86	**368-45684**	**54.51**	**40.88**	**36.79**
1.0	7.5	26	19.5	22.86	**368-45105**	**85.57**	**64.18**	**57.76**
1.5	8.5	30	20.5	27.94	**368-45155**	**108.41**	**81.31**	**73.18**
2.2	10.5	30	22.5	27.94	**368-45225**	**136.43**	**102.32**	**92.09**
400 VDC WORKING, METALLIZED POLYESTER FILM CAPACITORS (PETP ±10%)								
0.0010	4	12.5	12	10.16	**368-55102**	**19.18**	**14.39**	**12.95**
0.0022	4	12.5	12	10.16	**368-55222**	**16.14**	**12.10**	**10.89**
0.0033	4	12.5	12	10.16	**368-55332**	**16.14**	**12.10**	**10.89**
0.0047	4	12.5	12	10.16	**368-55472**	**16.14**	**12.10**	**10.89**
0.0068	4	12.5	12	10.16	**368-55682**	**16.14**	**12.10**	**10.89**
0.010	4	12.5	12	10.16	**368-55103**	**16.14**	**12.10**	**10.89**
0.015	4	12.5	12	10.16	**368-55153**	**16.14**	**12.10**	**10.89**
0.022	4	12.5	12	10.16	**368-55223**	**16.14**	**12.10**	**10.89**
0.033	4.5	12.5	12.5	10.16	**368-55333**	**17.66**	**13.25**	**11.92**
0.047	5	17.5	14	16.24	**368-55473**	**24.06**	**18.04**	**16.24**
0.068	5	17.5	14	16.24	**368-55683**	**24.06**	**18.04**	**16.24**
0.10	6	17.5	15	16.24	**368-55104**	**25.58**	**19.18**	**17.27**
0.15	7	17.5	16	16.24	**368-55154**	**35.33**	**26.49**	**23.84**
0.22	5.5	26	17.5	22.86	**368-55224**	**41.72**	**31.29**	**28.16**
0.33	6.5	26	18.5	22.86	**368-55334**	**48.12**	**36.09**	**32.48**
0.47	8	26	20	22.86	**368-55474**	**73.70**	**55.27**	**49.74**
0.68	8.5	30	20.5	27.94	**368-55684**	**90.45**	**67.83**	**61.05**
1.0	11	30	23	27.94	**368-55105**	**110.55**	**82.91**	**74.62**

Hammond Transformers

HAMMOND MANUFACTURING

L.V. RECTIFIER TRANSFORMERS
165, 166 & 167 SERIES

All primaries, 115 Volts, 60 Hertz
Primary leads 6-7" long
All models CSA certified

L.V. RECTIFIER TRANSFORMERS OPEN TYPE C or H, ENCLOSED TYPE X

MANU 195 CODE

165 SERIES

- Open horizontal frame.
- Mounting slots .22" x .44" H3-H15 28 x .56 (style H21-H27).
- Secondaries have 8-10" leads or copper tabs with holes.
- Insulation test 2000 volts.

166 SERIES

- Channel bracket mounting.
- 2-hole horizontal and vertical mounting as indicated.
- Secondary leads 6-7" long.
- Insulation test 2000 volts.

167 SERIES

- 4-hole "Type X" mounting.
- Secondary leads 6-8" long.
- Insulation test 2500 volts.
- Mounting slots

Secondary			Type H			Type C			Type X		
F.L. Volts	N.L. Volts	Amps.	Cat. No.	Dim. Ref.	Net Price	Cat. No.	Dim. Ref.	Net Price	Cat. No.	Dim. Ref.	Net Price
28.0 ct	30.8	1.0	—	—	—	**166 J28**	C12H	**$16.15**	**167 J28**	X2	**$19.62**
28.0 ct	29.8	2.0	**165 L28**	H4	**$25.55**	—	—	—	**167 L28**	X4	**24.03**
28.0 ct	29.6	3.0	—	—	—	—	—	—	**167 M28**	X6	**27.52**
30.0 ct	37.4	0.15	—	—	—	**166 E30**	C6H	**10.57**	—	—	—
30.0 ct	26.1	0.25	—	—	—	**166 F30**	C7H	**12.04**	—	—	—
30.0 ct	34.2	0.50	—	—	—	**166 G30**	C9H	**11.74**	—	—	—
30.0 ct	31.6	1.5	—	—	—	—	—	—	**167 K30**	X4	**26.79**
30.0 ct	32.0	3.0	—	—	—	—	—	—	**167 M30**	X7	**34.86**
30.0 ct	31.4	5.0	**165 P30**	H11	**39.71**	—	—	—	**167 P30**	X10	**42.18**
30.0 ct	31.4	10.0	**165 S30**	H21	**55.92**	—	—	—	**167 S30**	X16	**59.43**
33.0 ct	35.0	1.0	—	—	—	**166 J33**	C14H	**17.98**	**167 J33**	X3	**20.36**
35.0 ct	39.0	1.5	—	—	—	**166 K35**	C14H	**17.98**	—	—	—
36.0 ct	40.7	0.15	—	—	—	**166 E36**	C7H	**10.64**	—	—	—
36.0 ct	39.4	0.3	—	—	—	**166 F36**	C9H	**12.62**	—	—	—
36.0 ct	40.0	0.5	—	—	—	**166 G36**	C11H	**14.24**	—	—	—
36.0 ct	39.5	1.0	—	—	—	**166 J36**	C14H	**18.11**	**167 J36**	X2	**20.36**
36.0 ct	38.3	2.0	—	—	—	—	—	—	**167 L36**	X5	**25.13**
36.0 ct	37.9	3.0	—	—	—	—	—	—	**167 M36**	X8	**33.93**
36.0 ct	38.2	5.0	—	—	—	—	—	—	**167 P36**	X11	**41.83**
36.0 ct	37.4	8.0	—	—	—	—	—	—	**167 R36**	X17	**68.24**
36.0 ct	38.0	12.0	—	—	—	—	—	—	**167 T36**	X21	**100.53**
42.0 ct	43.9	2.0	**165 L42**	H7	**25.80**	—	—	—	—	—	—
44.0 ct	51.6	0.15	—	—	—	**166 E44**	C7H	**10.64**	—	—	—
44.0 ct	50.6	0.25	—	—	—	**166 F44**	C9H	**12.62**	—	—	—
44.0 ct	48.1	0.50	—	—	—	**166 G44**	C12H	**15.64**	—	—	—
44.0 ct	47.4	1.0	—	—	—	**166 J44**	C14H	**19.08**	—	—	—
44.0 ct	46.0	2.0	—	—	—	—	—	—	**167 L44**	X7	**28.80**
50.0 ct	62.0	0.075	—	—	—	**166 C50**	C5H	**9.98**	—	—	—

166 SERIES — HORIZONTAL CHANNEL BRACKET TYPE - C

Mtg. Style	DIMENSIONS A	B	C	D
C2H	1.63	1.13	0.81	1.38
C3H	2.06	1.25	1.19	1.75
C4H	2.06	1.38	1.19	1.75
C5H	2.38	1.38	1.38	2.00
C6H	2.38	1.50	1.38	2.00
C7H	2.81	1.88	1.69	2.38
C8H	2.81	2.00	1.69	2.38
C9H	3.25	2.00	2.00	2.81
C11H	3.25	2.25	2.00	2.81
C12H	3.69	2.13	2.31	3.13
C13H	3.69	2.25	2.63	3.56
C14H	4.00	2.50	2.63	3.56
C15H	4.00	3.25	2.63	3.56

NOTE: All dimensions are nominal and measured in inches, tolerance ± 0.06".

Page D10

Hammond Cases

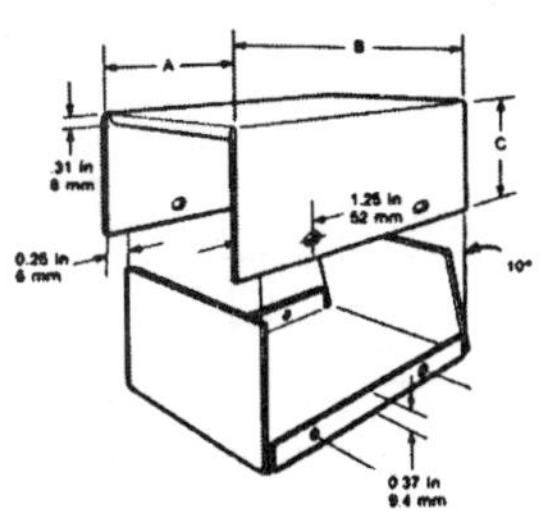

HAMMOND METAL INSTRUMENT CASES 1454 SERIES MINIATURE SLOPE FRONT INSTRUMENT CASE

20 ga. CRS. **Semi-Gloss Sand** colored panels with Textured **Black** covers. 6-32 Phillips, nickel plated screws and self-adhesive rubber feet included. Choice of sloping or vertical front panels by reversing case.

HAMMOND 1454 SERIES MINIATURE SLOPE FRONT INSTRUMENT CASES

Cat. No.	Dimensions mm (Inches)			Wt. Lbs.	Net Price
	A	B	C		
1454C	4.0(102)	2.5(64)	2.0(51)	#	**$9.74**
1454E	4.0(102)	5.0(127)	2.0(51)	#	**10.94**
1454F	4.0(102)	6.0(152)	3.0(76)	#	**14.82**
1454G	6.0(152)	2.5(64)	2.0(51)	#	**10.87**
1454H	5.0(127)	7.5(191)	2.5(64)	#	**16.66**
1454J	4.0(102)	3.0(76))	4.0(102)	#	**11.86**
1454K	8.0(203)	3.5(89)	2.5(64)	#	**13.56**
1454L	5.0(127)	6.0(152)	4.0(102)	#	**18.29**
1454M	4.0(102)	3.0(76)	6.0(152)	#	**14.11**
1454P	10.0(254)	5.0(127)	3.5(89)	#	**19.48**
1454R	5.0(127)	5.0(127)	7.5(191)	#	**21.88**

= Weight under 5 lbs.
NOTE: Models with **"B"** dimension less than 5 inches (127mm) have only one hole per side, centered in "B".

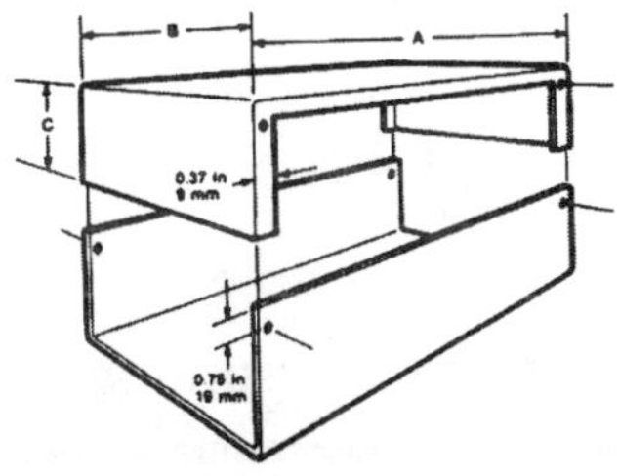

HAMMOND GENERAL PURPOSE CASES 1411 ALUMINUM UTILITY CASE

.040" thick aluminum with ASA 61 **Gray** baked enamel finish (outside only). 4 self-tapping screws are included.

HAMMOND 1411 SERIES ALUMINUM UTILITY CASES

Cat. No.	Dimensions mm (In.)			Wt. Lbs.	Net Price
	A	B	C		
1411B	69(2.7)	56(2.2)	41(1.6)	#	**$4.52**
1411C	81(3.2)	56(2.2)	28(1.1)	#	**4.38**
1411D	81(3.2)	56(2.2)	41(1.6)	#	**4.52**
1411F	102(4.0)	56(2.2)	76(3.0)	#	**4.66**
1411G	102(4.0)	56(2.2)	41(1.6)	#	**4.52**
1411H	102(4.0)	56(2.2)	56(2.2)	#	**4.87**
1411K	127(5.0)	56(2.2)	56(2.2)	#	**5.08**
1411L	127(5.0)	102(4.0)	76(3.0)	#	**6.06**
1411M	152(6.0)	76(3.0)	76(3.0)	#	**5.94**
1411N	127(5.0)	76(3.0)	56(2.2)	#	**5.50**
1411P	152(6.0)	127(5.0)	102(4.0)	#	**8.20**

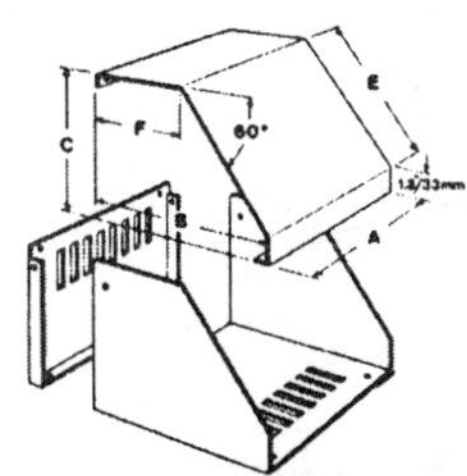

HAMMOND DESKTOP CONSOLES SCEM SERIES SLOPED FRONT, METAL INSTRUMENT CASES

An instrument case finished in **Textured Blue** with a choice of **Semi-Gloss Off-White** or finely **Textured Satin Black** panel. Convection ventilation is provided by slots in the bottom and removable rear panels. Complete with hardware and rubber feet.

HAMMOND SCEM SERIES SLOPED FRONT, METAL INSTRUMENT CASES

Cat. No.	Dimensions—mm (Inches)				
	A	B	C	E	F
SCEM 101110	102(4.0)	114(4.5)	102(4.0)	76(3.0)	74(2.9)
SCEM 161110	165(6.5)	114(4.5)	102(4.0)	76(3.0)	74(2.9)
SCEM 161414	165(6.5)	140(5.5)	140(5.5)	119(4.7)	79(3.1)
SCEM 251414	254(10.0)	140(5.5)	140(5.5)	119(4.7)	79(3.1)
SCEM 251619	254(10.0)	165(6.5)	191(7.5)	179(7.0)	74(2.9)
SCEM 351619	356(14.0)	165(6.5)	191(7.5)	179(7.0)	74(2.9)

HAMMOND SCEM SERIES SLOPED FRONT, METAL INSTRUMENT CASES

	Cat. No.	Wt. Lbs.	Net Price
	SCEM 101110 XX*	#	**$39.74**
	SCEM 161110 XX*	#	**44.05**
	SCEM 161414 XX*	#	**26.55**
	SCEM 251414 XX*	#	**56.16**
	SCEM 251619 XX*	#	**63.98**
	SCEM 351619 XX*	#	**69.97**
	1421J6	Extra Screws (pkg. of 25)	**3.15**

XX* Add suffix **WH** (white) or **BK** (black).
= Weight under 5 lbs.

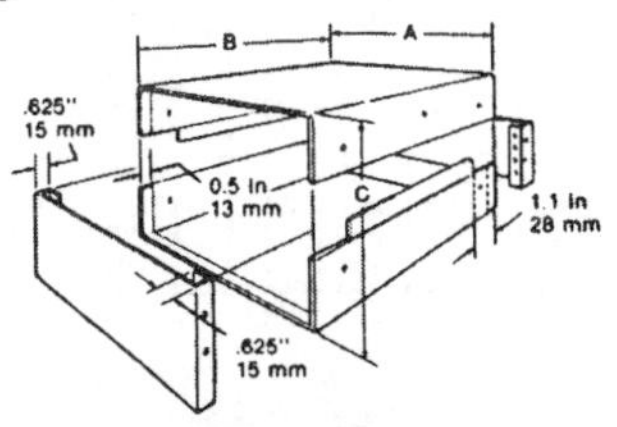

HAMMOND 1458 SERIES LOW PROFILE INSTRUMENT CASES

These symmetrically designed (either end can be front), 20 gauge CRS cases are finished in either **Black** or **Gulf Blue** textured baked enamel. Each case comes complete with; two .064" thick aluminum panels (one in **Satin Black** and one in **Smoke White** finish), 6-32 Phillips nickel plated screws and self-adhesive rubber feet. The case is split in two equal halves. Use **1448** Series internal rails (both top and bottom covers can then be removed without affecting component to panel connections). Also accommodates **1446** Series chassis.

HAMMOND 1458 SERIES LOW PROFILE INSTRUMENT CASES

Blacktex Cat. No.	Net Price	Bluetex Cat. No.	Net Price	Dimensions—Inches (mm)			Wt. Lbs.
				A	B	C	
1458A3	**$31.07**	**1458A3B**	**$31.07**	4(101)	4(101)	3(76.2)	#
1458B4	**32.48**	**1458B4B**	**32.48**	6(152)	4(101)	4(101)	#
1458C3	**33.89**	**1458C3B**	**33.89**	6(152)	6(152)	3(76.2)	#
1458C4	**36.72**	**1458C4B**	**36.72**	6(152)	6(152)	4(101)	#
1458D3	**39.53**	**1458D3B**	**39.53**	8(203)	8(203)	3(76.2)	#
1458D4	**44.48**	**1458D4B**	**44.48**	8(203)	8(203)	4(101)	#
1458D5	**49.42**	**1458D5B**	**49.42**	8(203)	8(203)	5(127)	#
1458E3	**40.94**	**1458E3B**	**40.94**	8(203)	10(254)	3(76.2)	#
1458E4	**46.59**	**1458E4B**	**46.59**	8(203)	10(254)	4(101)	#
1458E5	**52.24**	**1458E5B**	**52.24**	8(203)	10(254)	5(127)	#
1458G3	**40.94**	**1458G3B**	**40.94**	10(254)	8(203)	3(76.2)	#
1458G4	**46.59**	**1458G4B**	**46.59**	10(254)	8(203)	47(101)	#
1458G5	**52.24**	**1458G5B**	**52.24**	10(254)	8(203)	5(127)	#

= Weight under 5 lbs.

HAMMOND 1421 SERIES RUBBER MOUNTING FEET

Four different sizes of replacement rubber feet, 2 square and 1 round adhesive backed and 1 style of screw-on with self-tapping screws supplied.

Cat. No.	Description	Dimensions	Style	Pkg. Qty.	Net Price
RUBBER MOUNTING FEET					
1421T2	Self-adhesive	.37"×.12" high	Round	24	**$5.08**
1421T3	Self-adhesive	.5"×.5"×.25" high	Square	24	**7.50**
1421T4	Self-adhesive	.7"×.7"×.31" high	Square	24	**10.00**
1421T7*	Screw-on	.75" dia. x .5" high	Round	24	**18.40**

*1421T7 feet are c/w self-tapping screws.

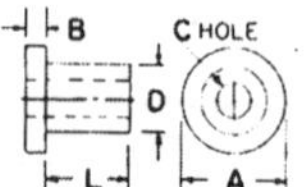

KEYSTONE NYLON SHOULDER BUSHINGS

Cat. No.	Screws	L	A	B	C	D	Net Price Per 1000 1-299	300-1499	1500-Up
7680	2	.025	.260	.030	.095	.154	**$109.18**	**$76.86**	**$70.31**
7681	4	.047	.250	.032	.115	.185	**64.93**	**45.71**	**41.81**
3053	4	.093	.235	.047	.115	.145	**75.78**	**53.35**	**48.80**
3102	4	.125	.235	.047	.115	.145	**82.05**	**57.76**	**52.84**
3103	4	.187	.235	.047	.115	.145	**82.05**	**57.76**	**52.84**
340	4	.250	.235	.047	.115	.145	**82.05**	**57.76**	**52.84**
7682	6	.032	.385	.062	.145	.240	**71.18**	**50.11**	**45.84**
7683	6	.032	.250	.062	.140	.187	**61.38**	**43.21**	**39.53**
7684	6	.032	.312	.062	.140	.187	**66.80**	**47.03**	**43.02**
3054	6	.125	.290	.047	.140	.170	**84.13**	**59.22**	**54.18**
3104	6	.187	.290	.047	.140	.170	**85.60**	**60.26**	**55.13**
341	6	.250	.290	.047	.140	.170	**82.05**	**57.76**	**52.84**
3055	6	.500	.290	.047	.140	.170	**90.60**	**63.78**	**58.35**
7685	8	.032	.375	.032	.171	.250	**71.18**	**50.11**	**45.84**
3056	8	.140	.344	.062	.171	.205	**97.08**	**68.34**	**62.52**
3105	8	.250	.344	.062	.171	.205	**87.88**	**61.86**	**56.59**
342	8	.375	.344	.062	.171	.205	**87.88**	**61.86**	**56.59**
3106	8	.500	.344	.062	.171	.205	**87.88**	**61.86**	**56.59**
7686	10	.032	.375	.062	.196	.308	**71.18**	**50.11**	**45.84**
3107	10	.093	.438	.062	.200	.260	**95.40**	**67.16**	**61.44**
3108	10	.250	.399	.062	.200	.260	**95.40**	**67.16**	**61.44**
3057	10	.500	.399	.062	.200	.260	**93.95**	**66.14**	**60.50**
3058	10	.750	.399	.062	.200	.260	**93.95**	**66.14**	**60.50**
7687	¼	.032	.437	.032	.253	.310	**76.40**	**53.79**	**49.20**
3059	¼	.187	.513	.062	.260	.312	**97.08**	**68.34**	**62.52**
3109	¼	.250	.513	.062	.260	.312	**97.08**	**68.34**	**62.52**
3110	¼	.375	.513	.062	.260	.312	**97.08**	**68.34**	**62.52**
3060	¼	.500	.513	.062	.260	.312	**97.08**	**68.34**	**62.52**

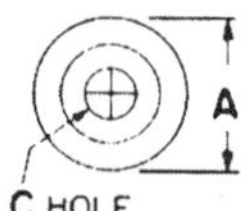

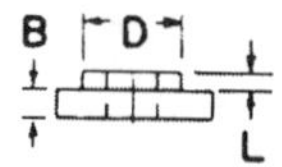

KEYSTONE FIBRE SHOULDER WASHERS

Cat. No.	Screws	L	A	B	C	D	Net Price Per 1000 1-299	300-1499	1500-Up
3062	4	.031	.250	.031	.115	.187	**$60.33**	**$42.47**	**$38.85**
3064	6	.031	.312	.062	.140	.187	**58.03**	**40.85**	**37.37**
3233	6	.031	.375	.062	.140	.240	**107.93**	**75.98**	**69.50**
3065	8	.031	.375	.062	.170	.246	**67.43**	**47.47**	**43.42**
3069	⅜	.031	.625	.062	.385	.498	**103.55**	**72.90**	**66.69**
3241	⅜	.031	.750	.062	.385	.498	**144.05**	**101.41**	**92.77**

Page D12

KEYSTONE

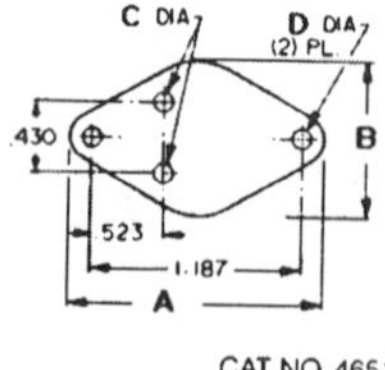

CAT. NO. 4651

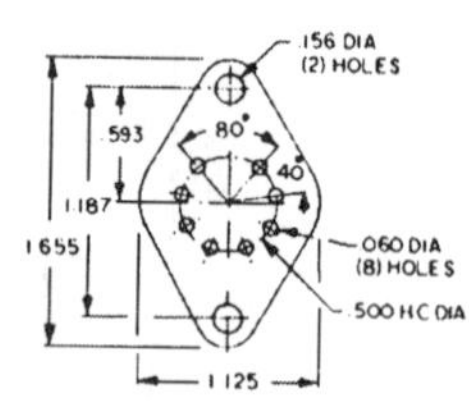

CAT. NO. 4658

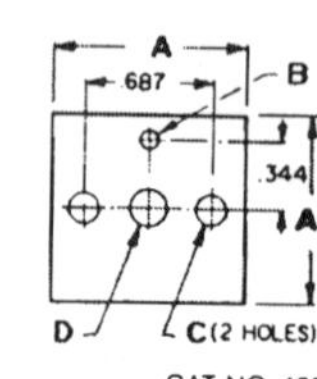

CAT. NO. 4655

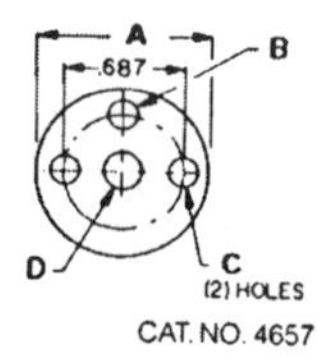

CAT. NO. 4660

CAT. NO. 4657

CAT. NO. 4678 TO 4683

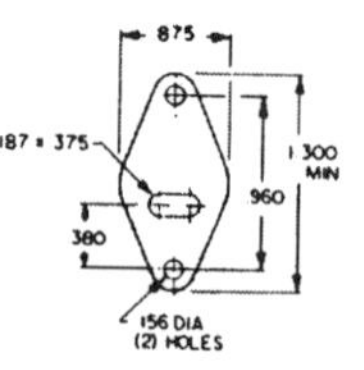

CAT. NO. 4665

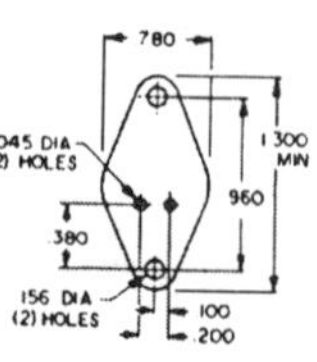

CAT. NO. 4668

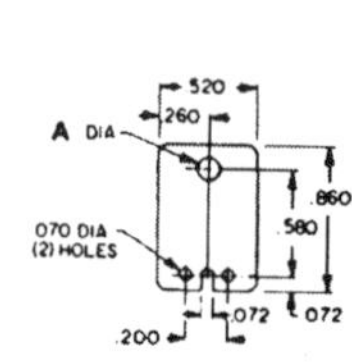

CAT. NO. 4672
CAT. NO. 4673

KEYSTONE MICA INSULATORS

These precision stamped mica insulators for power transistor mounting provide good thermal conductivity and will withstand high temperatures. High dielectric strength. Thermal degradation produces a non-conductive, non-toxic residue. All insulators are .003" thick.

Cat. No.	Dimensions A	B	C	D	Net Price Per 1000 1-299	300-1499	1500-Up
INSULATORS FOR TO-3 CASE							
4651	1.625	1.093	.093	.166	**$85.60**	**$60.26**	**$55.13**
4658	(For 8 Pin TO-3-see Diagram)				**85.60**	**60.26**	**55.13**
4660	(See Diagram)				**93.53**	**65.84**	**60.23**
INSULATORS FOR TO-36 CASE							
4655	1.125	.130	.187	.203	**85.60**	**60.26**	**55.13**
4656	1.250	.140	.156	.203	**108.35**	**76.28**	**69.78**
4657	1.062	.187	.187	.187	**85.60**	**60.26**	**55.13**
INSULATORS FOR TO-66 CASE							
4665	(See Diagram)				**89.35**	**62.90**	**57.54**
4666	(See Diagram)				**85.60**	**60.26**	**55.13**
INSULATORS FOR TO-220 CASE							
4672	.141	—	—	—	**34.85**	**24.53**	**22.44**
4673	.115	—	—	—	**34.85**	**24.53**	**22.44**

Cat. No.	Inside Dia.	Outside Dia.	Screw Size	Net Price Per 1000 1-299	300-1499	1500-Up
INSULATORS FOR THREADED STUD DEVICES						
4678	.120	.375	4	**$52.40**	**$36.89**	**$33.75**
4679	.140	.375	6	**54.08**	**38.07**	**34.82**
4680	.171	.500	8	**79.75**	**56.14**	**51.36**
4681	.187	.625	10	**52.83**	**37.19**	**34.02**
4682	.265	.750	¼	**69.53**	**48.95**	**44.77**
4683	.328	.875	5/16	**86.85**	**61.14**	**55.93**

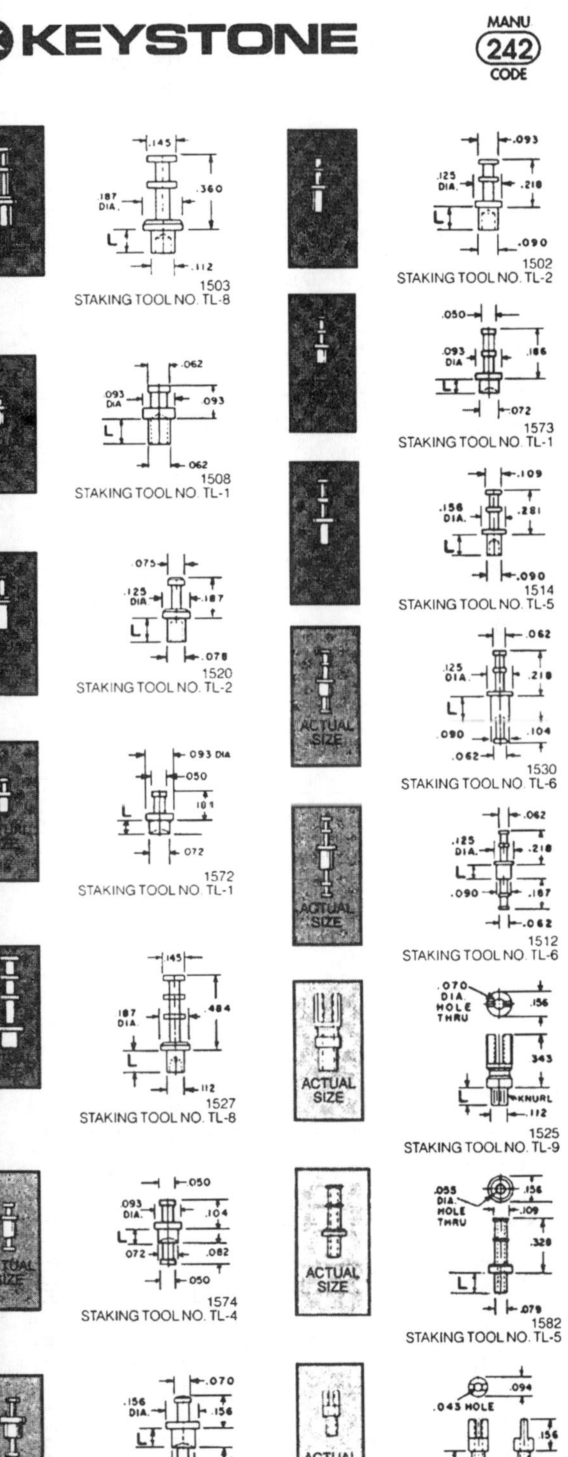

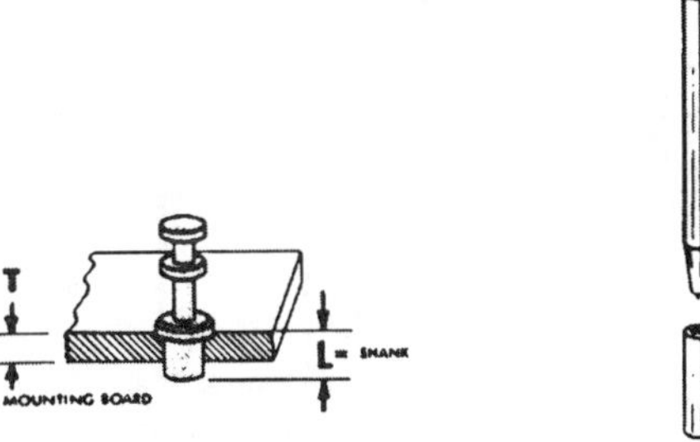

TERMINALS

Specify catalog number of terminal selected and add the dash number or mention the thickness of the insulator to be used, for example: **1503-1** for 1/32" insulator, **1503-2** for 1/16" insulator etc.—**1 for insulating board 1/32" thick—2 for insulating board 1/16" thick—3 for insulating board 3/32" thick—4 for insulating board 1/8" thick**

STAKING TOOLS

Catalog No. TL-series designates a set of tools, consisting of **TOP (staking)** and **BOTTOM (holder)**. Some sets of tools can be used for other terminals with similar dimensions. Tools are made of 1/4" dia. steel and **cadmium plated. Precision Machined** and carefully **hardened** to insure long wear.

Cat. No. TL-* Net Price **$17.20**

*Add number suffix for desired staking tool.

Cat. No.	L	T	Net Price Each		
			1-299	300-1499	1500-Up
1503-1	.078	1/32	**$.01**	**$.15**	**$.13**
1503-2	.109	1/16	**.21**	**.15**	**.13**
1503-3	.141	3/32	**.21**	**.15**	**.13**
1503-4	.172	1/8	**.21**	**.15**	**.13**
1503-6	.234	3/16	**.21**	**.15**	**.13**
1508-1	.053	1/32	**.09**	**.06**	**.05**
1508-2	.084	1/16	**.09**	**.06**	**.06**
1508-3	.115	3/32	**.09**	**.06**	**.05**
1520-1	.078	1/32	**.12**	**.08**	**.08**
1520-2	.109	1/16	**.12**	**.08**	**.08**
1520-3	.141	3/32	**.12**	**.08**	**.08**
1520-4	.172	1/8	**.12**	**.08**	**.08**
1572-1	.051	1/32	**.08**	**.06**	**.05**
1573-2	.082	1/16	**.09**	**.06**	**.06**
1572-3	.113	3/32	**.08**	**.06**	**.05**
1572-4	.145	1/8	**.08**	**.06**	**.05**
1527-1	.078	1/32	**.30**	**.21**	**.20**
1527-2	.109	1/16	**.30**	**.21**	**.20**
1527-3	.141	3/32	**.30**	**.21**	**.20**
1527-4	.172	1/8	**.30**	**.21**	**.20**
1574-1	.051	1/32	**.09**	**.06**	**.06**
1574-2	.082	1/16	**.09**	**.06**	**.06**
1574-3	.113	3/32	**.09**	**.06**	**.06**
1574-4	.145	1/8	**.09**	**.06**	**.06**
1522-1	.062	1/32	**.17**	**.12**	**.11**
1522-2	.094	1/16	**.17**	**.12**	**.11**
1522-3	.125	3/32	**.17**	**.12**	**.11**
1522-4	.156	1/8	**.17**	**.12**	**.11**
1502-1	.078	1/32	**.11**	**.08**	**.07**
1502-2	.109	1/16	**.11**	**.08**	**.07**
1502-3	.141	3/32	**.11**	**.08**	**.07**
1502-4	.172	1/8	**.11**	**.08**	**.07**
1573-1	.051	1/32	**.09**	**.06**	**.06**
1573-2	.082	1/16	**.09**	**.06**	**.06**
1573-3	.113	3/32	**.09**	**.06**	**.06**
1573-4	.145	1/8	**.09**	**.06**	**.06**
1514-1	.078	1/32	**.17**	**.12**	**.11**
1514-2	.109	1/16	**.17**	**.12**	**.11**
1514-3	.141	3/32	**.17**	**.12**	**.11**
1514-4	.172	1/8	**.17**	**.12**	**.11**
1530-1	.062	1/32	**.13**	**.09**	**.08**
1530-2	.094	1/16	**.13**	**.09**	**.08**
1530-3	.125	3/16	**.13**	**.09**	**.08**
1530-4	.156	1/8	**.13**	**.09**	**.08**
1512-1	.062	1/32	**.19**	**.13**	**.12**
1512-2	.094	1/16	**.19**	**.13**	**.12**
1512-3	.125	3/32	**.19**	**.13**	**.12**
1512-4	.156	1/8	**.19**	**.13**	**.12**
1525-2	.109	1/16	**.32**	**.22**	**.20**
1525-3	.141	3/32	**.32**	**.22**	**.20**
1525-4	.172	1/8	**.32**	**.22**	**.20**
1526-1	.054	1/32	**.15**	**.10**	**.10**
1526-2	.084	1/16	**.15**	**.10**	**.10**
1526-3	.115	3/32	**.15**	**.10**	**.10**
1526-4	.147	1/8	**.15**	**.10**	**.10**
1582-3	.135	3/32	**.23**	**.16**	**.15**

Page D13

KEYSTONE

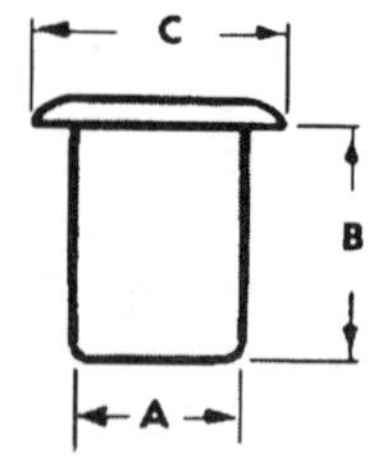

WIDE ROLL EYELETS

Cat. No.	Dimensions A	Dimensions B	Dimensions C	Hole Size	Net Price Per 1000: 1-299	300-1499	1500-Up
23	.059	.093	.105	.062 (1/16")	$31.73	$24.11	$22.21
44	.121	.125	.150	.093 (1/8")	35.50	26.98	24.85

Material: Brass. Plating: Bright Electro Tin, MIL-F-14072. Recommended for longer shelf life and better solderability.

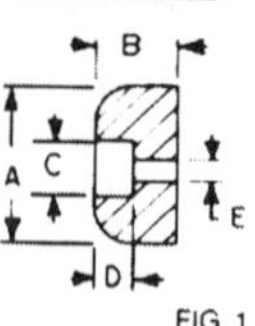

FIG. 1

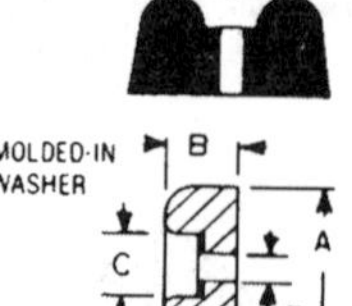

FIG. 2

RECESSED BUMPERS

Provides excellent traction and shock absorbing quantities. Available with or without Molded-In Washers.

Materials: Rubber, BUNA S, (figs. 1 & 2)

Cat. No.	Fig.	A	B	C	D	E	Screw Size	Net Price Per 1000: 1-299	300-1499	1500-Up
720	1	.500	.250	.250	.125	.125	#4	$158.03	$111.25	$101.77
721	1	.625	.312	.250	.187	.125	#4	250.50	176.35	161.32
722	1	.718	.437	.250	.218	.125	#4	502.45	353.72	323.58
723	2	.937	.375	.375	.187	.187	#8	481.80	339.19	310.28

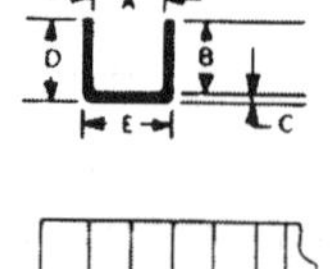

FLEXIBLE GROMMETING

Ideal for maintenance, short-run production, R&D or prototyping. Supplied in 25 ft. rolls. Make your own grommets to fit any size or shape required.

Material: Black Nylon Type 6/6

CAT. NO.	A	B	C	D	E	Net Price Each: 1-19	20-99	100-Up
8615	.052	.150	.040	.190	.125	$34.92	$30.93	$29.93
8616	.085	.155	.045	.200	.162	39.73	35.19	34.06
8617	.125	.155	.045	.220	.206	40.45	35.82	34.67

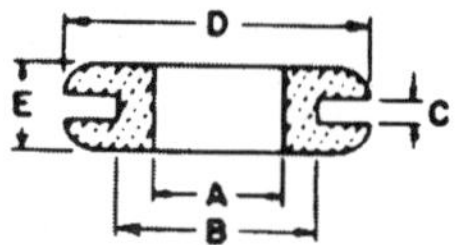

RUBBER GROMMETS

Ideal for electrical insulation and mechanical damping. Made from BUNA S.

Cat. No.	A	B	C	D	E	Net Price Per 1000: 1-299	300-1499	1500-Up
739	.250	.375	.062	.500	.218	$124.63	$87.74	$80.26
752	.625	.875	.062	1.125	.312	367.83	258.95	236.88

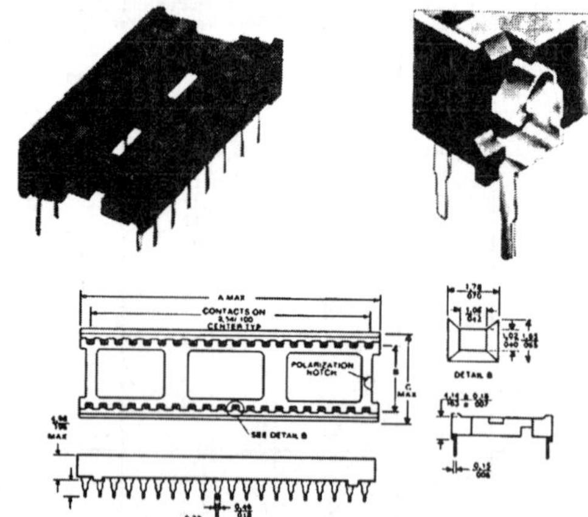

AUGAT 200C SERIES STAMPED SINGLE BEAM CONTACT CLOSED BOTTOM DIP SOCKETS

Augat's lowest cost socket provides maximum normal force and consistent mechanical performance. Non-wicking, closed bottom design gives protection against flux and solder contamination. Protected open entry for easy hand or automatic insertion. Anti-overstress design protects against oversize leads. High normal force results in gas tight connections. Thermoplastic polyester insulator rated 94V-0. Phosphor bronze contacts per QQ-B-750. Tin/lead plating per MIL-P-81728.

Cat. No.	No. of Contacts	Dimensions mm (Inches) Spacing "B"	Length "A"	Net Price Each 1-499	500-999	1000-Up
STAMPED PHOSPHOR BRONZE CONTACTS, TIN/LEAD PLATING, .125-.135" PC TAILS						
208-AG19DC	8	7.62(.300)	10.16(.400)	$.15	$.14	$.12
214-AG19DC	14	7.62(.300)	17.78(.700)	.17	.16	.13
216-AG19DC	16	7.62(.300)	20.32(.800)	.19	.18	.15
218-AG19DC	18	7.62(.300)	22.86(.900)	.21	.20	.17
220-AG19DC	20	7.62(.300)	25.40(1.00)	.24	.22	.19
224-AG10DC	24	7.62(.300)	30.48(1.20)	.29	.27	.23
224-AG19DC	24	15.24(.600)	30.48(1.20)	.29	.27	.23
228-AG19DC	28	15.24(.600)	35.56(1.40)	.33	.31	.26
232-AG19DC	32	15.24(.600)	40.64(1.60)	.38	.35	.30
240-AG19DC	40	15.24(.600)	50.80(2.00)	.48	.44	.37

electro sonic

Page D14

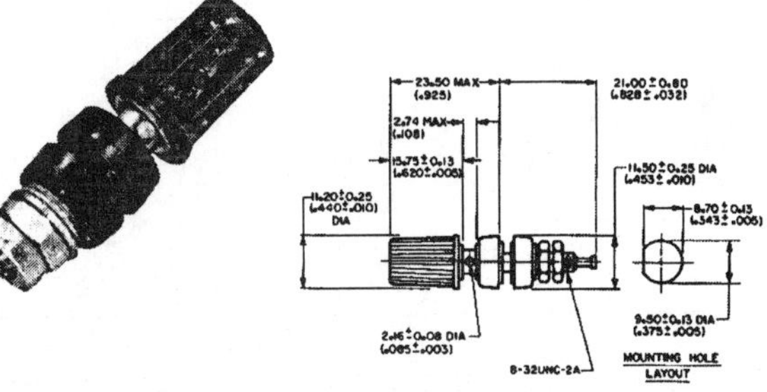

STANDARD UL TYPE BINDING POSTS WITH KEYED INSULATOR

esigned to meet UL-1244, IEC 348.P and C39.5-1974 specifications for dis- nce of thumb nut openings. Keyed insulators prevent turning of installation. For anel thickness from .062" to .250" Insulating material: Nylon. Stud: Brass. Fin- h: Gold or Bright Tin. Current: 15 amps. Breakdown voltage: 1000 Vrms min. tud-to-Panel: 3.0pF nominal. Contact Resistance: 0.010 ohm max.

at. No.	Color	Net Price Per 100	Cat. No.	Color	Net Price Per 100
IN PLATED			GOLD PLATED		
11-0741-003	White	$309.20	111-0741-002	White	$456.61
11-0742-003	Red	309.20	111-0742-002	Red	456.61
11-0743-003	Black	309.20	111-0743-002	Black	456.61
11-0744-003	Green	309.20	111-0744-002	Green	456.61
11-0747-003	Yellow	309.20	111-0747-002	Yellow	456.61
11-0750-003	Blue	309.20	111-0750-002	Blue	456.61

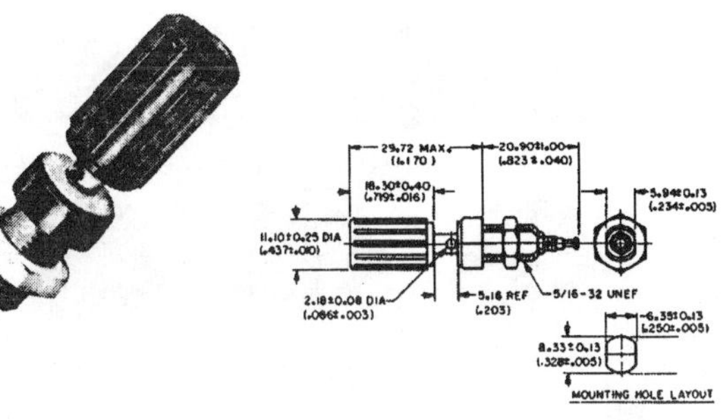

INSULATED STANDARD BINDING POSTS

ully insulated nylon binding post solves many common design problems. Nylon flatted to prevent turning. Metal body is silver plated with turret terminal. hreaded nylon body withstands 6.0 lb. in. (7 kg. cm.) of thread stripping force. or panel thickness up to .281" (7.15mm). Current rating: 15 amps. Breakdown oltage 5700 Vrms. Capacitance 3.3pF.

at. No.	Color	Net Price Per 100	Cat. No.	Color	Net Price Per 100
11-0101-001	White	$179.77	111-0107-001	Yellow	$179.77
11-0102-001	Red	179.77	111-0108-001	Brown	179.77
11-0103-001	Black	179.77	111-0108-001	Blue	179.77
11-0104-001	Green	179.68	111-0112-001	Violet	179.77
11-0106-001	Orange	179.77	111-0113-001	Gray	179.77

Page D15

Cinch Terminal Strips

NO. 2000

NO. 2000 TERMINAL STRIPS

These compact and sturdy junction terminal strips are useful in assembling radio chassis, wiring etc. Terminals: .019" brass, pre tin plate spaced on 5/16" centers. Insulation: Bakelite. Brackets: Cadmium-plated steel mounting holes: .140 diameter. Center of mounting hole to end of bracket: 5/32".

Cat. No.	No. of Terminals	Mounting Centers	Net Price Each 1-99	100-499	500-999	1000-Up
2002	2	1"	$.74	$.66	$.57	$.49
2003	3	1 5/16"	.85	.75	.66	.56
2004	4	1 5/8"	.94	.83	.73	.62
2005	5	1 15/16"	.90	.80	.70	.60
2006	6	2 1/4"	1.06	.94	.83	.71
2007	7	2 9/16"	1.37	1.22	1.06	.91
2008	8	2 7/8"	1.51	1.34	1.18	1.01
2009	9	3 3/16"	1.66	1.47	1.29	1.10
2010	10	3 1/2"	1.80	1.60	1.40	1.20
2011	11	3 13/16"	1.94	1.73	1.51	1.30
2012	12	4 1/8"	2.11	1.87	1.64	1.40
2013	13	4 7/16"	2.23	1.98	1.74	1.49

51 51A 51B 51C 51F 51R 51L 52 52A 52B 52C

52R 53 53A 53B 53C 53E

53F 54 54A 54B 54C

55 55A 55B 55C

56 56A 56B 56C

LUG TYPE TERMINAL STRIPS

These low-cost terminal strips are recommended for commercial applications where it is necessary to provide support for small electronic components in places where existing tie points are inadequate. Solder coated steel lugs are placed on 3/8" centers with .140" diameter mounting holes. Chocolate colored bakelite straps are 1/16" thick and 3/8" wide.

Cat. No.	Mtg Centers	Net Price Each 1-99	100-499	500-999	1000-Up
51	—	$.36	$.32	$.28	$.24
51A		.36	.32	.28	.24
51B		.36	.32	.28	.24
51C		.36	.32	.28	.24
51F	—	.61	.54	.48	.41
51L		.61	.54	.48	.41
51R		.61	.54	.48	.41
52	—	.36	.32	.28	.24
52A		.47	.42	.36	.31
52B		.36	.32	.28	.24
52C		.47	.42	.36	.31
52R	—	.61	.54	.48	.41
53	1 1/2"	.61	.54	.48	.41
53A	1 1/2"	.61	.54	.48	.41
53B	3/4"	.61	.54	.48	.41
53C	3/4"	.61	.54	.48	.41
53E	—	.56	.50	.43	.37
53F		.61	.54	.48	.41
54	1 7/8"	.68	.61	.53	.46
54A		.77	.69	.60	.52
54B	—	.61	.54	.48	.41
54C		.68	.61	.53	.46
55	2 1/4"	.88	.78	.69	.59
55A	2 1/4"	1.03	.91	.80	.68
55B	1 1/2"	.88	.78	.69	.59
55C	1 1/2"	1.03	.91	.80	.68
56	2 5/8"	.94	.83	.73	.62
56A	2 5/8"	.94	.83	.73	.62
56B	1 7/8"	.94	.83	.73	.62
56C	1 7/8"	.94	.83	.73	.62

Eaton Switches

SPDT (FORMERLY T01 SERIES)

DPDT

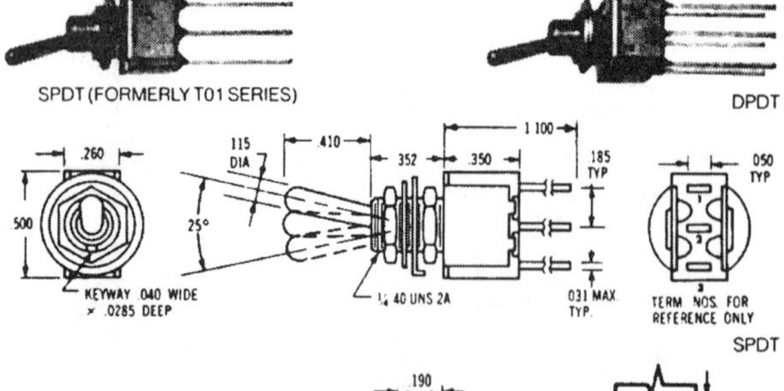

SPDT

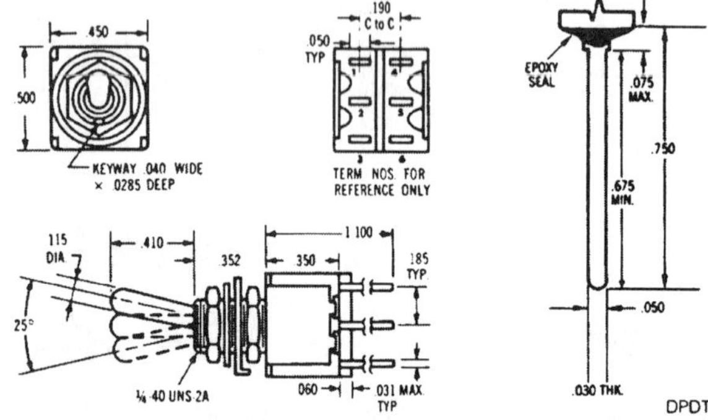

DPDT

Note	Contact Sequence		
	Lever Left	Centre	Right (Keyway)
(A)	2-3 5-6	2-3 5-4	2-1 5-4
(B)	2-3, 8-9 5-6, 11-12	2-3, 8-9 5-4, 11-10	2-1, 8-7 5-4, 11-10

BAT LEVER, WIRE-WRAP TERMINALS

Cat. No.	Lever Up	Lever Centre	Lever Down	VA Max. @ 28V AC-DC	Net Price Each 1-24	25-99	100-499	500-Up
SPDT—3 WIRE-WRAP TERMINALS								
A121S1YWB	On	Off	On	0.5	**$5.75**	**$4.87**	**$4.27**	**$3.77**
A123S1YWB	On	None	On	0.5	**5.03**	**4.26**	**3.74**	**3.34**
A126S1YWB	On	None	On*	0.5	**5.75**	**4.87**	**4.27**	**3.77**
A127S1YWB	On*	Off	On*	0.5	**5.75**	**4.87**	**4.27**	**3.77**
A131S1YWB	On	Off	On*	0.5	**5.75**	**4.87**	**4.27**	**3.77**

*Momentary

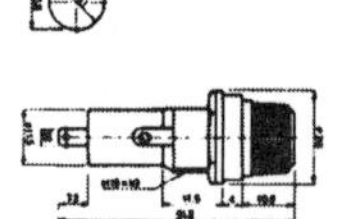

CSA/UL approved panel mount fuseholder for 3AG size fuses (¼" × 1¼"). Screw type cap.

Cat. No. 55-802-0 Net Price Per 1000 **$1760.00**

ADHESIVE RUBBER FEET

Cat. No.	Size	Thickness	Net Price
54-880-0	½" round	.232"	**$.19**
54-885-0	½" square	.232"	**.22**

Page D16

MODE ELECTRONICS

MANU 279 CODE

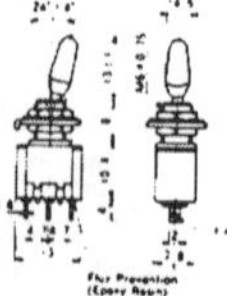

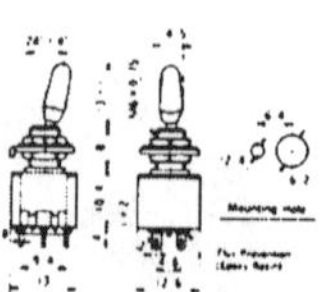

HIGH CURRENT TOGGLE SWITCH

Sub-miniature toggle. 10 Amp @ 125VAC. With solder terminals.

Cat. No.	Description	Net Price
41-210-0	SPST On-Off	**$3.20**
41-213-0	SPDT On-On	**3.44**
41-223-0	DPDT On-On	**4.72**

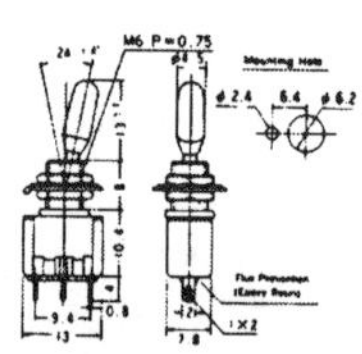

1 POLE

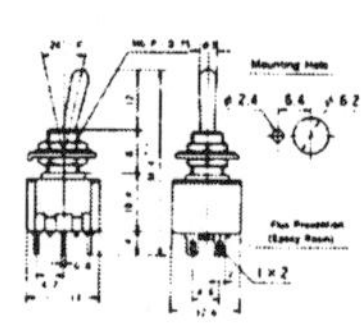

2 POLE

SUB-MINIATURE TOGGLE SWITCHES

UL/CSA approved sub-miniature toggles. 6 Amp @ 125VAC with solder terminals.

Cat. No.	Description	Net Price
41-230-0	SPST On-Off	**$2.88**
41-233-0	SPDT On-On	**2.96**
41-243-0	DPDT On-On	**4.16**
41-245-0	DPDT On-Off-On	**4.64**

54-122-0

54-217-0

54-218-0

54-227-0

54-234-0

54-232-0

54-279-0

54-278-0

54-228-0

54-374-0

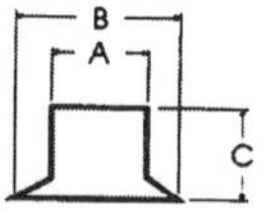

KNOBS

All knobs have brass bushings with set screw.

Cat. No.	A	B	C	Shaft	Net Price
54-122-0	15mm	15mm	12mm	6.35mm	**$1.44**
54-132-0	20mm	22mm	13mm	6.35mm	**1.16**
54-217-0	18mm	23mm	15mm	6.35mm	**1.26**
54-218-0	27mm	33mm	17mm	6.35mm	**1.26**
54-227-0	25mm	31mm	15mm	6.35mm	**1.31**
54-228-0	31mm	37mm	15mm	6.35mm	**1.42**
54-232-0	16mm	20mm	12mm	6.35mm	**.80**
54-234-0	24mm	27mm	16mm	6.35mm	**1.18**
54-278-0	23mm	37mm	15mm	6.35mm	**1.76**
54-279-0	23mm	37mm	15mm	6.35mm	**1.57**
54-374-0	30mm	Pointer knob		6.35mm	**.87**

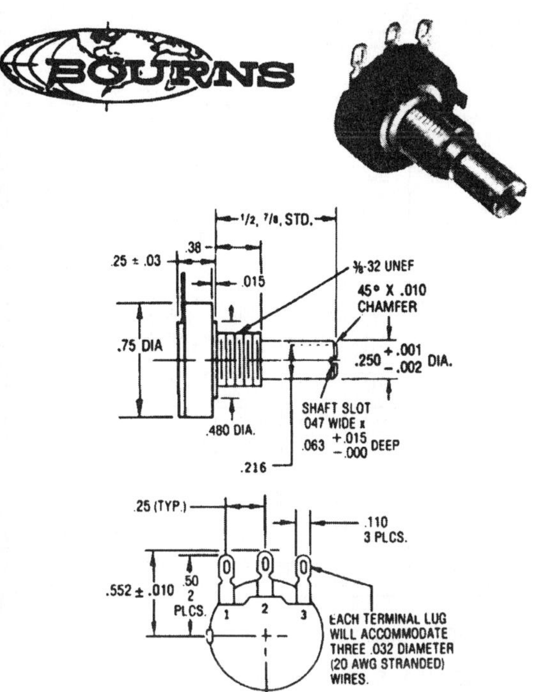

BOURNS MODEL 3852
¾″ DIA. CONTROL
SINGLE-TURN, BUSHING MOUNT, CERMET ELEMENT

One model replaces several styles including RVC5. Meets or exceeds the applicable requirements of MIL-R-23285. Cermet infinite resolution element offers stable performance in all environments. The highest power rating available in a thin ¾″ diameter case: 2 watts at +70°C. Unique one-piece molded-in terminations eliminate bonded connections and stress problems. Operating temperature range: −65° to +150°C. Temperature coefficient: ±150 ppm/°C. Rotational life: 50,000 cycles.

Ohms ±10% Tol.	⅞″ Slotted Shaft* Solder Lugs	½″ Slotted Shaft* Solder Lugs	Resolution
NON-LOCKING ⅜″ DIA. BUSHING, ¼″ DIA. SHAFT, LINEAR TAPER			
50	**3852A-282-500A**	**3852A-162-500A**	Infinite
100	**3852A-282-101A**	**3852A-162-101A**	Infinite
250	**3852A-282-251A**	**3852A-162-251A**	Infinite
500	**3852A-282-501A**	**3852A-162-501A**	Infinite
1,000	**3852A-282-102A**	**3852A-162-102A**	Infinite
2,500	**3852A-282-252A**	**3852A-162-252A**	Infinite
5,000	**3852A-282-502A**	**3852A-162-502A**	Infinite
10K	**3852A-282-103A**	**3852A-162-103A**	Infinite
25K	**3852A-282-253A**	**3852A-162-253A**	Infinite
50K	**3852A-282-503A**	**3852A-162-503A**	Infinite
100K	**3852A-282-104A**	**3852A-162-104A**	Infinite
250K	**3852A-282-254A**	**3852A-162-254A**	Infinite
500K	**3852A-282-504A**	**3852A-162-504A**	Infinite
1 Meg	**3852A-282-105A**	**3852A-162-105A**	Infinite
2.5 Meg	**3852A-282-255A**	**3852A-162-255A**	Infinite

Series	Net Price Each			
	1-24	25-49	50-99	100-Up
3852A	**$9.88**	**$7.65**	**$6.17**	**$4.63**

*Shaft length is measured from mounting surface.

Wire-Wrap

CooperTools

WIRE-WRAP® CUT-STRIP-WRAP BITS AND SLEEVES

The cut-strip-wrap bit and sleeve assembly is an economical means for making solderless wrapped connections. The one-step operation results in time savings for terminating the loose ends of wire harness and cables when compatible combinations of wire and insulation types are used. Features: Wire loaded conventionally. No exposed moving parts—wrapping bit rotates inside stationary sleeve. Bare wire-turns on terminals are constant because of fixed window cut-off design. Close terminal spacing—only slightly larger than conventional bits and sleeves. Highly reliable modified wrap. No retraining of operators.

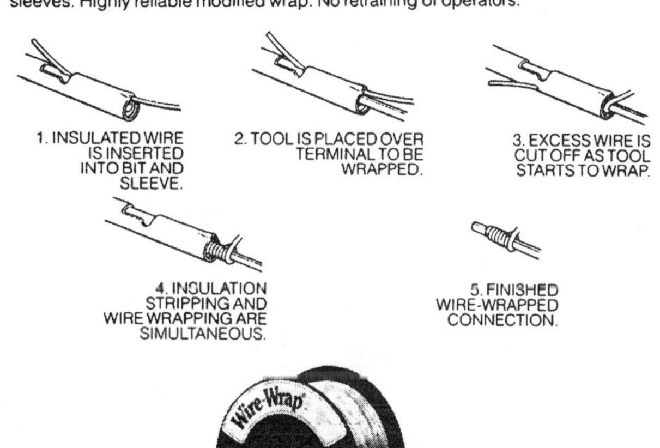

WIRE-WRAP® CUT-STRIP-WRAP INSULATED WIRE

Wire-Wrap® insulated wire is guaranteed to work with your Wire-Wrap® CSW bits and sleeves. That means good wraps that meet specifications every time. The wire comes on 1000 ft. (304mm) spools in 24, 26 and 30 AWG sizes.

Cat. No.	Wire Size AWG	Insulation O.D. Inches	Color	Net Price per 1000 ft. Roll
990222	30	.0195	Black	**$109.11**
990223	30	.0195	Green	**109.11**
990224	30	.0195	Orange	**109.11**
990080	30	.0195	White	**78.80**
990081	30	.0195	Yellow	**78.80**
990082	30	.0195	Red	**78.80**
990083	30	.0195	Blue	**78.80**
990084	26	.0275	White	**127.28**
990085	26	.0275	Yellow	**127.28**
990086	26	.0275	Red	**133.38**
990087	26	.0275	Blue	**133.38**
990092	24	.034	White	**133.38**
990093	24	.034	Yellow	**195.03**
990094	24	.034	Red	**195.03**
990095	24	.034	Blue	**195.03**

Wrapping Bit Cat. No.	Net Price	Sleeve Cat. No.	Net Price	Wire Size AWG	Terminal Diagonal Max. (Inches)	Terminal Diagonal Min. (Inches)	Insulation Range Max. (Inches)	Insulation Range Min. (Inches)	Insul. Turns	Bare Wire Turns	Effective Radius (Inches)
9066	**$185.66**	**519067**	**$113.91**	24	.070	.059	.045	.040	1/2	6	.133
8910	**185.66**	**518911**	**113.91**	24	.072	.059	.045	.040	1/2	6	.133
8910	**185.66**	**519929**	**106.97**	24	.072	.059	.037	.032	1/2	5	.133
9926	**185.66**	**519927**	**113.91**	26	.058	.054	.029	.026	3/4	6-1/2	.111
2202	**215.30**	**522201**	**113.91**	26	.072	.059	.035	.031	1/2	7	.128
7812	**202.16**	**527813**	**113.91**	26	.034	.030	.029	.025	1-1/4	7	.082
0764	**202.77**	**990765**	**111.86**	30	.034	.030	.0215	.019	1	7	.0705

Page D17

Belden

Hook-Up Wire

Description U.L. Style C.S.A. Type	Cat. No.	Standard Lengths ft.	Standard Lengths m	Std. Unit Lbs. ea.	AWG (stranding)	Insulation Thickness Inch	Insulation Thickness mm	Nominal O.D. Inch	Nominal O.D. mm	Available Colors See Color Chart No. 10
STYLE 1007 300V—80°C HOOK-UP WIRE—(CSA TYPE TR-64—90°C)										
Tinned copper, PVC insulated. Rated -10°C to 105°C, 300V. Rated 600V peak for electronic circuits, and internal wiring of electronic and electrical equipment. UL 1007 and 1569 CSA	**9918†**	100 1000	30.5 304.8	.8 7.6	18(16×30)	.016	.41	.079	2.01	1-5, 7-10, 13
STYLE 1015 600V—105°C HOOK-UP WIRE—(CSA TYPE TEW)										
Tinned copper, PVC insulated. Rated -40°C 105°C, 600V. Rated 2500V peak for electronic circuits, and internal wiring of electronic and electrical equipment.	**9924†**	100 1000	30.5 304.8	.6 5.3	24(7×32)	.031	.79	.088	2.24	1-5, 9, 10, 13
	8920†	100 1000	30.5 304.8	.7 6.4	22(7×30)	.031	.79	.93	2.36	1-5, 9, 10, 13
	8919†	100 1000	30.5 304.8	.8 7.8	20(10×30)	.031	.79	.100	2.54	1-5, 9, 10, 13
	8918†	100 1000	30.5 304.8	1.0 10.3	18(16×30)	.031	.79	.110	2.79	1-5, 9, 10, 13
	8917†	100 500 1000	30.5 152.4 304.8	1.7 7.4 14.3	16(26×30)	.031	.79	.123	3.12	1-5, 9, 10, 13
UL 1015 CSA	**8916†**	100 500	30.5 152.4	2.3 10.3	14(41×30)	.031	.79	.138	3.51	1-5, 9, 10, 13
	9912†	100 250	30.5 76.2	3.1 7.5	12(65×30)	.031	.79	.158	4.01	1-5, 9, 10, 13
	9910†	100 250	30.5 76.2	4.5 11.0	10(65×28)	.031	.79	.183	4.65	2, 4, 9, 10

UNSHIELD AND SHIELDED SPECIAL PURPOSE CORDS
COMBICORD☆
POWER CORDS FOR PERMANENT CONNECTIONS

Combicord solves EMC problems when internal filtering, due to design difficulties is not possible or on products in service that were produced before the EMC problem was fully recognized. Brings a valid solution to these problems through the use of an external filter. The Combicord consists of a shielded supply cord and molded plug that incorporates the filter components. Available as a detachable or permanently wired appliance cord with a NEMA plug and IEC320 appliance plug on the detachable version. The shielded supply cord is a SVT3 × 18AWG. For hard-wiring it is recommended to add suitable Y-capacitors (Cy=2200pF) between hot (Brown) resp. neutral (Blue) and ground (Yellow-Green), connected with the flex wire of the cord-shield.

Description	Cat. No. Plug	Cat. No. Conn.	Length of Cord ft.	Length of Cord m	Pack-aging	Color Coding	AWG stranding [Dia. in mm]	Nominal O.D. Inch	Nominal O.D. mm	Rating	Type of Cordage	Jacket Color
Molded PVC Grounding Plug PH-290B	**17237**		6'	1.83	B-50 S-10	NA	18(41×34) [1.19]	.246	6.25	1250 Watts 10A-125V	SVT	Gray
	290B		60°C									
							N.S.N. 6150-01-057-0941††			6150-01		
Outer Jacket Stripped 1 1/2" Conductors Stripped and Twisted 5/8"	**17239**		8'	2.44	B-50 S-10	NA	18(41×34) [1.19]	.246	6.25	1250 Watts 10A-125V	SVT	Gray
	290B		60°C									
							N.S.N. 6150-00-433-1104†					
UL CSA E N L Type SVT PVC Jacketed NEMA 5-15P												

AGC, MTH, GLH, & MGB TYPE FUSES
NON-TIME DELAY

These fast acting glass tube, visual indicating fuses are designed for instruments, electronic and small appliance circuits. Most common American size glass fuse.

PANEL MOUNTED FUSEHOLDERS

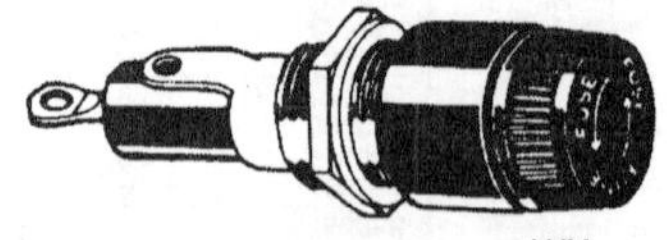

HJM
HKP

SERIES HCM

PANEL MOUNTED FUSEHOLDERS

Panel mounted non-indicating fuseholders. With model HCM knob grips 9/32" fuse tightly and pulls it from holder when knob is removed. 1/4" fuse is not held by knob but extends from holder when knob is removed. HJM, HJM-CC for 1/4"× 1" fuses, HKP, HKP-CC, HKP-HH for 1/4"× 1 1/4" fuses. -CC is 3/32" shorter behind panel. HKP-HH, quick connect terminals.

Buss No.	Amps/Volts	Net Price Each 1-99	Net Price Each 100-Up
FOR 1/4"× 1" FUSE			
HJM	5/125	**$2.12**	**$1.91**
HJM-CC	5/125	**3.79**	**3.41**
FOR 1/4"× 1 1/4" FUSE			
HKP	30/250	**1.91**	**1.72**
HKP-CC	30/250	**1.83**	**1.65**
HKP-HH	15/250	**2.26**	**2.03**
HKP-L	2,250V	**4.42**	**3.98**
FOR 1/4"× 1 1/4" AND 9/32"× 1-1/4" FUSE			
HCM	18/250	**3.38**	**3.04**
REPLACEMENT KNOB FOR HKP HOLDER			
9435½	Knob only	**1.45**	**1.30**

Page D18

appendix E

Practice Sessions

EXERCISE 1 A PRACTICE SESSION USING UNDO AND ERASE

To practice using the Undo and Erase commands, we must first draw some lines to work with. We'll create Figure 1E-1A. Set:

Limits = 100,100 units
Grid = 5 units
Snap = 5 units
Zoom = A (All)
Coordinates = ON (F6)

Developing the Drawing and Using Undo

TRAP: Undo can get you into a few problems if you do not fully understand its use. There are two methods to use Undo. First, while you are still in a command, such as drawing lines (the line command), and secondly when you have *terminated* a command.

If you are in a command, such as the Line command, selecting the Undo option or command will undo or remove the last entity drawn on the screen. Continuous selection of Undo will remove subsequent entities, one at a time.

If you have terminated a command and you Undo a line on an object, every line you drew to create that object becomes undone. This is the problem. You could erase everything you did over the past ten minutes. If this occurs, type REDO to bring the objects back. Redo is like an antidote to Undo.

To draw Figure 1E-1A, select Draw from the AutoCAD Root Menu and then Line from the Draw Menu.

```
Command: SELECT LINE
Command: LINE From point: select 25.000,25.000
```

Watch the coordinates area and using the input device, select the following:

```
    To point: 15.0000 <90       Line A
    To point: 10.0000 <180       "   B
    To point:  25.0000 <90       "   C
    To point: 21.2132 <45        "   D
    To point:  5.0000 <90        "   E
    To point: 30.0000 <0         "   F
    To point: 20.0000 <270       "   G
    To point: 25.0000 <0         "   H
    To point: 35.3553 <225       "   I
    To point: 15.0000 <270       "   J
    To point: 35.0000 <180       "   M
(Do not select CLOSE or TERMINATE the LINE command)
```

Note: We are still in the Line command. Now to select Undo from the Line Menu.

```
To point: SELECT UNDO(FROM THE LINE MENU).
```

Note: The last line (M) is removed from the drawing, and the cursor moves back to location 15.0000 <270, (60,25).

Select Undo again.

```
To point: SELECT UNDO
```

The cursor moves back one more step, and another line (J) is removed.

Now reselect point 15.0000 <270, and reestablish Line J.

```
To point: 15.0000 <270 Line J
To point: 21.2132 <135 Line K
To point: CLOSE (FROM THE LINE MENU) Line L
```

The line now closes to the starting point. You should now have a drawing that looks similar to Fig. 1E-1A.

UNDO AND REDO

Now to see how Redo works.

```
Command: SELECT UNDO FROM THE LINE COMMAND
```

As expected, the complete Figure 1E-1A disappeared! This is because the Line command was terminated, using Close, and Figure 1E-1A now looks like a single entity to the Undo command. To resolve this problem, use the Redo command, now! Don't try Oops, it won't work. Redo should be used immediately after an Undo mistake is created, else it may not work.

```
Command: REDO (RETURN)
```

If all went well, Figure 1E-1A should reappear.

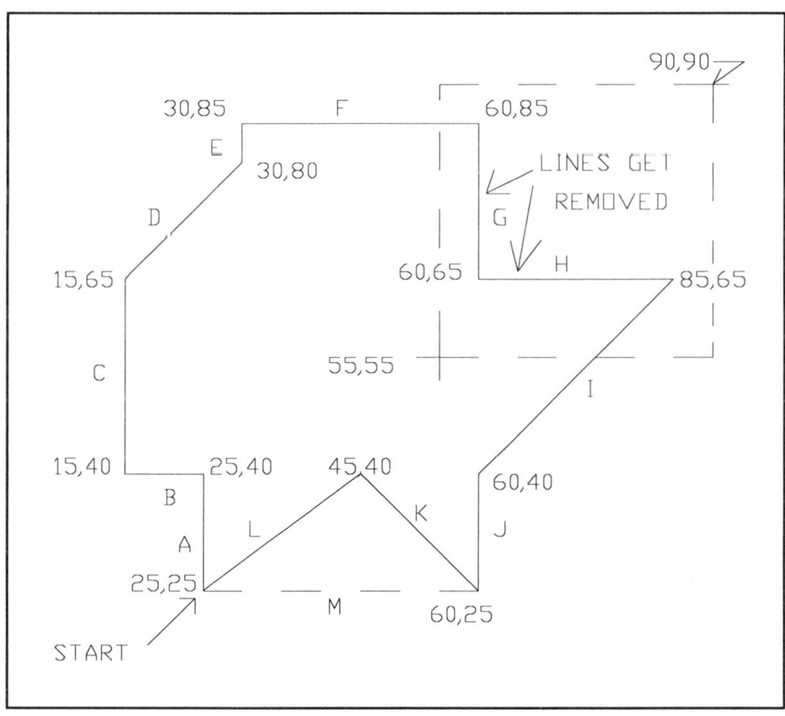

Figure 1E-1A Erase Editing Diagram

Playing with Erase

LAST (L)

We will now investigate other Erase commands. Select Edit, using the input device from the Line menu, then Erase.

The Command line requests:

```
Command: SELECT ERASE
Select objects:
```

Select E Last from the Erase menu.

```
Command: SELECT E LAST FROM ERASE MENU
```

Upon pressing the device select button, the last line (L) disappears. Now choose Oops;

```
Command: SELECT OOPS
```

the last line reappears.

Selecting Objects by Pointing (P)

Select Erase again, but this time we will point to several lines before removing them.

```
Command: SELECT ERASE
Select objects:
```

Move the square cursor over to a line on our drawing and press the device select button. The line selected becomes dotted but it is not yet removed; all we have done is select it.

```
Select objects: SELECT A LINE
```

Move the cursor to another line and press the device select button; it also becomes dotted (selected).

```
Select objects: SELECT A LINE
Select objects: (RETURN)
```

Pressing the Return key on the keyboard, causes both lines to disappear.

Move your cursor to Oops and select it;

```
Command: SELECT OOPS
```

the lines reappear.

Now we'll play with the Window Object Selection method found under the Erase command.

Playing with Window (W)

Using the input device, select ERASE and then type W (WINDOW) at the command line:

```
Command: Select ERASE
Select objects: W (RETURN)
First point: select 55.0000,55.0000
Second point: select 90.0000,90.0000
CLICK THE MOUSE SELECTION BUTTON (LEFT BUTTON)
```

Note that only the two lines (G and H), those totally enclosed by the window, became dotted when the input device select button was pressed.

Now press the Return key and the two dotted lines disappear. Select Oops, and they reappear.

Let's Erase the whole object by using a Window. By now *you* should be able to erase it and bring it back without my help. When you have completed this practice session, return to Exercise 3-3c.

EXERCISE 2a: A PRACTICE SESSION ON DIMENSIONING USING AutoCAD VERSION 10 THROUGH VERSION 12 (Supported by V13)

Discussion

The following exercises, 2A and 2B, are based on AutoCAD versions up to and including version 12. In these versions, the Command prompt to change to "DIM" while doing dimensioning. Exercise 3, which follows, demonstrates how AutoCAD version 13 differs from version 12, and shows the new command where the dimension variables are used directly from the Command prompt. *Note:* Version 13 also supports the DIM command structure.

In this exercise we will develop a house and dimension the outside using the Linear Dimensioning options of Horizontal, Vertical, Aligned, Rotated, Baseline, and Continue. We will also set the default values for the arrowhead size (DIMASZ) to .1 and text height variable (DIMTXT) to .25.

Remember, to get out of the Dimension mode, type EXIT or press the CTRL-C Keys.

TIP: To save you time in developing the HOUSE1 drawing, you will find it on the disk that comes with this tutorial as HOUSE1.DWG; or, if you are having fun, you can develop it now using the following procedure.

Housekeeping for the Drawing HOUSE1

Set up the following housekeeping:

Drawing name = HOUSE1
Limits = 14,10
Grid = .5
Snap = .1
Zoom = (All)
Coordinates = ON

Setting the Dimension Variables (Version 12)

Now to set the arrowhead size to 0.1 and the text height to 0.25.

```
Command: SELECT DIM IN THE AUTOCAD ROOT MENU
Dim: SELECT DIM VARS FROM MENU
```

A series of Dimension Variables are displayed in the menu area.

Note: The Command line is now DIM.

```
Dim: SELECT DIMASZ
Dim: DIMASZ Current value <default>
New value: .1 (RETURN)
```

Next set the text height.

```
Dim: SELECT DIMTXT
Dim: DIMTXT Current value <default>
New value: .25 (RETURN)
```

Now exit the Dimension Command menu.

```
Dim: EXIT (RETURN)
Command:
```

The Command line reappears.

Building the House1 Drawing

With the Coordinates ON, draw a line between the following coordinate points, starting at location (7,8). Use Fig. 2E-1A as a reference.

```
Command: LINE (RETURN)
From point: 7,8 (RETURN)     ...Starting point (7, 8)
To point: 3.9051 <310        ...Line C          (9.5, 5)
To point: 2.0000 <270        ...Line D          (9.5, 3)
To point: 2.0000 <180        ...Line E          (7.5, 3)
```

```
To point: 1.5000 <90        ...Line F        (7.5, 4.5)
To point: 1.0000 <180       ...Line G        (6.5, 4.5)
To point: 1.5000 <270       ...Line H        (6.5, 3)
To point: 2.0000 <180       ...Line I        (4.5, 3)
To point: 2.0000 <90        ...Line A        (4.5, 5)
Close:                      ...Line B        (7, 8)
```

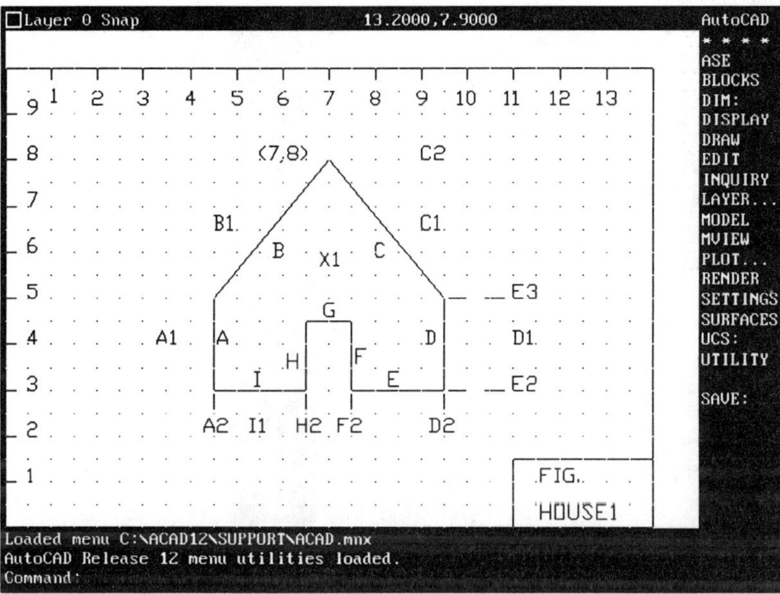

Figure 2E-1A House 1

Your house should look similar to that of Fig. 2E-1A. If not, recheck your coordinates.

Dimensioning the House

In this exercise we will dimension all of the lines on the outside of the house, starting with Line A (Fig. 2E-1A).

From the AutoCAD Root Menu, use the input device to select "Dim:." Notice the command line changes to DIM:. We will next choose the Linear Dimensioning mode.

```
Command: SELECT DIM FROM ROOT MENU
Dim: SELECT LINEAR
```

Using the Vertical Dimensioning Option

Now we are ready to dimension line A of Fig. 2E-1A, using the "automatic" dimensioning feature of AutoCAD. Select line A by pointing to it with the input device, when requested.

```
Dim: SELECT VERTICAL (VRTICAL)
Dim: VERT (appears on the DIM line)
First extension line origin or RETURN to select: (RETURN)
Select line, arc, or circle:  SELECT LINE "A"'
```

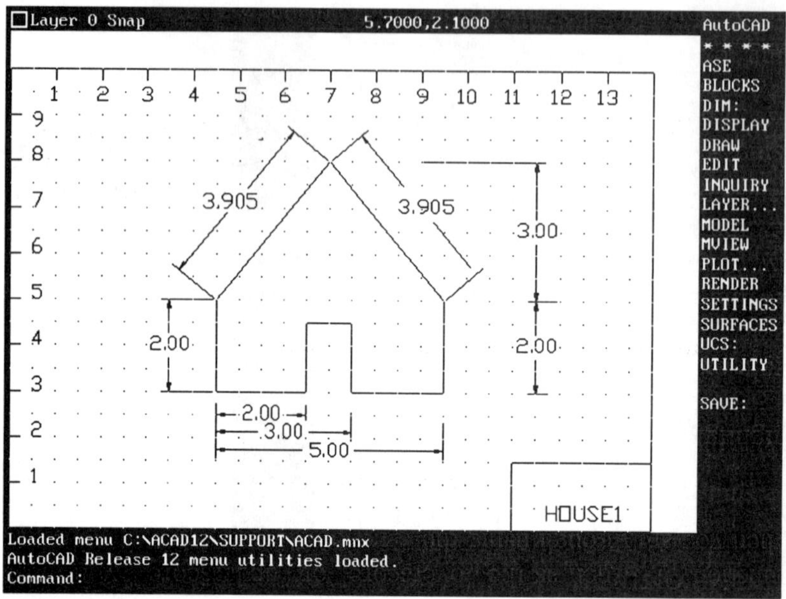

Figure 2E-1B Completed House1.

AutoCAD now asks where the dimension line is to be located; select location A1.

```
Dimension line location: SELECT LOCATION A1
```

AutoCAD now informs us of the length of Line A and allows us to change it. We will keep it the same.

```
Dimension text <2.0000>: (RETURN)
```

If all went well, AutoCAD drew a vertical line at A1 and automatically dimensioned it (see Fig. 2E-1B).

Using the Aligned Dimensioning Option

Now we will dimension line B, using "Aligned"

```
Dim: SELECT ALIGNED
Dim: ALI (appears at the DIM line)
First extension line origin or RETURN to select: (RETURN)
Select line, arc, or circle: SELECT LINE "B"
```

```
Dimension line location: SELECT LINE B1
Dimension text <3.9051>: (RETURN)
```

If everything is again running smoothly, AutoCAD now draws and dimensions the ALIGNED dimension line B1.

Using the Rotated Dimensioning Option

To dimension line C using Rotated, we must inform AutoCAD of the rotation angle of the dimension line. If you choose a wrong angle, the dimension line will not be parallel to the drawn line. If this occurs, Undo it and try again. It was found that 130 degrees was the correct angle for this line.

```
Dim: SELECT ROTATED
Dim: Rotate dimension line angle <0>: 130 (RETURN)
First extension line origin or RETURN to select: (RETURN)
Select line, arc, or circle: SELECT LINE "C"
Dimension line location: SELECT LINE C1
Dimension text <3.9051>: (RETURN)
```

AutoCAD now draws and dimensions the rotated dimension line C1.

Some Fun Using BASELINE

Let's dimension the bottom length of our house, lines I, G, E, using the Baseline option. *Note:* When using Baseline or Continue, you must first establish the initial dimension line by using either the Horizontal, Vertical, Aligned, or the Rotated options. You then enter the BASELINE option, which employs only the second extension line.

The following text will walk you through this procedure.

```
Dim: SELECT HORIZONTAL
First extension line origin or RETURN to Select: SELECT
EXTENSION LINE A2
Second extension line origin: SELECT EXTENSION H2
Dimension line location: SELECT I1
Dimension line text <2.0000>: (RETURN)
```

The first line I1 is automatically drawn and labelled. Now select the Baseline option. *Note:* Only the second extension line is called for.

```
Dim: SELECT BASELINE OPTION
Dim: BASE (appears at the DIM line)
Second extension line origin: SELECT "F2"
Dimension text <3.0000>: (RETURN)
```

The second BASELINE line, F2, is automatically inserted and labelled.

Now select the Baseline option again.

```
Dim: SELECT BASELINE OPTION
Dim: BASE (again appears in the DIM line)
Second extension line origin: SELECT "D2"
Dimension text <5.0000>: (RETURN)
```

The third BASELINE, D2, is drawn and labelled.

Working with CONTINUe

Now let's dimension the overall height of the house, using Continu. First we will dimension line D, using the Vertical option, and then use Continu for the height of the roof, line C (Fig. 2E-1A).

```
Dim: SELECT VERTICAL (VRTICAL)
Dim: VERT (appears at the DIM line)
First extension line origin or RETURN to select: SELECT POINT E2
Second extension line origin: SELECT E3
```

TIP: Make sure that the location you choose for the drawing dimension line D1 will not interfere with the dimension line at C1.

```
Dimension line location: SELECT D1
Dimension text <2.0000>: (RETURN)
```

The first vertical height dimension line D1 is drawn and dimensioned. See Fig. 2E-1B. Now for the continuation of this vertical dimensioning, select Continu.

```
Dim: SELECT CONTINU
```

In the selection of the "second extension line origin" (point C2), move it far enough to the right of the roof line so that, again, it does not interfere with the C1 dimension line.

```
Second extension line origin: SELECT C2 POINT
Dimension text <3.0000>: (RETURN)
```

Now the vertical height of the building is dimensioned. Your drawing should look similar to Fig. 2E-1B.

Let's Get Back to the "COMMAND" Line

Exit back to AutoCAD'S command line.

```
Dim: EXIT (RETURN)
Command:
```

At this point we are back in AutoCAD and ready to continue.

EXERCISE 2b: A PRACTICE SESSION USING ANGULAR, DIAMETER, AND RADIUS DIMENSIONING (AutoCAD V12, Supported by V13).

Discussion

In Exercise 2A, you were introduced to the various Linear dimensioning methods. In this exercise we will use Angular, Diameter, and Radius dimensioning methods, which draw Arcs and Leader lines to create the dimension line.

Angular Dimensioning

Using Fig. 2E-1A as a reference, we will have AutoCAD automatically measure the angle between the two roof lines B and C, and then insert an Angular dimension line between them. From the Root Menu, select DIM.; then follow the commands below:

```
Command: SELECT DIM:
Dim: SELECT ANGULAR
Dim: ANG Select first line: SELECT LINE "B"
Second line: SELECT LINE "C"
```

AutoCAD requests which side of the lines B and C we want to measure the arc from, that is, the top side or bottom. We'll use the bottom side; select X1.

```
Enter dimension line arc location: SELECT X1
```

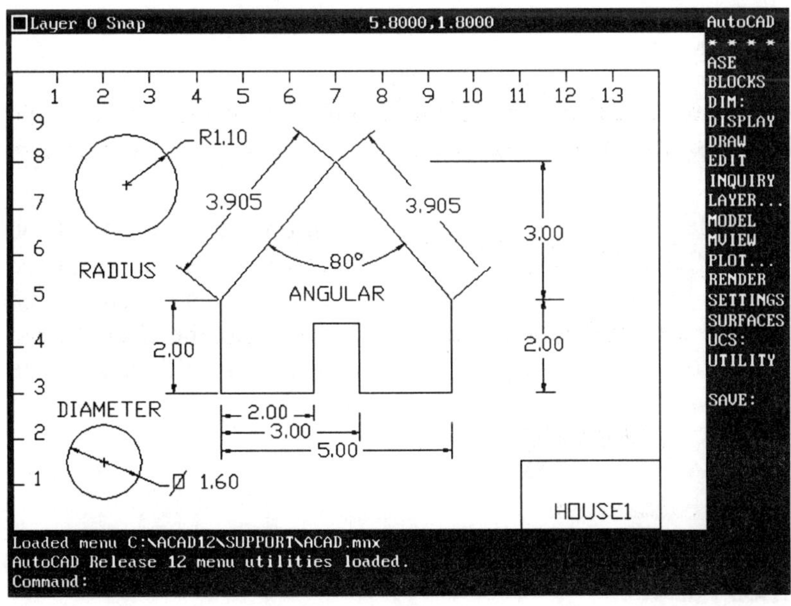

Figure 2E-1C House1 dimensioned.

AutoCAD informs us that the angle between lines B and C is 80 degrees, or thereabout, and gives you the opportunity to change the text value. If you are satisfied, press RETURN.

```
Dimension Text <80>: (RETURN)
```

You now have the opportunity to show AutoCAD where the text is to be placed. Select the X1 location again.

```
Enter text location: SELECT X1
```

An arc is drawn between the roof lines B and C at location X1 and is automatically labelled as 80 degrees. See Fig. 2E-1C.

<u>TIP:</u> Had we selected the top side of the two lines B and C as the dimension line arc location, the arc would have been drawn on the outside of the roof and labelled as 280 degrees. Your results should look similar to that of Fig. 2E-1C.

Diameter Dimensioning

To measure the diameter of a circle or arc, we will draw a "small" circle on our work sheet, using the Center, radius command from the Circle menu. Don't forget to Exit the dimensioning menu before starting.

```
Command: SELECT CIRCLE
Command: SELECT CEN,RAD:
Command: Circle 3P/2P/<center point>:SELECT
Command: Diameter/<radius>: DRAG OUT THE CIRCLE
```

The circle is now drawn. Now to dimension it.

When AutoCAD asks "select arc or circle," select a point on the perimeter of the small circle. *Note:* Select this location carefully, as it becomes the point where the Diameter Dimension Line terminates when the line is drawn. It also becomes the location from which a Leader line will extend should the area inside the circle be too small to insert the text. If you have drawn your circle small enough, this will be the case.

From the AutoCAD Root Menu select DIM.

```
Command: SELECT DIM
Dim: SELECT DIAMETER
Dim: Diam select arc or circle: SELECT CIRCLE
```

AutoCAD next informs us of the measured diameter and the value it will use as text.

If the area inside the circle is too small, you will get a "Text does not fit message," and AutoCAD will ask "Enter Leader length for text." Reply with a length appropriate for the space available.

```
Dimension text <value found>: (RETURN)
Text does not fit
Enter Leader Length for text: .5 (RETURN)
```

The diameter of the circle will be drawn at the end of a Leader line if your circle size was small enough. If AutoCAD could squeeze the text inside the circle, it did so.

Redraw the circle to a "large" size and repeat the exercise. Your results should look similar to those of Figs. 2E-1D and 2E-1E.

Figure 2E-1D Small circle with Leader Diameter line.

Figure 2E-1E Large Circle. (Diameter Dimensioned)

Radius Dimensioning

In this exercise we will use the Radius command. Redraw the "small" circle again.

```
Dim: SELECT RADIUS
DIM: RAD Select Arc or Circle: SELECT CIRCLE
```

Again, the point selected on the perimeter of the circle should be chosen with care as it becomes the location from which the Leader line starts, if needed.

```
Dimension text <value found>: (RETURN)
Text does not fit
Enter Leader length for text: .5 (RETURN)
```

A Radius Dimension line is automatically drawn from the center of the circle to the point selected on the perimeter of the circle; then a leader line 0.5 units long is drawn beyond this point. Note how all lines, arrowheads, and text are inserted automatically.

Figure 2E-1F Small circle with Leader Radius line.

Figure 2E-1G Large circle (Radius Dimensioned)

Repeat the exercise, using a large circle. Your results should look similar to Figs. 2E-1F and 2E-1G.

EXERCISE 3: A PRACTICE SESSION ON DIMENSIONING USING AutoCAD VERSION 13

Discussion

In practice exercise 2A and 2B, you played with dimensioning a house using the DIM command prompt feature of AutoCAD version 12. While in this mode, you were restricted from using a number of AutoCAD commands such as Erase. In version 13, this all changed with the DIM series of commands such as Dimlinear which is used directly at the command prompt. If you have version 13 installed, try the following exercise.

Housekeeping for Exercise 3

Limits = 100, 100
Grid = 5
Snap = 5

Using Fig. 3E-1 as a reference, create a triangle that is 40 units long on side C and 80 units high on side B. The angle of C and A will become 63 degrees.

In this exercise we will play with the Dimlinear command to show you the basic difference between version 13 commands and those of the DIM series of dimensioning commands in version 12. Version 13 now allows you to make use of all the editing features of AutoCAD while dimensioning, where as the DIM command series restricted you in this area..

Using Figure 3E-1 as a reference, we will explore a number of AutoCAD version 13 dimensioning methods.

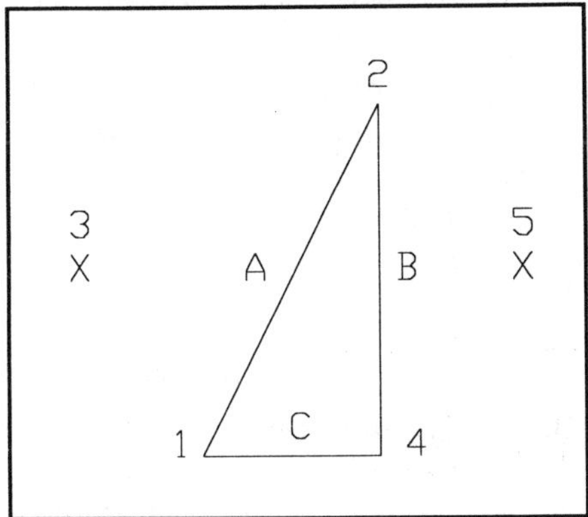

Figure 3E-1 Triangle.

Playing With Version 13

To start, we'll set the number of decimal places in the dimension text to two, then the arrowhead size to .2 and finally the Dimension text height to 2 units.

```
Command: DIMDEC (RETURN)
New value for DIMDEC <4>: 2 (RETURN)
Command: DIMASZ (RETURN)
New value for DIMASZ <0.1800>: .2 (RETURN)
Command: DIMTXT (RETURN)
New value for DIMTXT <0.1800>: .2 (RETURN)
```

Our next step will be to automatically dimension line A of Fig.3E-1 using Dimlinear and the "Select Object" format. At the command prompt type:

```
Command: DIMLINEAR (RETURN)
First extension line origin or RETURN to select:
```

We'll use the Select method, press the Return key

```
First extension line origin or RETURN to select: (RETURN)
Select object to dimension:
```

You are now asked to select the object line (Line A). AutoCAD automatically determines the origin point of the line selected and the two extension lines. *Note:* the extension line offset is determined by the DIMEXO system variable.

```
Select object to dimension: SELECT LINE "A"
Dimension line location (Text/Angle/Horizontal/Vertical/Rotated):
SELECT POINT 3
```

Take note that AutoCAD is smart enough to know that the extension lines will be either Horizontal or Vertical, depending upon which direction you drag the mouse. When location point 3 is selected, you should also note that AutoCAD has already determined the vertical distance between point 1 and point 2 of line A, and places the dimension lines and extension lines automatically.

TEXT Option Of The DIMLINEAR Command

There are a number of options to the Dimlinear command which we will now explore, such as modifying the dimensional text with the Text option.

If the text that is displayed needs to be modified, select the "T" option at the Dimension Line Location request.

```
Dimension line location (Text/Angle/Horizontal/Vertical/Rotate?: T
(RETURN)
```

This brings up the Edit MText Dialogue Box. See Fig. 3-9d on page 81. The dimensional text is referenced by the < > arrows in this editor. Use the Delete key to remove these arrows and type in the new text; then select the OK button. You are immediately taken back to the drawing editor and the new text is in place.

VERTICAL Option Of The DIMLINEAR Command

The default option of the Dimlinear command is Horizontal. To change the text to a vertical display, we'll use the Vertical option. See Figure 3E-2.

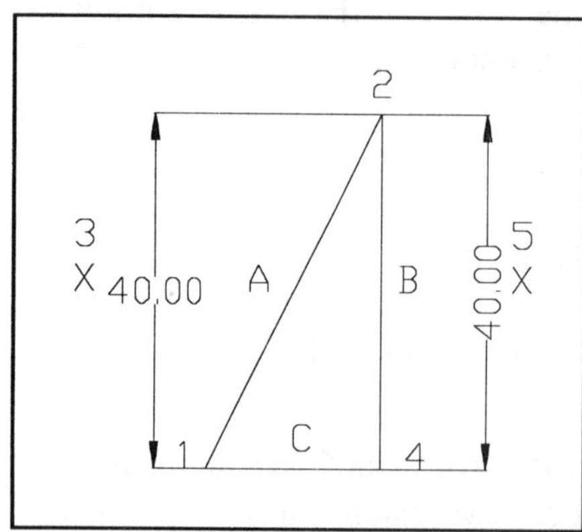

Figure 3E-2 Version 13 Horizontal and Vertical text.

```
Command: DIMLINEAR (RETURN)
First extension line origin or RETURN to select: (RETURN)
Select object to dimension: SELECT LINE "B"
Dimension line location (Text/Angle/Horizontal/Vertical/Rotate): V
(RETURN)
```

```
Dimension line location (Text/Angle): A (RETURN)
Enter text angle: 90 (RETURN)
```

Now drag the extension line to the desired location, at point 5, and click the mouse select button. Your drawing should now look similar to Figure 3E-2.

Selecting The "End Points Of A Line" Option

At the Dimlinear command, you are given the option to select the first extension line by pointing the mouse to the end of the line. We'll play with this selection method and place the text and the dimension line on an angle of 63 degrees, the slope (hypotenuse) of a triangle 4 units by 8 units (Figure 3E-1).

Erase the old dimension lines on Figure 3E-2 to make room for this next exercise.

```
Command: DIMLINEAR (RETURN)
First extension line origin or RETURN to select:
```

We'll select point 1, but first let's use the Osnap End option to select our point. Type:

```
First extension line origin or RETURN to select: END (RETURN)
SELECT POINT 1
Second extension line: END (RETURN)
SELECT POINT 2
Dimension line location (Text/Angle/Horizontal/Vertical/Rotated): A
(RETURN)
Enter text angle: 63 (RETURN)
Dimension line location (Text/Angle): DRAG THE LINE TO POINT 3, AND
CLICK IN PLACE.
```

Your dimension line and text are now parallel the line A, on a 63 degree angle, similar to Figure 3E-3.

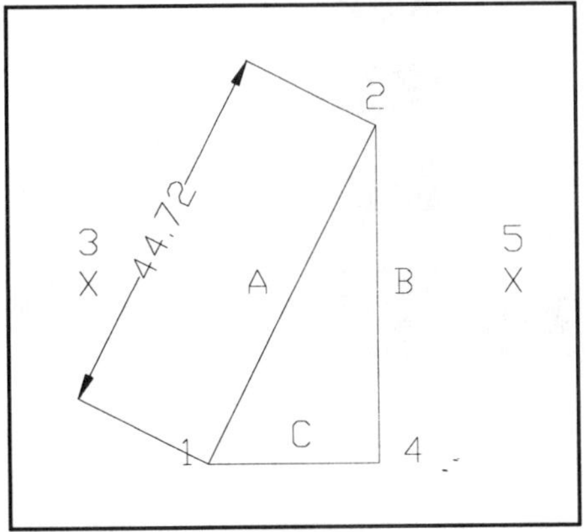

Figure 3E-3 Rotated Text and Dimension Line.

EXERCISE 4a CREATING AND USING TEXT STYLES

Discussion

AutoCAD, when first used, provides only one text style, called STANDARD. It is based on the text font TXT, which is quick and simple for AutoCAD to draw on the screen.

In this exercise we will learn to create other text styles, and make use of the text fonts supplied with AutoCAD. See Fig. 4E-1 and Appendix - A1.

First, let's examine the fonts and see how they differ.

TXT Font. The TXT font is a simple stick-like font with no rounded corners. It is drawn from short lines and thus is simple for AutoCAD to redraw when time is of the essence.

SIMPLEX Font. Simplex font produces a more natural-looking character with rounded corners. The "O" looks round instead of box-shaped, thus giving the character a smooth shape.

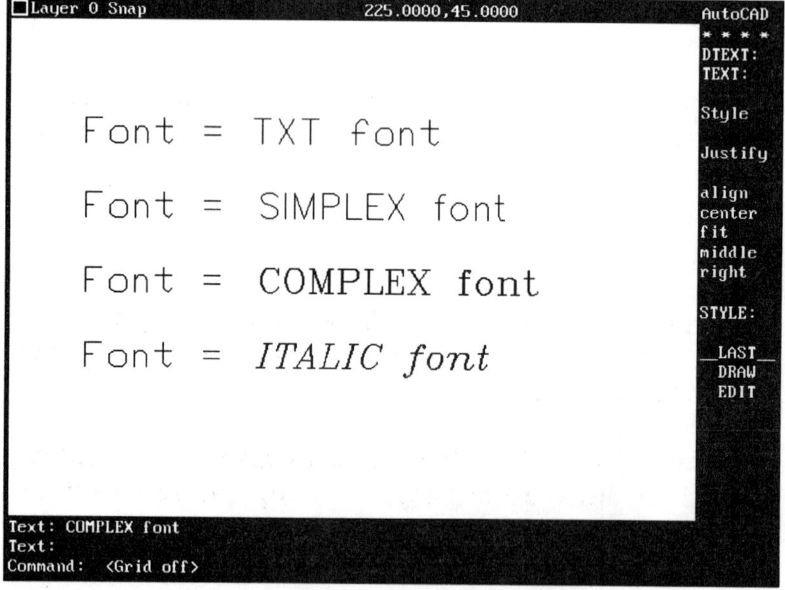

Figure 4E-1 AutoCAD Font Styles.

COMPLEX Font. Complex font looks similar to "letter-quality" characters found on a typewriter. There are small strokes at the ends of the characters to finish the letters off. Note the characters "T" and "E" of a typed character set. This extra touch in forming the character is paid for in time when you are plotting.

ITALIC Font. The Italic font has a slight forward slant (15 degrees), which makes it useful for highlighting purposes. This added touch, unfortunately, is paid for in time when you are plotting.

Creating Our First Text-Style "ITAL-1."

We will now create a simple text style named "ITAL-1," based on the Italic font.

From the keyboard, at the command line, type STYLE to initiate the style command. Do not select Style from the Text Menu.

```
Command: STYLE (RETURN)
Text style name (or ?)<STANDARD>: ITAL-1 (RETURN)

The Select Font File Dialogue Box may appear. If so, click on TYPE
IT and continue.

Font file <default>: ITALIC (RETURN)
```

Next thc height of the character is requested. If you reply with "0," AutoCAD will prompt you for the height when the text is written, during the Text command. If you specify a height here, this will become the height used forever. I suggest you respond with "0."

Next, AutoCAD requires a width factor. A factor of 1 produces a normal proportioned character, greater than 1 a wide character, and less than 1 a tall character.

```
Height <default>: 0 (RETURN)
Width factor <1.00>: 1 (RETURN)
```

The oblique angle determines the slant of the character. Zero degrees is a straight-up-and-down character. A positive angle slants the character to the right, and a negative angle slants the character to the left.

The Italic character already has a slant to the right of about 15 degrees. If we add an angle here, it will be added to the already existing angle of 15 degrees. This we don't want. Therefore, our obliquing angle will be kept to 0 degrees.

```
Obliquing angle <default>: 0 (RETURN)
Backwards? <Y/N>: N (RETURN)
Upside-Down? <Y/N>: N (RETURN)
```

In special cases, you may want characters that are printed backwards or upside down or both. In this case, we responded with No.

```
Vertical? <Y/N>: N (RETURN)
ITALIC-1 is now the current text style.
Command:
```

We now have a new character style called ITAL-1, which is based on the Italic font. In fact, it is the Italic font given the name ITAL-1. This character style is stored in the ACAD file, so make sure that the default drive has this file available when you go to use it.

By using the above procedure, the remaining text fonts may be converted into character styles that are useful for your drawing applications. See Appendix - A1.

Using Character Styles in Your Drawing

We now have two character styles available for us to use on the drawing, STANDARD and ITAL-1. You invoke the desired style by selecting the Style option "S", under the Text menu.

```
Command: DTEXT (RETURN)
Justify/Style<Start point>: S (RETURN)
Style name (or ?) <default>: ITAL-1 (RETURN)
```

The style is now set to **ITAL-1.** Continue with the text command.

```
Justify/Style<Start point>: SELECT START POINT
Height <default>: .25 (RETURN)
Rotation angle <default>: 0 (RETURN)
Text: THIS IS TEXT (RETURN) (RETURN)
```

The text is written in Italic font. If the rotation angle is set to 90 degrees, the text is written in a vertical plane.

What Styles Are Available?

To determine what styles are presently available, type STYLE at the Command prompt. This will display the Text style prompt. Reply with the question mark symbol, "?." AutoCAD then displays a list of all styles saved in the ACAD file.

```
Command: STYLE (RETURN)
Text style name (or ?) <ITAL-1>: ? (RETURN)
Text style(s) to list <*> (RETURN)
```

AutoCAD now displays the Text Style table showing the status of all options. See Table 4E-1.

TABLE 4E-1 Text styles

```
Style name: STANDARD          Font file: TXT
   Height: 0.0000   Width factor: 1.00   Obliquing angle: 0
   Generation:  Normal

Style name: ITAL-1            Font file: ITALIC
   Height: 0.0000   Width factor: 1.00   Obliquing angle:  0
   Generation:  Normal
```

EXERCISE 4b SPECIAL CHARACTER TEXT

Discussion

There are some circumstances when you may want to print special characters that are not found on the standard computer keyboard, such as the degree symbol, the small plus and minus sign, or the underscore or overscore of a character. AutoCAD provides a special character function to allow you to do this.

The special character mode is entered by placing two percent signs (%%) as a control function before the special character to be inserted.

When you toggle a function ON, such as the "overscore" or "underscore," you must also toggle it OFF when you are through using the function. If you don't, it will stay on forever.

The following list highlights the characters available:

%%o	Toggle overscore mode on/off.
%%u	Toggle underscore mode on/off.
%%d	Draw the degree symbol.
%%p	Draw plus/minus tolerance symbol.
%%c	Draw the circle diameter dimension symbol.
%%%	Create a percent sign.
%%nnn	Draw ASCII character nnn where nnn is ASCII number from 032 to 129.

To print the following text, you would use the special character set as shown.

Text:
80% OF THE PEOPLE, +/- 3, LIKE THEIR **STEAK**
COOKED AT 450°F.

Special character to achieve the text:

```
80%%% OF THE PEOPLE, %%p3, LIKE THEIR %%uSTEAK%%u
COOKED AT 450%%dF.
```

Note the toggle On and Off underscore symbol around STEAK.

EXERCISE 5 ISOMETRIC DRAWING

Discussion

To see the details on three sides of an object, a "pictorial view" such as an Isometric, a Diametric or a Trimetric view may be used. By far, the Isometric view is the most commonly used view because of its simplicity to draw. AutoCAD supports the drawing of Isometric views through the Snap Style option.

Isometric pictorials are always drawn at 30 degrees on receding sides with a 35-degree, 16-minute tilt forward. See Fig. 5E-1.

Playing with the Isometric Grid

To become familiar with the Isometric view, we'll create a simple drawing. Set your Limits to 12,0000, 9.0000 and use a Grid and Snap spacing of 0.5 unit.

Housekeeping

Limits = 12,0000, 9.0000
Grid = .5
Snap = .5

A Grid of 0.5 center is displayed on the screen. Now to switch to the Isometric view by reusing the Snap command.

```
Command: SNAP (RETURN)
Snap spacing or On/OFF/Aspect/Rotate/Style<default>: S (RETURN)
Standard/Isometric <.5>: I (RETURN)
Vertical spacing <0.5000>: .5 (RETURN)
```

The Grid is regenerated with a series of dots at 30 degrees on receding sides.

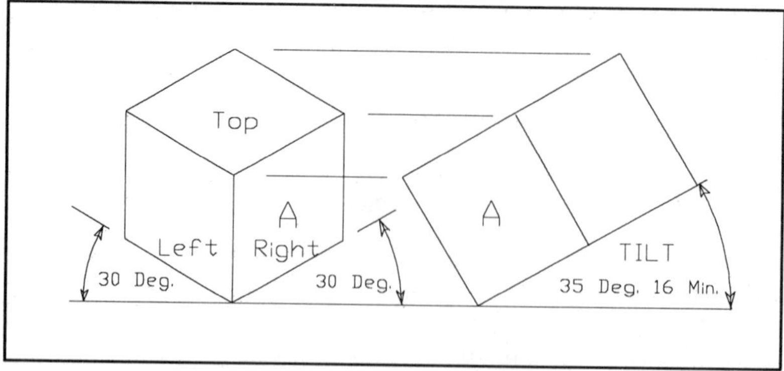

Figure 5E-1 Isometric tilt view.

As you move the input device (mouse or tablet pen), you will notice that the axis of the cursor lines is viewed from a Right side perspective. This may be changed to a Left side view, then a Top view, and back to a Right side view by continuously pressing the CTRL-E keys.

```
Command: CTRL-E
Command: <Isoplane Right>
```

A Right side view is seen.

```
Command: CTRL-E
Command: <Isoplane Left>
```

And a Left side view is seen.

```
Command: CTRL-E
Command: <Isoplane Top>
```

And a Top view is seen again.

Notice that you do not have to press the Return key.

To switch back to the Standard view, access the Snap command, and set the Styles to Standard again.

```
Command: SNAP (RETURN)
Snap spacing or On/OFF/Aspect/Rotate/Style <default>: S (RETURN)
Standard/Isometric <I>: S (RETURN)
Spacing aspect <0.5000>: (RETURN)
```

Drawing In The Isometric Mode

We will now draw the cube of Fig. 5E-2, using the Isometric option of the Snap command. The Left side will be drawn first, followed by the Top, and then the Right side. Switch back to the Isometric view through the Snap command.

Toggle the Coordinates to ON with the F6 function key to allow us to see the cursor locations and press the CTRL-E key until you are in the *Isoplane Left* mode.

Drawing the Left side.

```
Command: <Isoplane Left>
Command: SELECT DRAW
Command: SELECT LINE
Line from point: SELECT POINT A (See Fig. 5E-2)
To point: SELECT POINT B, 3 UNITS @ 330<
To point: SELECT POINT C, 3 UNITS @ 90<
To point: SELECT POINT D, 3 UNITS @ 150<
To point: SELECT CLOSE IN ROOT MENU
To point: (RETURN)
```

Drawing the Top

```
Command: CTRL-E (Select Isoplane Top)
```

Starting at point D, draw points E and F, and back to C.

```
Command: SELECT LINE
Line from point: SELECT POINT D
To point: SELECT POINT E, 3 UNITS @ <30
To point: SELECT POINT F, 3 UNITS @ <330
To point: SELECT POINT C, 3 UNITS @ <210
To point: (RETURN)
```

Our last drawing will complete the right side of the cube. It is a line starting at point F, to G, and finishing back at B.

Drawing the Right side

```
Command: CTRL-E (Isoplane Right)
Command: SELECT LINE
Line from point: SELECT POINT F
To point: SELECT POINT G, 3 UNITS @ <270
To point: SELECT POINT B, 3 UNITS @ <210
To point: (RETURN)
```

The cube is now complete!

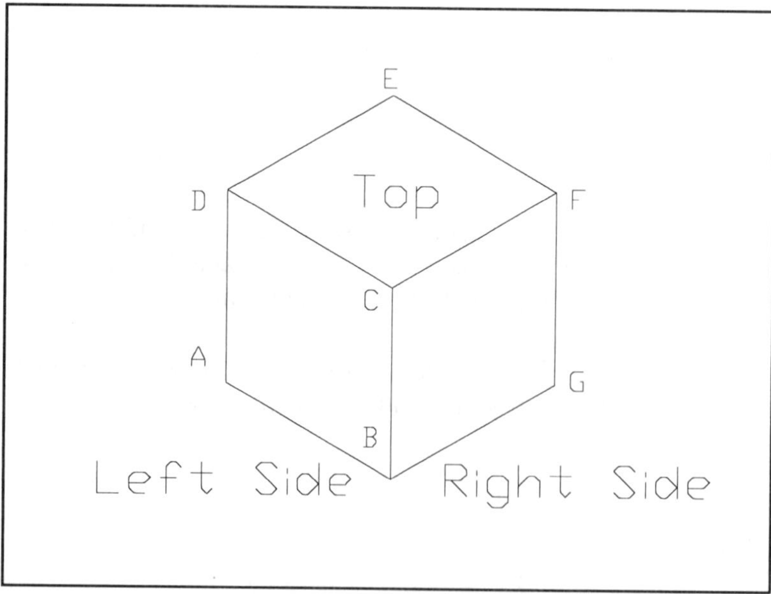

Figure 5E-2 Isometric cube.

Adding Text in Isometric Mode

We will add text to our cube to show you some of the limitations of the Isometric option by placing the labels "Left Side," "Top" and "Right Side" on the drawing.

Adding Text to the Left Side

```
Command: CTRL-E (Select Isoplane Left)
Command: SELECT DTEXT
DText Justify/Style <start point>: SELECT A POINT TO LEFT OF CUBE
Height <.2000>: .5 (RETURN)
Rotation angle <0>: 330 (RETURN)
Text: LEFT SIDE (RETURN)(RETURN)
```

Adding Text to the Top

```
Command: CTRL-E (Select Isoplane Top)
Command: SELECT DTEXT
DText Justify/Style <start point>: SELECT A POINT ON TOP OF CUBE
Height <.5000>: .5 (RETURN)
Rotation angel <330>: 0 (RETURN)
Text: TOP (RETURN)(RETURN)
```

Adding Text to the Right Side

```
Command: CTRL-E (Select Isoplane Right)
Command: SELECT DTEXT
DText Justify/Style <start point>: SELECT A POINT ON RIGHT OF CUBE
Height <.5000>: .5 (RETURN)
Rotation angle <0>: 30 (RETURN)
Text: RIGHT SIDE (RETURN)(RETURN)
```

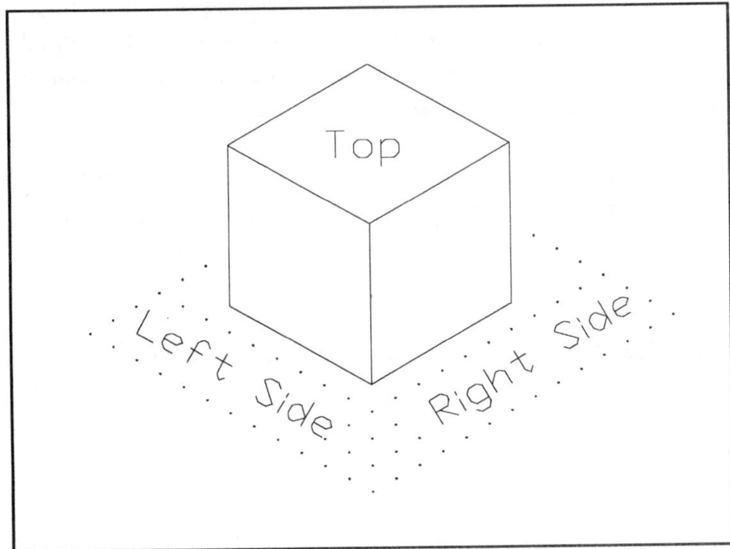

Figure 5E-3 Isometric view with text.

Your drawing should be similar to Fig. 5E-3. Note that the text does not lie flat on the surface of the three Isometric planes. Neither does it lie flat on the base elevation of the grid. It is for this reason that Isometric views may best be left as "Pictorial" views without lettering.

Having an appreciation for Isometric Pictorials, let us now examine the Three-Dimensional (3-D) command of AutoCAD.

EXERCISE 6 3-D DRAWINGS AND MODELLING

Discussion

The most exciting drawing feature of AutoCAD is its ability to create three-dimensional modelling. This is accomplished by changing a two-dimensional "plan" view (top down view) of an object into a three-dimensional "wire frame" viewing port through the use of the VPOINT or viewpoint command. This feature is a function of the ADE-3 software option of AutoCAD.

Three-dimensional models may be created from an existing plan view of an object by using the Change command, or while making a new drawing. In either case, you must inform AutoCAD about (1) the base elevation of the part and (2) the thickness. The base elevation can be thought of as the elevation where the bottom of the part lies.

In the following exercises, we will explore both methods of creating 3-D drawings. In the first part, we will examine a much simpler method to draw the cube developed in Exercise 5. Our first step in this drawing is to establish a thickness for a square 3 units by 3 units. To create the cube, we will make the thickness also 3 units.

In the second exercise, we will create a simple plan view of the eyelet, Fig. 6AE-1, and convert it, using the Change command, into the wire frame 3D drawing of Fig. 6AE-2.

EXERCISE 6a THREE-DIMENSIONAL MODELLING OF A CUBE

Housekeeping

Limits = 12.0000, 9.0000
Grid = .5
Snap = .5

To establish the base elevation of our cube, select Elev from the Modes menu found under the Root Menu. The base elevation of our cube will be the surface of our drawing paper and thus set to "0." The thickness of the cube will be "3" units high.

```
Command: ELEV (RETURN)
New current elevation <0.0000>: 0 (RETURN)
New current thickness <0.0000>: 3 (RETURN)
```

Toggle the coordinates to ON with the F6 function key so that we can see the cursor location.

In the center of the screen draw a square 3 units by 3 units. (See Fig. 6AE-1)

```
Command: SELECT DRAW
Command: SELECT LINE
Line From point: SELECT POINT A
To point: SELECT SECOND POINT
To point: SELECT THIRD POINT
To point: SELECT FOURTH POINT
To point: CLOSE
```

We now have a square drawn similar to Fig. 6AE-1.

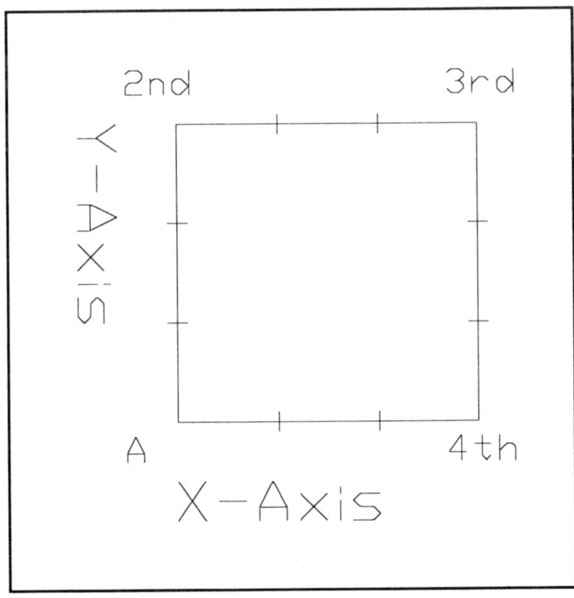

Figure 6AE-1 Three-dimensional plan view of cube (square).

Now to view our square as a 3-D cube by using the Vpoint command. In AutoCAD version 12, Vpoint is selected from the Display and Vpoint submenu.

The Vpoint menu displays three selections: Axis, Plan, and Rotate.

AXIS displays a visual Target and shows the X-Y-Z axis on which the object will appear. See Fig. 6AE-2A. Through the use of the input device, such as a mouse, you can position the view to display any side, top, or bottom elevation., See Fig. 6AE-2B, which shows X-Y-Z axis with the cursor placed in the lower, inner quadrant. When you are satisfied with the viewpoint, press the input device selection button, and the cube will appear as selected.

PLAN will cause the view to switch back to the original PLAN view of the square (Fig. 6AE-1). You may have to Zoom to All to see the drawing.

ROTATE will allow you to rotate the view of the cube in all three axis. Be careful when viewing the results as not everyone can think in 3D views easily. It takes time and much thought as to what you are seeing.

Let's look at our cube.

```
Command: VPOINT (RETURN)
Enter view point <0.0000, 0.0000, 1.0000>: (RETURN)
```

By pressing Return or selecting Axis at this point, the "Target and Axis" of Fig. 6AE-2A is displayed.

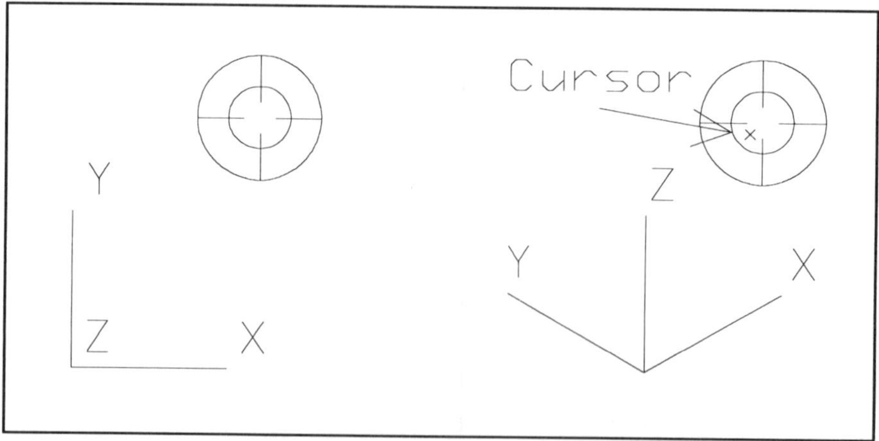

Figure 6AE-2A Three-dimensional target and axis.

Figure 6AE-2B Target and new axis.

Move the cursor to the Target area, and watch how the X-Y-Z axis changes to show you how the cube will eventually be placed. When satisfied with the viewpoint, fix it in position by pressing the input device select button.

Our 3 x 3 x 3 cube appears. Notice that the Grid is on an Isometric ratio, and that the UCSICON shows the X-Y plane you have chosen.

By selecting VPOINT and Axis again, from the Root Menu, you can change the position of the cube to that which best represents the pictorial view desired.

Adding Text to a 3-D View

To add text to our cube, it is best to switch back to a Plan view of the cube. Select Plan from the Vpoint menu.

```
Command: SELECT VPOINT
Command: SELECT
Command: Zoom to All (RETURN)
```

Our cube should now be seen as a top-down or plan view similar to Fig. 6AE-1.

We want our text to have no thickness. It should appear as if it were a pencil line on our drawing. Remember, the thickness of lines drawn to this point are three units thick. Therefore; we must first reduce the thickness to 0 so that any "new" lines will appear to be paper thick.

```
Command: SELECT ELEVation
New current elevation <0.0000>: 0 (RETURN)
New current thickness <3.0000>: 0 (RETURN)
```

Now to add the X-Axis text to the plan view.

```
Command: SELECT DRAW
Command: SELECT DTEXT
Text start point: SELECT X AXIS POINT
Height <.2000>: .5 (RETURN)
Rotation angle: 0 (RETURN)
Text: X-Axis (RETURN)(RETURN)
```

Now the Y-Axis text.

```
Command: SELECT DTEXT
Text start point: SELECT Y AXIS POINT AS SHOWN
Height <.5000>: (RETURN)
Rotation angle: 270 (RETURN)
Text: Y-Axis (RETURN)(RETURN)
```

Now let's view our cube with this text, in 3-D.

```
Command: SELECT VPOINT
Command: SELECT AXIS
```

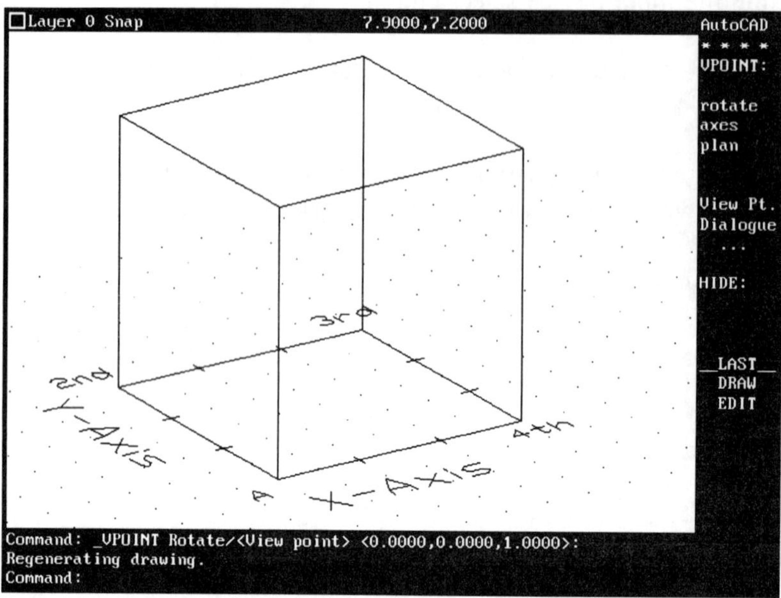

Figure 6AE-3 Three-dimensional view of cube and text.

Using the input device, select a location on the Target, such that the viewpoint will show our text from the best perspective. When satisfied, press the input device select button. *Note:* The UCS Icon represents the plane you have chosen.

Our cube reappears and should be similar to Fig. 6AE-3, complete with text of zero height, lying "flat" on the base elevation of the Isometric plane. Note that the text appears to follow the X and Y planes, giving it a true 3D look when compared with that found in the Isometric view of Exercise 5.

EXERCISE 6b THREE-DIMENSIONAL MODELLING OF AN EYELET

Discussion

In this exercise we will use the Change command to modify an existing drawing. To begin, draw the plan view of the eyelet shown in Fig. 6BE-1 and label the X-Axis and Y-Axis, but first set up the housekeeping criteria.

Housekeeping

Limits = 12.0000, 9.0000
Grid = .5
Snap = .5

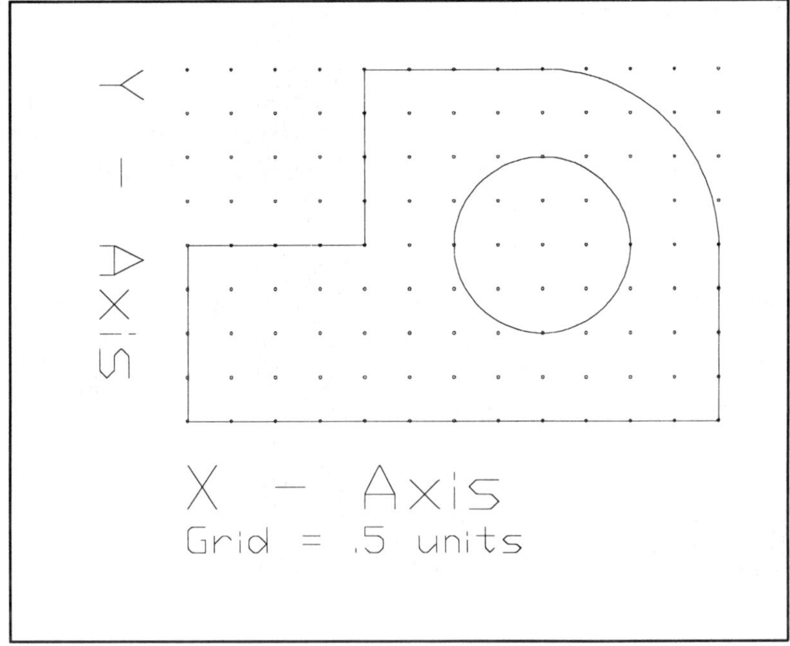

Figure 6BE-1 Plan view of eyelet.

Having drawn the plan view of the eyelet, we are now ready to give it a depth of 1 unit by using the Change command. When Change requests us to select the objects to change, use a Window and select only the eyelet portion of the drawing, not the text.

```
Command: CHANGE (RETURN)
Select objects or Window or Last: W (RETURN)
First point: SELECT FIRST POINT (RETURN)
Second point: SELECT SECOND POINT (RETURN)
```

Now we will give the eyelet a thickness of 1 unit, but keep its base elevation flat to the surface of the paper.

```
Change point (or Layer or Elevation): E (RETURN)
New elevation <0.0000>: 0 (RETURN)
New thickness <0.0000>: 1 (RETURN)
```

You may notice the object flicker as AutoCAD updates the memory.

We can now view the eyelet as a 3-D part. Selecting Vpoint and the appropriate axis on the Target will display the eyelet as a three-dimensional object.

```
Command: SELECT VPOINT
Command: SELECT AXIS.
Command: SELECT TARGET AXIS USING THE INPUT DEVICE
```

The eyelet now appears as a 3-D Isometric "pictorial" wire frame drawing similar to Fig. 6BE-2. For practice purposes, select various views of the eyelet to understand how AutoCAD handles the various positions on the Target. The X and Y text will help you with the orientation of the eyelet.

Now let's examine the Hide command of the Vpoint Menu. Select a "new" Axis viewpoint such that the circle of the eyelet falls behind the front edge (X side) of the eyelet. Next, select the Hide command. This causes the screen to go blank for a short period while AutoCAD is regenerating the drawing. When the drawing reappears, the eyelet has all of the hidden lines removed from view.

Note: It appears at present that AutoCAD does not support the option to save, print, or plot the Hide view. It can only be seen on the screen.

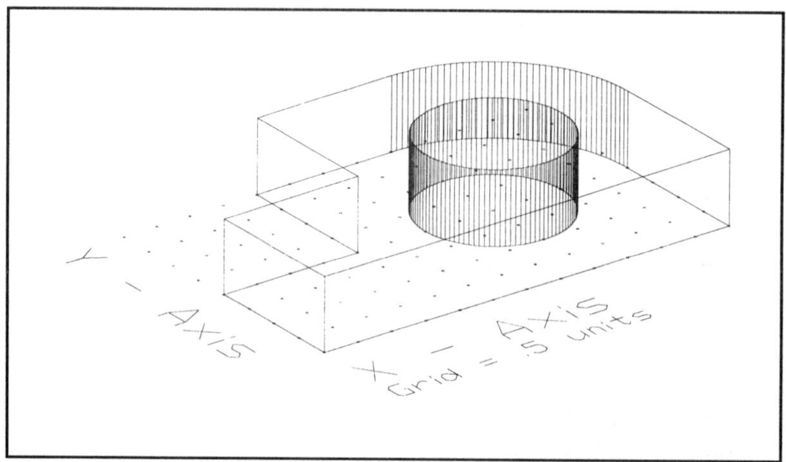

Figure 6BE-2 Three-dimensional view of eyelet.

EXERCISE 7 MODIFYING LINES WITH "STRETCH" AND "GRIPS"

Discussion

There are countless times when developing a printed circuit board drawing you may want to move a line a few millimetres, shift a component, reposition a land area or move a PCB trace farther away from an adjacent track. In this next exercise, you will learn some of the tricks to do this with ease and elegance.

In the past, this meant redrawing the line, polyline, donut or component. Today, through the use of the Stretch command and the newly introduced Grips in version 12, and our old favorite, the Move command, we are able to reposition just about anything, anywhere.

In Chapter three, we had an opportunity to play with Break, Trim and the Extend command. These commands differ from the above three, in that they modify the line or component by changing its actual length, either by adding to it, or by removing something from it. For example; Extend allowed us to increase the length of a line, where Stretch allows us to move or stretch a line or component to a new position without redrawing it. The rubber-band effect.

The STRETCH Command

First, we'll play with the Stretch command. Using the Polyline and Donut command, create a diagram similar to Fig. 7E-1. This will be our basic building block for the next exercises, so create four copies of it using the Copy command.

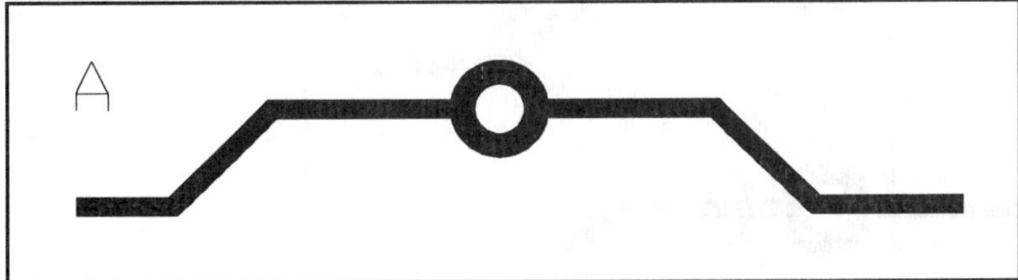

Figure 7E-1 Tracks and Donut.

In printed circuit applications, we may want to adjust where a land-area (Donut) is positioned on a trace, with respect to a component. In our first use of the Stretch command, we'll slide the Donut farther to the right on the trace, using Fig. 7E-2 as a reference.

To understand the Stretch command, we need to review how objects are selected. In "Window object selection", we have not made much use of the Crossing option.

This option basically says, *"any object which crosses into the window will be selected as part of the entities captured by the window."* This becomes a very useful tool when using Stretch, for it allows additional objects associated with the object to be moved, to also be captured and come along for the ride. We'll see how this works in this next exercise.

Exercise 7a Moving The Donut

Now to slide the Donut (Fig. 7E-2 A) to the right on the trace, making use of the Windows object selection option of Crossing C. Use Fig. 7E-2 as a reference and type the following:

```
Command: STRETCH (RETURN)
Select objects to stretch by window or ploygon...
Select object: C (RETURN)
First corner: WRAP A WINDOW AROUND THE DONUT (See Fig. 7E-2 B)
Other corner: SELECT (RETURN)
```

Use the center of the Donut as a base point for selecting the Donut.

```
Base point or displacement: SELECT CENTER OF DONUT
New point or displacement: DRAG THE DONUT, HORIZONTALLY TO THE RIGHT
                           AND CLICK IN PLACE.
Command:
```

When you have finished, your drawing will look similar to Fig. 7E-2 C.

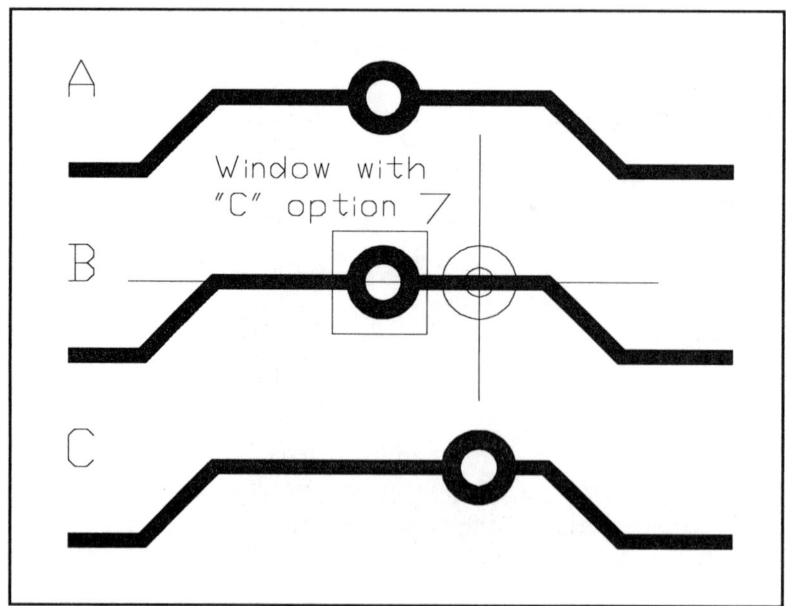

FIGURE 7E-2 Sliding the Donut right, using Stretch.

TIP: *What did we observe?* We should note that when the window was wrapped around the Donut, which included the left and right ends of the Trace (Polylines) lines, they too were captured by the window. This is the window's Crossing option in action.

Exercise 7b Stretching The Donut and Tracks (Rubber-Banding)

In this exercise, we will take the Stretch command one step further and rubber-band the Donut and attached Traces to a new location. During this exercise, note that the rubber-band effect only stretches the Trace back to the last bend in the Trace line. It does not reposition the entire trace line. I'll show you how to do this additional step in Exercise 7d.

Bring up another view of your test building block (Fig. 7E-1). Now, using Figure 7E-3 as a reference, we'll rubber-band the Donut and Traces just slightly above its present position. Type:

```
Command: STRETCH (RETURN)
Select object to stretch by window or polygon...
Select object: C (RETURN)
First corner: SELECT WINDOW AS IN FIG. 7E-3 B
Other corner: SELECT (RETURN)
Base point or displacement: SELECT CENTER OF DONUT
New point or displacement: DRAG THE DONUT AND TRACES VERTICALLY AND
                           CLICK IN PLACE.
Command:
```

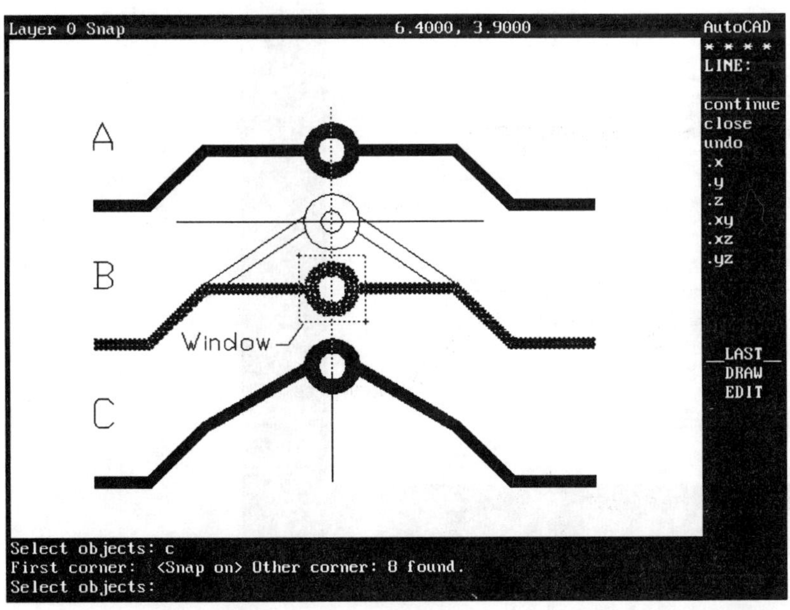

FIGURE 7E-3 Rubber-Banding the Donut and Trace through Stretch command.

What did we observe? a) We saw both the Donut and the two Trace lines were dragged together as the Donut moved to the new location. b) It was the donut base point that became the center for the objects being moved.

TIP 1: Select only that portion of the object you want to Stretch. If you were to wrap a window around the entire object (Fig. 7E-1), then Stretch would work like the Move command.

TIP 2: The Stretch command works well when Ortho is on. This allows the objects to be moved exactly horizontally or vertically.

Exercise 7c Getting a Grip on GRIPS

Discussion

Grips were introduced in version 12. They allow you to increase your productivity by providing a unique editing tool so you can immediately Stretch, Move, Rotate, Scale and Mirror objects. They are, by default, turned ON when AutoCAD starts. The Grip Selection Box, attached to the cursor, is always displayed when the Command prompt is empty and disappears when a command is being executed.

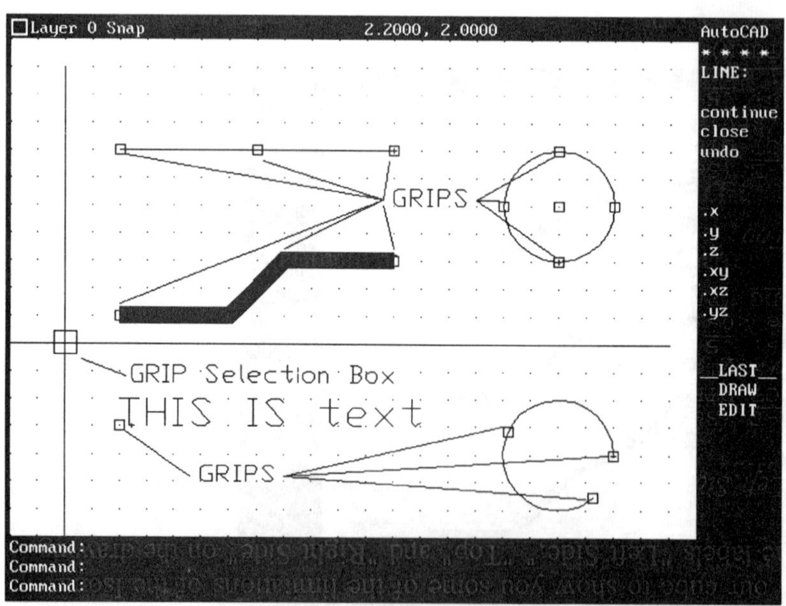

FIGURE 7E-4 Grips

Grips are small rectangles that are placed on objects that can be used to grab hold of an object. They are usually placed at Object Snap locations on the object and can be activated by simply clicking on the object (as shown in Figure 7E-4) or by wrapping a window around them with the Grip Selection Box, using the left mouse button. Once selected, you can manipulate them using the graphics cursor and key words.

There are usually a minimum of three grip points on every object. On a straight line for example (see Fig. 7E-4), there is one at either end of the line and one in the middle. When the middle Grip is selected, by clicking on it which causes it to turn red, you can move the entire line to a new location. By clicking on a circle and then choosing the center Grip point, you can move the entire circle to a new location. This is similar to using the Move command.

A Grip point is also positioned at the starting point of text. This allows you to easily move the entire line of text around your drawing. Likewise, polylines have Grips at each joint in the polyline, again making it easy to edit the entity.

The Grip points on the various objects are either turned On or Off from the Grips command or by the Ddgrips command which displays the Grips dialogue box. To use these, type:

```
Command: GRIPS (RETURN)
New value for Grips <1>: 1 (RETURN) (Will keep them active)
```

A value of 1 will turn Grips On and 0 will turn Grips Off.

To evoke the Grips dialogue box, type:

```
Command: DDGRIPS (RETURN)
```

The Grips dialogue box is immediately displayed. See Fig. 7E-5

Grips are selected by clicking on the Enable Grips Check Box and then OK.

Now to play. *Create a diagram similar to that of Fig. 7E-4.*

Moving a line. Through the Grips Selection Box on the cursor, select the line by clicking on it and then choose the center of the line. You will notice that the center Grip becomes red, a hot point. Upon choosing the hot point and holding the left mouse (select) button down, you can drag the line to any location on the screen and click it in place. To lock it there, press the Return key.

Moving the circle. Through the Grips Selection Box, select the circle. This causes five Grip points to appear. See Fig. 7E-4. Choosing the center point causes it to become a hot point. With the mouse select button held down, drag the circle to a new location. When complete, press the Return key to lock it in place.

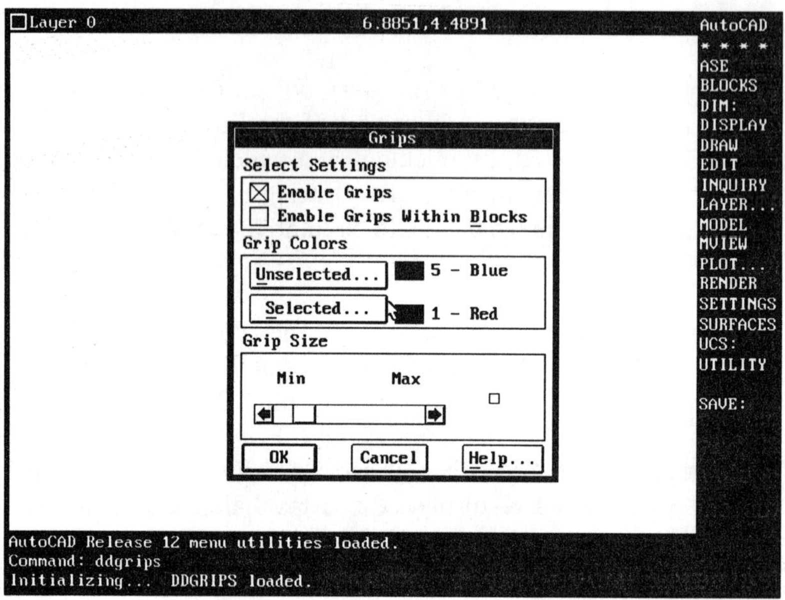

FIGURE 7E-5 GRIPS Dialogue Box.

TRAP: Be aware that the Grips Selection Box is always active when no command is being executed. This could get you into some trouble if you are not careful. The Grips Selection Box cannot be turned Off like the UCSICON, it is always present.

TIP: After each move using Grips for editing, terminate the action by pressing the Return key.

Exercise 7d USING STRETCH and GRIPS To Move DONUTS and TRACES

Discussion

Using Fig. 7E-6 A as a reference, we'll change the Donut and Trace lines so they fall on a straight line, as shown in Fig. 7E-6 D. To do this, first the Stretch command is used to reposition the Donut, and then the Grips command will move the remaining Trace lines. To move the Donut, type:

```
Command: STRETCH (RETURN)
Select objects to stretch by window or polyline...
Select object: C (RETURN)
First corner: SELECT A WINDOW AS SHOWN IN FIG. 7E-6 B
Second corner: Select (RETURN)
Base point: SELECT CENTER OF DONUT
New point: DRAG THE DONUT PARALLEL TO THE LOWER LINE
```

Your diagram should appear similar to Fig. 7E-6 B.

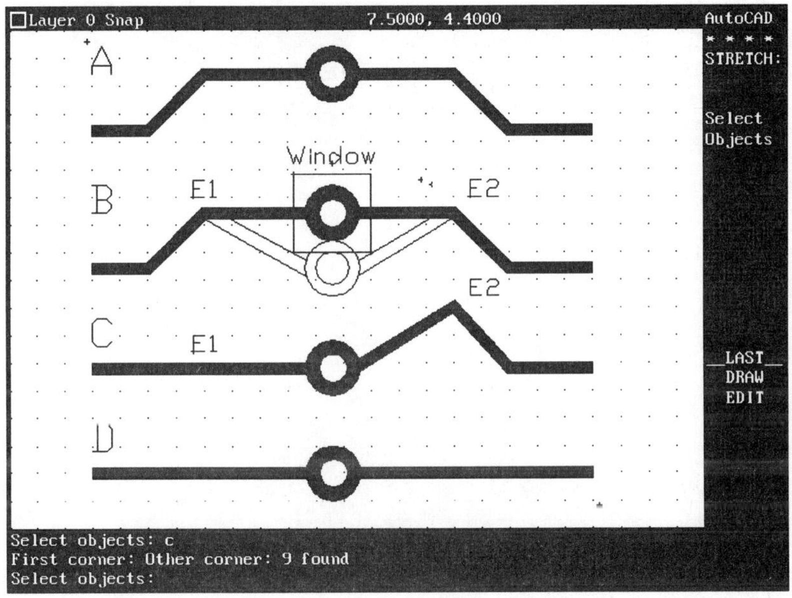

FIGURE 7E-6 Stretching and Gripping to straighten a Donut and Trace.

Now to grip the polylines and drag them in place.

```
Command: WITH THE GRIPS SELECTION BOX, SELECT GRIP E1 (FIG. 7E-6 B)
AND DRAG IT TO THE LOCATION SHOWN IN FIG. 7E-6 C.

Command: REPEAT THE SAME PROCEDURE WITH GRIP E2.
```

Your drawing should now look similar to Fig. 7E-6 D.

TIP: In this exercise, we used the combination of the Stretch command and the Grips command to move the Donut and Trace lines in Figure 7E-6 A to Figure D. They were not deleted and then redrawn. Redrawing could be very time consuming if the Trace line threaded itself across the entire printed circuit board.

Notes:

appendix F

Answers to Quizzes

CHAPTER 2 QUIZ: THE COMPUTER ENVIRONMENT

Circle the most correct answer or fill in the blanks.

1. Which of the following should be done at regular intervals?
   a. Save your drawing.
   b. Use CHKDSK.
   c. Make a backup of the complete disk.
   (d.) All of the above.

2. List four things you can do to ensure extended media life of your floppy disk.
   1. *Do not touch exposed media surface.*
      Keep disks away from magnetic fields.
   2. *Do not bend or fold disks.*
      Insert disk into disk-drive carefully.
   3. *Use disk cover around disk when not in use.*
      Store at 10 to 52 C (50 to 125 F).
   4. *Do not write on disk.*
      Protect disk from moisture
      Do not insert/remove disk when drive light is on.

3. Syntax refers to:
   a. How a disk is formatted.
   b. A copy program.
   (c.) Following the prescribed command structure.
   d. A program under DOS.

4. DOS is short for:
   Answer: *DISK OPERATING SYSTEM*

5. Which of the following commands will display the files on your disk?
   a. A:\DOS
   b. A:\FORMAT
   c. A:\CHKDSK
   (d.) A:\DIR

6. The purpose of formatting a disk is to:
   (a.) Prepare it to receive information.
   b. Scratch lines on it.
   c. Verify the files on the disk.
   d. Copy information from one disk to another.

7. Before formatting an old disk, what should one check for?
   Answer: *That it does not contain useful files.*

8. In a two-drive DOS system, what is the command structure to copy the file ELECTEMP.DWG in the A drive onto a formatted disk in the B drive?
   Answer: A: > *COPY A:ELECTEMP.DWG B:ELECTEMP.DWG*
   or *A: COPY ELECTEMP.DWG B:ELECTEMP.DWG*

9. What command is used to monitor the status of your disk?
   a. A:\ DIR
   b. A:\ DOS
   c. A:\ DISKCOPY
   (d.) A:\ CHKDSK

10. What is the DOS command structure to DISKCOPY the complete disk in the A drive onto a formatted disk in drive B?
    Answer: *A: DISKCOPY A: B:*
    or *A: DISKCOPY B:*

11. When using Windows, what is an easy method to copy a file from one drive to another?
    Answer: *By the "Drag and Drop" method.*

12. In the latest versions of DOS, what command is used to allow you to recover the information on a disk that has been formatted by mistake?
    Answer: *UNFORMAT*

13. What do you type at the C:\ prompt to start Windows running?
    Answer: *WIN*

14. What command is used to recover a file that was erased by mistake?
    Answer: *UNDELETE*

15. The MOVE command under windows:
    a. Makes a backup copy of a file.
    b. Changes places of two files.
    (c.) Removes a file from one place and relocates it.
    d. None of the above.

CHAPTER 3 QUIZ: GETTING AutoCAD RUNNING

Circle the most correct answer or fill in the blanks.

1. Which of the following commands will start AutoCAD running?
   (a.) ACAD or ACADR12 or ACADR13.
   b. RUN AutoCAD.
   c. START.
   d. None of the above.

2. After most keyboard commands, you should press:
   a. The RETURN key.
   b. The ENTER key.
   c. The carriage Return key.
   (d.) All of the above.

3. "Housekeeping" in AutoCAD refers to:
   a. Doing a "chkdsk".
   b. Making sure your Disk is cleaned once a week.
   (c.) The process of setting up an AutoCAD drawing before any work is performed on it.
   d. Formatting a disk.

4. To Cancel a Command, which of the following keys are pressed together?
   a. SHIFT + C
   (b.) CTRL + C
   c. ALT + C
   d. CTRL + ALT + DELete

5. What are four things that are usually set up at the beginning of most drawings, before any lines are drawn?
   a. *LIMITS* d. *SNAP*
   b. *GRID* e. *ZOOM (All)*
   c. *UNITS* f. *Drawing Name*

6. What are the two steps that must be done in order to view a portion of your drawing?
   a. *SELECT ZOOM*
   b. *SELECT WINDOW*

7. What command is used to bring the cursor directly to a grid line or mark?
   Answer: *SNAP*

8. What command restores an object when it disappears from the screen during a Block or Erase command?
   Answer: *OOPS*

9. What two simple methods can be used to restore or repeat an AutoCAD command?
   a. *PRESS THE RETURN KEY*
   b. *PRESS THE SPACE BAR*
   c. *PRESS THE RIGHT MOUSE BUTTON*

10. What two commands must be issued in order to save an object from your drawing to disk?
    a. *BLOCK*
    b. *WBLOCK*

CHAPTER 4 QUIZ: THE SCHEMATIC

Circle the most correct answer or fill in blanks.

1. To save destroying the prototype drawing called C-Sht, you should:
   (a.) Duplicate it under a new name such as SCHEMA.
   b. Draw on it, but use a layer called SCHEMA.
   c. Redraw it from scratch.
   d. None of the above.

2. Which of the following commands will COPY a single file from one disk to another, so it can later be used by AutoCAD?
   a. COPY A:*.* B:*.*
   b. DISKCOPY A: B:
   c. COPY A:ELECTEMP B:ELECTEMP
   (d.) COPY A:ELECTEMP.DWG B:ELECTEMP.DWG

3. To capture an "entity" from your drawing, you should use the command sequence:
   a. COPY A:*.* and B:*.*
   (b.) BLOCK and WBLOCK

c. SAVE and BLOCK
d. WBLOCK and BLOCK

4. Before starting a drawing, you should:
   a. Examine the hand sketches to establish sheet size.
   b. Perform a "rough" block layout of the drawing.
   c. Give some thought as to where to begin.
   (d.) All of the above.

5. "DOTS" and "NO-DOTS" refer to:
   a. The small marks plotter pens make.
   b. Reside on the drawing paper.
   (c.) A drawing standard.
   d. None of the above.

6. What command is used to place a BLOCK or WBLOCK into your drawing?
   ANSWER: *INSERT*

7. To latch to the end of a line previously drawn, you should use the command:
   (a.) OSNAP.
   b. SNAP.
   c. TARGET.
   d. All of the above.

8. Briefly describe the process in Question 7.
   Choose "LINE"; then select or type "END."

9. In AutoCAD, an ARC may be drawn through:
   (a.) All degrees up to 359.99.
   b. 360 Degrees.
   c. Only to 180 Degrees.
   d. None of the above.

10. Where is a quick reference place on the screen to see if a LAYER is "current"?
    ANSWER: *The top left corner of the screen, in the mode area.*

11. Layers in AutoCAD may be thought of as:
    a. New drawings.
    b. Colored areas.
    c. Sheets of paper.
    (d.) Sheets of clear cellulose.

CHAPTER 5 QUIZ: PRINTED CIRCUIT BOARDS

Circle the most correct answer or fill in the blanks.

1. A "TRACE" line may:
   (a.) Be drawn solid or filled in.
   b. Be drawn in an arc.
   c. Have different starting and ending widths.
   d. All of the above.

2. A "POLYLINE" (PLINE) may:
   a. Be drawn solid or filled in.
   b. Be drawn in an arc.
   c. Have different starting and ending widths.
   (d.) All of the above.

3. A TRACE or PLINE, when meeting a donut, should:
   a. Just touch the outer edge of the donut.
   b. Cover the complete pad.
   (c.) Fall completely within the pad area but not cover the center hole.
   d. All of the above.

4. The "FILL" command:
   a. May be turned on and off at will.
   b. Should be left off until you plot.
   c. Allows solid polylines and traces.
   (d.) All of the above.

5. What are the three types or groups of PCBs?
   a. *Single-sided*
   b. *Double-sided*
   c. *Multilayer*

6. Hatching:
   a. Is the process of filling in an area with dotted and/or dashed lines.
   b. Has at least 53 patterns available in AutoCAD.
   c. Allows you to construct your own pattern.
   (d.) All of the above.

7. Before designing in a component, you should:
   (a.) Consult the manufacturer's data sheet.
   b. Test its resistance.
   c. Get the cheapest unit.
   d. All of the above.

8. Some components are polarized and should:
   a. Never be used.
   (b.) Have their orientation verified.
   c. Be mounted vertically.
   d. All of the above.

9. Electrolytic capacitors:
   (a.) Will explode if mounted backwards.
   b. Are nonpolarized.
   c. Come with only axial leads.
   d. All of the above.

10. The input leads on the LM317 and LM337 regulator:
   a. Are the same.
   (b.) Are different.
   c. Have axial leads.
   d. Should be heat-sinked.

11. To manually lay out a PCB, you should:
   a. Observe the schematic.
   b. Check component orientation.
   c. Interconnect components as in the schematic.
   (d.) All of the above.

CHAPTER 6 QUIZ: ASSEMBLY DRAWINGS

Circle the most correct answer or fill in the blanks.

1. Dimensional Information:
   a. Must be precise and clearly understood.
   b. Use Dimension Lines and Extension Lines.
   c. Always use tolerances.
   (d.) All of the above.

2. The UCS system uses:
   a. The Cartesian Coordinate System.
   b. Only the World Coordinate System.
   c. The X, Y coordinate system.
   (d.) A method that allows you to work more easily with three-dimensional objects.

3. The WCS system uses:
   (a.) The Cartesian Coordinate System.
   b. Only the World Coordinate System.

c. The X, Y coordinate system.
d. A method that allows you to work more easily with three-dimensional objects.

4. Tolerances use:
   a. The plus "+" and minus "-" symbols.
   b. The Limit Dimensional System.
   c. Upper and Lower Limit Dimensions.
   (d.) All of the above.

5. In version 12, the Prompt when the Dimension Command is active, is:
   Answer: *DIM*

6. The two methods to get out of the Dimensional mode are:
   Answer: *Type EXIT of press the CTRL-C keys.*

7. The Extension Line is also known as:
   (a.) Witness Line
   b. Leader Line
   c. Center Line
   d. Dimension Line

8. The DIM1 Command:
   a. Is now obsolete.
   b. Provides a Horizontal Array.
   c. Only uses the "+" and "-" symbols.
   (d.) Allows you to draw a single Dimension Line.

9. How many types of Dimensioning methods are there in AutoCAD?
   a. 1 type of Dimensioning.
   b. 3 types of Dimensioning methods.
   (c.) 5 types of Dimensioning methods.
   d. 7 types of Dimensioning methods.

10. The term "DIMTXT" changes what Dimensional Variable?
    Answer: *Text Height*

11. The term "DIMASZ" changes what Dimensional Variable?
    Answer: *Arrowhead size*

12. What AutoCAD Command is used to remove a segment from a circle?
    Answer: *BREAK*

appendix G

Dimension Variables and Default Settings

Dimension Variables And Default Settings For AutoCAD Version 12

Variable Name	*Default Value*	*Functional Description*
DIMALT	Off	When ON enables an alternate units dimension sysetm such as metric
DIMALTD	2	Decimal places used by DIMALT
DIMALTF	25.4	Multiplier factor for DIMALT (Inches/mm conversion)
DIMAPOST	<"">	Alternate units text suffix
DIMASO	On	Associative dimensioning
DIMASZ	0.1800	Arrowhead size
DIMBLK	None	Customized arrowhead block name
DIMBLK1	None	Separate arrow block 1
DIMBLK2	None	Separate arrow block 2
DIMCEN	0.0900	Center mark size
DIMCLRD	Byblock	Dimension line color (Bylayer = 0 to 256 colors)
DIMCLRE	Byblock	Extension line color (Bylayer = 0 to 256 colors)
DIMCLRT	Byblock	Dimension text color (Bylayer = 0 to 256 colors)
DIMDLE	0.0	Dimension line extension
DIMDLI	0.38	Dimension line continuation increment
DIMEXE	0.1800	Distance extension line is beyond dimension line
DIMEXO	0.0625	Extension line origin offset
DIMGAP	0.0900	Gap from dimension line to text

DIMLFAC	1.0000	Linear unit scale factor
DIMLIM*	Off	Add limit dimension tolerance
DIMPOST	None	Default suffix for dimension text
DIMRND	0.0000	Dimension rounding tolerance
DIMSAH	Off	Separate arrow blocks
DIMSCALE	1.0000	Overall scale factor
DIMSE1	Off	Suppress the first extension line
DIMSE2	Off	Suppress the second extension line
DIMSHO	On	Update dimensions while dragging
DIMSOXD	Off	Suppress outside extension dimension
DIMSTYLE	UNNAMED	Current dimension style (read-only)
DIMTAD	OFF	Place text above the dimension line
DIMTFAC	1.0000	Tolerance text height scaling factor
DIMTIH	On	Text inside extension is horizontal
DIMTIX	OFF	Place text inside extension lines
DIMTM	0.0000	Minus sign tolerance
DIMTOFL	0.0000	Text outside, force line inside
DIMTOH	1.0000	Horizontal text outside extension lines
DIMTOL*	Off	Adds dimension tolerance text
DIMTP	0.0000	Plus sign tolerance
DIMTSZ	0.0000	Tick size
DIMTVP	0.0000	Text vertical position
DIMTXT	0.1800	Text height
DIMZIN	Off	Suppress zero inches after feet

* *DIMTOL and DIMLIM cannot both be on at the same time.*

Other useful system variables:

FILEDIA	When On (1) allows Dialogue Boxes to be visible.
DDEDIT	Modify text dialogue box.
DDEDITMT	Modify text attributes dialogue box.
DDINSERT	Insert dialogue box.
DDOSNAP	OSNAP dialogue box.

Dimension Variables And Default Settings For AutoCAD Version 13

Variable Name	*Default Value*	*Functional Description*
DIMALT	Off	When ON enables an alternate units dimension system such as metric
DIMALTD	2	Decimal places used by DIMALT
DIMALTF	25.4	Multiplier factor for DIMALT (Inches/mm conversion)
DIMALTTD	2	Decimal places for DIMALT tolerance values
DIMALTTZ	0	Toggles suppression of zeros for tolerance values
DIMALTU	2(1-5)	Sets units format: 1 Scientific, 2 Decimal, 3 Eng. 4 Architectural, 5 Fractional
DIMALTZ	0= (Off)	Toggles suppression of zeros for DIMALTU
DIMAPOST	<"">	Alternate units text suffix
DIMASO	On	Associative dimensioning (Off/On toggle)
DIMASZ	0.1800	Arrowhead size
DIMAUNIT	0(0-4)	Sets angle format of angle dimensions: 1Decimal degrees, 2 Deg/min/sec, 3 Gradians, 4 Radians, 5 Surveyor's units
DIMBLK	None	Customized arrowhead block name
DIMBLK1	None	Separate arrow block 1
DIMBLK2	None	Separate arrow block 2
DIMCEN	0.0900	Center mark size (0 = none, <0 = Center Lines drawn, >0 =Center Marks drawn)
DIMCLRD	Byblock	Dimension line color (Bylayer = 0 to 256 colors)
DIMCLRE	Byblock	Extension line color (Bylayer = 0 to 256 colors)
DIMCLRT	Byblock	Dimension text color (Bylayer = 0 to 256 colors)
DIMDEC	4	Decimal places for tolerance values
DIMDLE	0.0	Dimension line extension
DIMDLI	0.3800	Dimension line continuation increment
DIMEXE	0.1800	Distance extension line is beyond dimension line
DIMEXO	0.0625	Extension line origin offset
DIMFIT	3(0-4)	Controls placement of Dimtext and arrowheads
DIMGAP	0.0900	Gap from dimension line to text
DIMJUST	0	Controls horizontal dimension text position
DIMLFAC	1.0000	Linear unit scale factor
DIMLIM	Off	Add limit dimension tolerance
DIMPOST	None	Default suffix for dimension text
DIMRND	0.0000	Dimension rounding tolerance
DIMSAH	Off	Separate arrow blocks (Off = DIMBLK settings, On = User-defined arrowhead blocks are used)
DIMSCALE	1.0000	Overall scale factor
DIMSD1	Off	When ON suppresses the first dimension line
DIMSD2	Off	When ON suppresses the second dimension line
DIMSE1	Off	When ON suppresses the first extension line
DIMSE2	Off	When ON suppresses the second extension line
DIMSHO	On	Update dimensions while dragging
DIMSOXD	Off	Suppress outside extension dimension
DIMSTYLE	UNNAMED	Current dimension style read-only <STANDARD>

DIMTAD	0 (0-3)	Controls text placement above the dimension line
DIMTDEC	4	Decimal places of primary units dimensions.
DIMTFAC	1.0000	Tolerance text height scaling factor
DIMTIH	On	ON, Draws text inside extension horizontally.OFF, Aligns text with dimension line
DIMTIX	OFF	Place text between extension lines
DIMTM	0.0000	Minus sign tolerance
DIMTOFL	0.0000	Text outside, force line inside
DIMTOH	1.0000	Horizontal text outside extension lines
DIMTOL	Off	Adds dimension tolerance text
DIMTOLJ	1(0-2)	Sets vertical tolerance text position
DIMTP	0.0000	Plus sign tolerance
DIMTSZ	0.0000	Arrow/Tick size (0 Draws Arrow, >0 Draws Tick)
DIMTVP	0.0000	Text vertical position
DIMTXSTY	STANDARD	Sets text style for dimension text
DIMTXT	0.1800	Text height for dimension text
DINTZIN	0	Toggles suppression of zeros for tolerance values
DIMUNIT	2(1-5)	Sets units format: 1 Scientific, 2 Dec., 3 Eng., 4 Architectural, 5 Fractional
DIMUPT	0 (0-1)	Cursor control for User Positioned Text
DIMZIN	0 (0-3)	Controls zero feet/inches relationship

appendix H

Hatch Patterns

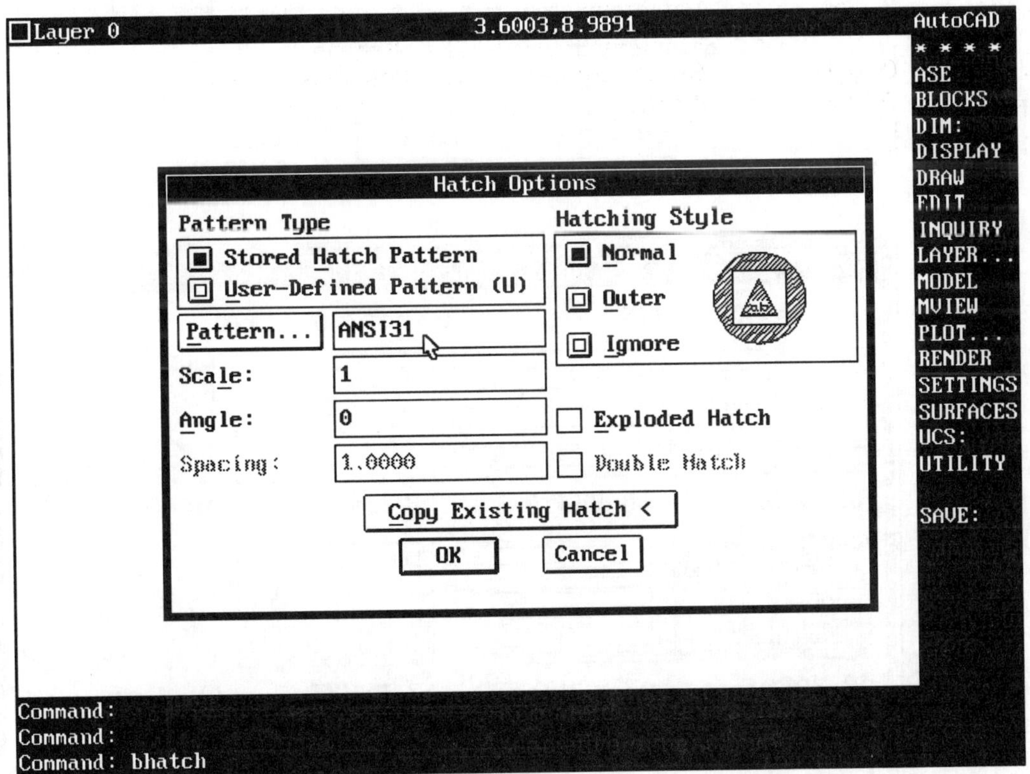

Figure H-1. Hatch Option dialogue box.

From the Hatch Option dialogue box, the following hatch patterns are available.

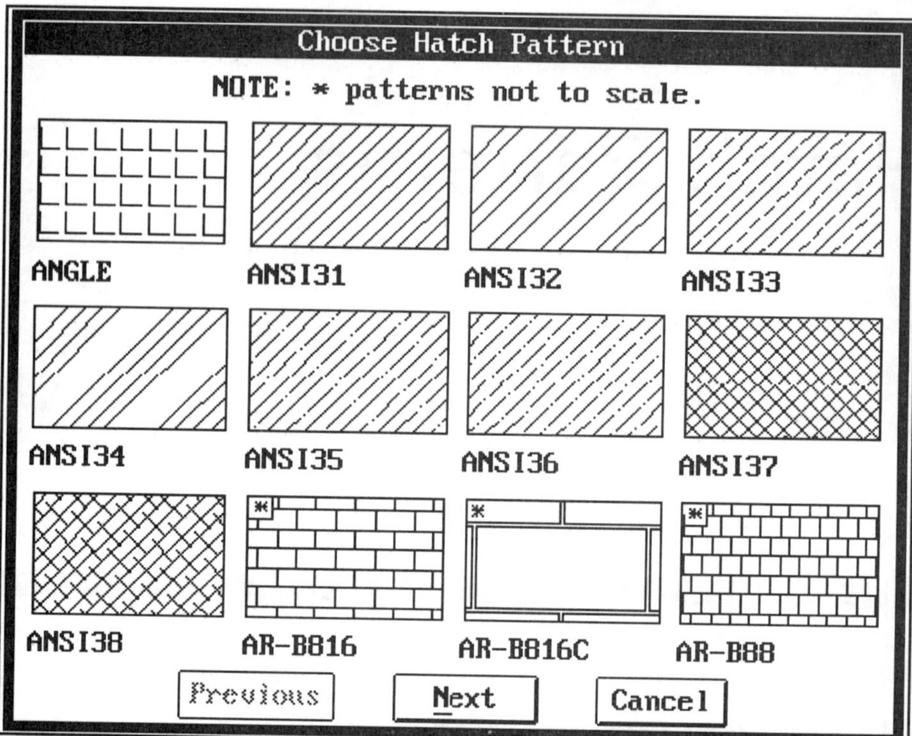

Figure H-2 Hatch Patterns (Part 1).

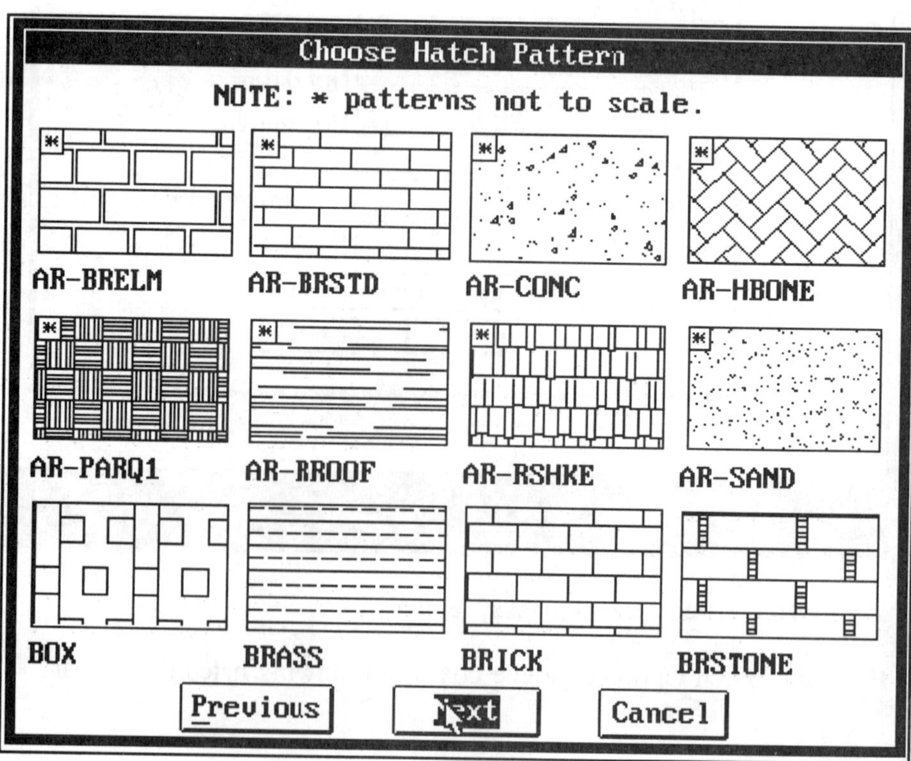

Figure H-3 Hatch Patterns (Part 2).

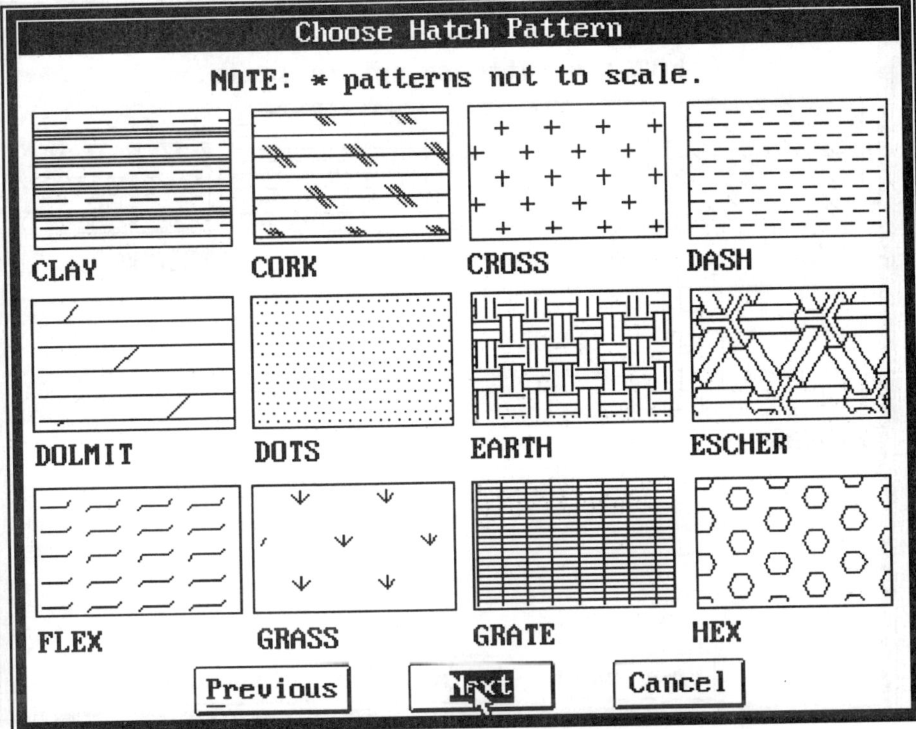

Figure H-4 Hatch Patterns (Part 3).

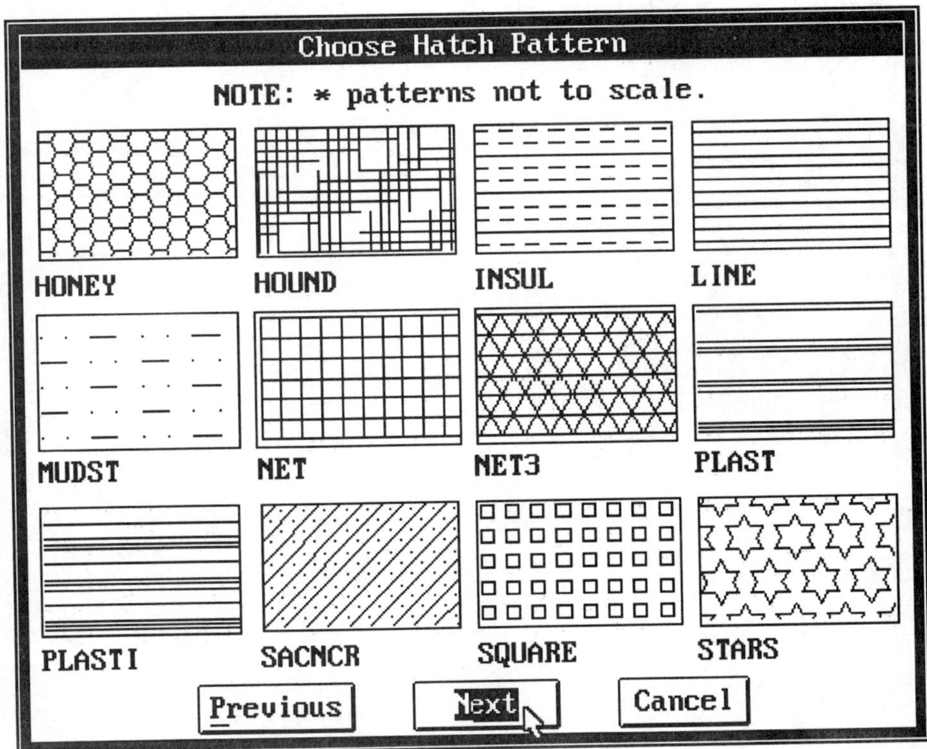

Figure H-5 Hatch Patterns (Part 4).

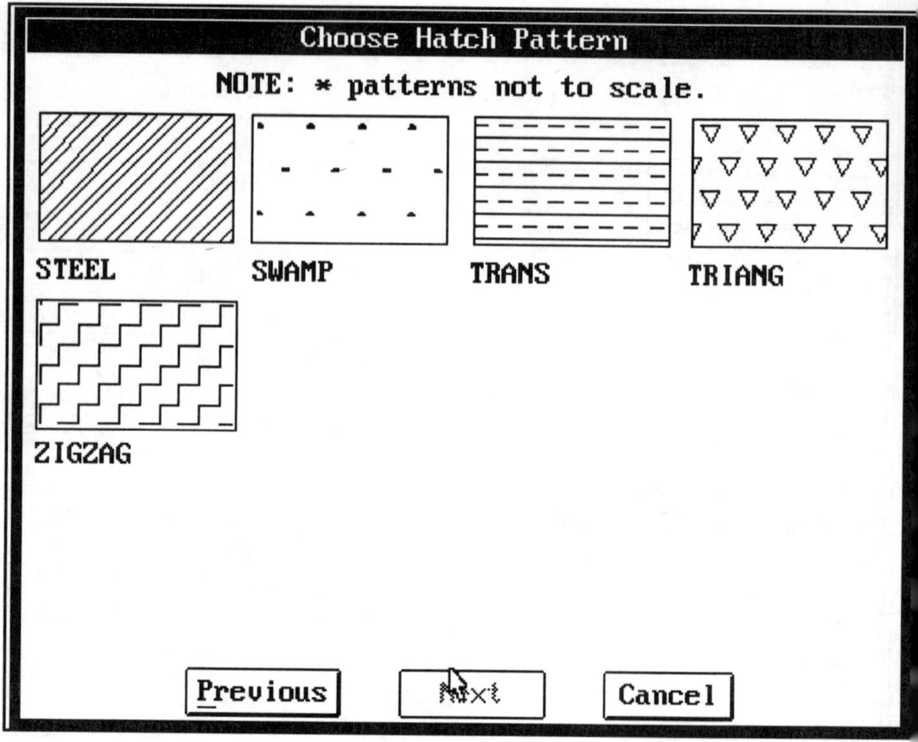

Figure H-6 Hatch Patterns (Part 5).

Index

A

B

C

E

F

G

H

I

K

L

Q

R

S

T

U

V

W

Z